Mathematical Preparation for Physical Chemistry

The beginner with a little background in algebra and trigonometry will find this book a useful adjunct to his interest in chemistry. It was written for those who lack the time, or the courage, to master higher mathematics in the regular way, and may be used either in class or independently.

Farrington Daniels is Professor-emeritus of Chemistry at the University of Wisconsin.

McGRAW-HILL PAPERBACKS IN SCIENCE, MATHEMATICS AND ENGINEERING

Mathematical Preparation for Physical Chemistry

Farrington Daniels

McGraw-Hill Book Company, Inc.

New York Toronto London

Printed in the United States of America

121314151617 MUMU 7

ISBN 07-015301-9

PREFACE

Science is best expressed when it is mathematically expressed; and yet most students of natural science cringe at the prospect of higher mathematics. It is because the author has seen so many students struggling against the handicap of an inadequate preparation in mathematics—spending more effort in avoiding mathematics than needs to be spent in learning it—that this book has been written.

The work is based on a one-semester course of three hours, given to sophomores in preparation for physical chemistry. Its only prerequisites are a little knowledge of algebra, a speaking acquaintance with trigonometry and an interest in chemistry.

Mathematics is a deductive science while chemistry is essentially an inductive science. Since the chemist's mind finds it easier to reason from specific facts to general propositions, there will be found in this book more of the inductive approach than is usual. As a consequence, the mathematical derivations suffer in rigor of proof. Many other things have been sacrificed, also, in the attempt to cover in a small book, those portions of higher mathematics which are of special value to the natural scientist.

There must be no mistake concerning the purpose of this book. It is *not* equivalent to the standard courses in analytical geometry, differential calculus and integral calculus. It is designed for those who lack the time or the courage to master these subjects in the regular way. It is meant either for class use, or for individual students taking physical chemistry, or for workers in science who have escaped into larger fields without calculus in college. The material of this book is sufficient mathematical preparation for a first course in physical chemistry and for some advanced courses, but a fully prepared *physical chemist* should have a better foundation than can be given in these few pages.

The ability to state a problem in mathematical language is invaluable, but the mathematical view point and the technique necessary for carrying out the operations come only with earnest practice. For this reason several problems are included at the end of each chapter and answers are given for alternate problems. It is hoped that the book may help students of chemistry and other natural sciences to appreciate and to use one of their most valuable tools.

If the student is satisfied with this book the author will be disappointed, for it is his hope that many who read these pages will resolve to go farther in the study of mathematics.

FARRINGTON DANIELS.

MADISON, WISCONSIN.

1928

CONTENTS

CHAPTER I

LARGE AND SMALL NUMBERS

When numbers get beyond the thousands or below the thousandths it is difficult to use them in the ordinary way. Physical chemistry has to deal with very large and very small numbers, such, for example, as the number of molecules in a gram molecule (six hundred thousand billion billion) or the wave length of light expressed in millionths of a millimeter.

The best way to express a large or small number is *to write it down as an integer, and multiply it by* 10 *with the proper exponent.* Thus, $123 = 1.23 \times 10^2$, and $1{,}230 = 1.23 \times 10^3$, and $0.000{,}000{,}001{,}23 = 1.23 \times 10^{-9}$. In this way there is always one figure at the left of the decimal point and the magnitude of the number is obtained at once by a glance at the exponent.

The exponent represents the number of times which 10 must be multiplied by 10. For example: 10^1 = tens. $10^3 = 10 \times 10 \times 10$ = thousands. 10^6 = millions. $10^{-2} = \dfrac{1}{10 \times 10}$ = hundredths. 10^{-6} = millionths.

There are many advantages in this procedure. Economy of space is effected, a point of particular importance in tabulating data. The labor of counting figures to the right or the left of the decimal is eliminated. Since the exponent of ten is identical with the characteristic of the corresponding logarithm the use of logarithms or a slide rule is simplified. Thus the logarithm of 5×10^{-19} is $\overline{19}.699$ and the antilogarithm of 12.699 is 5×10^{12}.

The accuracy with which a quantity is known is readily given by the number of figures retained after the decimal point. When this system of exponents is not used it is sometimes necessary to retain several figures which are not significant.

The pointing off of decimals is tedious when large or small numbers are multiplied or divided, but when all numbers are expressed as units multiplied by powers of ten the process is simplified. It is easy to keep track of the decimal point when only units are involved, and the exponents are simply added together or subtracted, thus

$$\frac{500{,}000 \times 12{,}000{,}000 \times 0.006}{30{,}000 \times 0.02 \times 10} = \frac{5 \times 10^5 \times 1.2 \times 10^7 \times 6 \times 10^{-3}}{3 \times 10^4 \times 2 \times 10^{-2} \times 10^1} =$$

$$\frac{5 \times 1.2 \times 6}{3 \times 2} \times 10^{+5+7-3-4+2-1} = 6 \times 10^6.$$

It will be remembered that in multiplication the exponents of a number are added, and that in division the exponents in the denominator are subtracted from those of the same number in the numerator (p. 266). Exponents applying to the same number may be transferred from the denominator to the numerator or vice versa by merely changing their sign.

The practice of expressing numbers always with only one figure at the left of the decimal point has not been universally adopted, but it is hoped that it will soon become universal, for it simplifies any calculation. Thus although some authors write 60.6×10^{22}, it is better to write 6.06×10^{23}.

The decimal point may be shifted at will in this way, if a compensating alteration is made in the exponent. For example $36. \times 10^{-10}$ may be changed to 3.6×10^{-9} by dividing and multiplying by 10 for this is an operation which does not change the value of the number.

$$36 \times 10^{-10} = {}^{36}\!/_{10} \times 10^{-10} \times 10 = 3.6 \times 10^{-10+1} = 3.6 \times 10^{-9}.$$

Also,

$$0.0036 \times 10^5 = 1{,}000 \times 0.0036 \times \frac{10^5}{1{,}000} = 3.6 \times 10^2.$$

There are two exceptions to this rule of "one figure at the left of the the decimal point." If in a subsequent operation it is desired to extract a root of the number it is advantageous to have the exponent of ten exactly divisible, regardless of the number of figures at the left of the decimal. Thus if the square root of 3.6×10^{-9} is to be extracted. it is written as 36×10^{-10}, the square root of which is 6×10^{-5}. Also, when numbers are to be added or subtracted they must have the same exponents. Thus

$$3 \times 10^{-6} + 5 \times 10^{-4} = 3 \times 10^{-6} + 500 \times 10^{-6} = 503 \times 10^{-6} = 5.03 \times 10^{-4}.$$

A different convention is sometimes used for small numbers. According to this convention, the number of zeros before the first significant figure is denoted by a subscript. Thus 0.000,0023 is written 0.0_523. For mathematical operations the previous method is better, *i.e.*, 2.3×10^{-6}.

Considerable progress has been made by visualizing the small units of physical chemistry and assigning to each its numerical value. The atomic weights no longer need to be intangible relative numbers. They may be turned into actualities by simply dividing by 6.06×10^{23}. For example, 1 atom of oxygen with an "atomic weight" of 16 weighs $\frac{16}{6.06 \times 10^{23}} = 2.64 \times 10^{-23}$ g. In the past it was known that an atom of gold is 197 times as heavy as an atom of hydrogen but it is known now that an atom of gold weighs $\frac{197}{6.06 \times 10^{23}} = 3.25 \times 10^{-22}$ g. Likewise the final unit of electricity is the electron and the unit of radiant energy is the quantum.

In the problems which follow, and in others which the student should set for himself, the following numerical values may be helpful. 1 in. = 2.540 cm. 1 lb. = 453.6 g. $1\mu = 10^{-3}$ mm. 1 $m\mu = 10^{-6}$ mm. The Avogadro number, the number of molecules in a gram molecule, is 6.06×10^{23}. The volume of a gram molecule of a gas at 0° C. and 760-mm. pressure is 22.4 liters. The charge on an electron is 1.59×10^{-19} coulombs. The velocity of light is 3×10^{10} cm. per second. These and other constants may be found on page 271. A gram molecule is a weight in grams equal to the molecular weight. For example, the molecular weight of sodium chloride is 58.5 and a gram molecule of sodium chloride is 58.5 g.

Exercises

1. $25{,}000 \times 2 \times 63{,}000{,}000 = 2.5 \times 10^4 \times 2 \times 6.3 \times 10^7 = ?$

   Ans.: $31.5 \times 10^{11} = 3.15 \times 10^{12}$.

2. $\frac{3 \times 10^{-9} \times 6 \times 10^{21} \times 200}{9 \times 10^3} = ?$

3. $0.00004 \times 50 \times 3{,}600 \times 0.000{,}000{,}00001 = ?$ *Ans.:* 7.2×10^{-11}

4. $\dfrac{0.0012 \times 200 \times 400{,}000{,}000}{0.000{,}006 \times 1{,}000} = ?$

5. $\dfrac{1{,}234 \times 0.000{,}567 \times 78{,}900{,}000}{1{,}500 \times 0.000{,}000{,}735} = ?$ *Ans.:* 5.007×10^{10}.

6. $\dfrac{1.2 \times 10^{-10} \times 3 \times 10^{6}}{1 \times 10^{-9} \times 6 \times 10^{2}} + 30{,}000 - \dfrac{5 \times 10^{3}}{2} = ?$

7. $\dfrac{235{,}000 \times 3{,}600 \times 0.000{,}421}{429 \times 0.00096} - \dfrac{129{,}000 \times 23{,}600}{98{,}000{,}000 \times 0.023} = ?$

8. How many seconds are there in a lifetime of 75 years?

9. How many milligrams are there in a ton? *Ans.:* 9.07×10^{8}.

10. How many dimes can be spread out in a single layer on an acre of flat land?

11. A cube of gold 1 cm. on an edge has an area of 6 sq. cm. If this is cut up into cubes 1 mm. on an edge, what is the total surface area? If cut up into cubes 1μ on a side? One $m\mu$ on a side?

Ans.: 60. 6×10^{4}. 6×10^{7} sq. cm.

12. What is the surface area of 1 g. of glass wool, whose threads are 0.01 mm. in diameter? The threads may be considered cylindrical and the area of the ends may be neglected. Density of glass = 2.6.

13. A certain star is so far away that it takes light traveling at the rate of 3×10^{10} cm. per second, 10 years to reach the earth. How many centimeters is it to the star?

Ans.: 9.467×10^{18} cm.

14. The limits of astronomical observation have now been pushed to a distance of a million light-years. What is this distance in centimeters; in $m\mu$; in miles?

15. The product of the wave length of light and its frequency gives its velocity. (*a*) If the velocity of light is 3×10^{10} cm. per second what is the frequency of blue light which has a wave length of 400 $m\mu$? (*b*) What is the frequency of red light, if its wave length is 750 $m\mu$?

Ans.: (*a*) 7.5×10^{14} (*b*) 4.00×10^{14}.

16. In a representative volumetric analysis, the titration may be made with an accuracy of 1 drop, or $\frac{1}{20}$ cc. of a $\frac{1}{10}$ molar solution. How sensitive is this test in terms of numbers of molecules of dissolved reagent?

Ans.: $\left(\dfrac{1}{20} \times \dfrac{1}{1{,}000} \times \dfrac{1}{10}\right) \times (6 \times 10^{23}) = 3 \times 10^{18}$.

17. *a.* Certain colorimetric tests will detect one-millionth of a gram molecule of material. How many molecules are necessary to give a detectable test? *Ans.:* 6.06×10^{17}.

b. How many molecules of sodium chloride are necessary to give a spectroscopic test, sensitive to 0.000,000,001 gram molecule? *Ans.:* 6.06×10^{14}.

c. How many molecules of radium emanation (radon) are necessary to give a test with an electroscope if 10^{-15} gram molecule can be detected? *Ans.:* 6.06×10^{8}.

d. With the mass spectograph 10^8 molecules of material can be detected. What fraction of a gram molecule is this?

Ans.: 1.65×10^{-16}.

18. By experimental observation it was found that a certain quantity of radium emitted 2.8×10^{13} alpha particles in a year. During the same time 0.001 cu. mm. of helium was formed, measured at 0° and 760 mm. Making the reasonable assumption that the alpha particles turned into helium molecules, what is the number of molecules of helium in 22.4 l. at 0° and 760 mm.? This is a measure of the Avogadro number, 6×10^{23}.

19. A certain thermopile is so sensitive that it will register a definite deflection when light is falling on it at the rate of 1×10^{-8} cal. per second. How long would it take this light to melt a pound of ice? (The heat of the fusion of ice is 80 cal. per gram.) *Ans.:* 3.63×10^{12} sec.

20. During a certain storm 2.54 çm. (1 in.) of rain fell.

a. How many cubic centimeters of water fell on a square mile of land?

b. How many grams? (Density of water = 1).

c. How many moles? (Molecular weight of water = 18).

d. How many molecules?

e. How many drops, assuming each drop was 0.1 cc.?

21. How many molecules are there in a liter vessel which has been evacuated to 0.1 mm. at 0° C.?

$$\left(\frac{0.1}{760} \times \frac{1}{22.4} \times 6.06 \times 10^{23} = ?\right)$$

22. How many years would it take the entire population of the earth (one billion) to count the gas molecules in 1 cc. of air at room temperature $\left(\frac{1}{24{,}000} \text{ of a gram molecule of gas}\right)$, each person counting 3 molecules per second?

23. How many cubic centimeters are there in a circular lake 4 miles in diameter and 6.1 m. (20 ft.) in depth?

a. How many gram molecules in the lake? 1 g. molecule of water weighs 18 g.; and occupies 18 cc.

b. How many molecules? *Ans.:* 6.68×10^{36} molecules.

24. *a.* How many molecules are there in a liter of ethyl alcohol? Density = 0.78 and molecular weight = 46.

b. The liter of alcohol is poured into the lake described in Prob. 23 and completely mixed.

How many molecules of alcohol would be recovered if a liter sample of the lake water was then taken?

25. If the molecules of a gram molecule of hydrogen (2 g.) could be strung out in a single row what would be the length of the row in miles? The diameter of the hydrogen molecule is 2×10^{-8} cm. *Ans.:* 75 billion miles.

26. An electron carries a charge of 1.59×10^{-19} coulombs. An ampere is a current of such strength that a quantity of electricity equal to 1 coulomb passes a given point in the circuit every second.

a. How many electrons are involved per second in the passage of a current of 1 amp.?

b. If a sensitive galvanometer can detect a current of 10^{-10} amp., how many electrons must flow each second to be detected?

27. *a.* How many ampere-hours would be required to convert a liter of water into hydrogen and oxygen by electrolysis? Two electrons are involved in the decomposition of each molecule of water, and each electron carries a charge of 1.59×10^{-19} coulombs. An ampere-hour is equivalent to 3,600 coulombs.

b. How long would it take a current of 100 amp. to decompose all the water in the lake described in Prob. 23?

Ans.: a. 2,974 ampere-hours. *b.* 6.70×10^{8} years.

28. During electrolysis silver is deposited one atom at a time, each ion taking up an electron as it deposits (Prob. 26).

a. How many atoms are deposited each second by a current of 1 amp.?

b. How many coulombs of electricity are required to deposit a gram atom of silver (6.06×10^{23} atoms)? This quantity of electricity is known as the Faraday.

CHAPTER II

LOGARITHMS[1]

THEORY

Logarithms are Exponents.—The equation $10^2 = 100$ may be written $2 = \log_{10} 100$. Expressed in words, the first equation states that ten raised to the second power is equal to 100 while the second expresses the same fact by stating that 2 is the logarithm to the base 10 of 100. In the first equation 2 is an exponent while in the second it is a logarithm.

In general if $10^x = y$, then $x = \log_{10} y$. If y is given, its logarithm, x, may be found by reference to ordinary logarithm tables. Likewise if x is given, its antilogarithm, y, may be found by looking up in the table the number which has x for its logarithm. The antilogarithm, y, is the number obtained by raising 10 to the x power. In the numerical example given above, 100 is the antilogarithm of 2.

Natural logarithms have e, or 2.71828 . . . for a base instead of 10. Thus if $e^2 = 7.389$, $2 = \log_e 7.389$. In general if $e^x = y$, $x = \log_e y$. The full significance of e will become apparent later (p. 126) but it is sufficient now to state that it is extremely important in mathematics and in chemistry. It is so important, that if the base is not specified a mathematician assumes that $\log_e$ is meant. In chemistry, however, as in algebra, if the base is not specified, the base 10 is understood. Log means $\log_{10}$ in chemistry, while natural logarithms are designated by *ln* (logarithm natural) or by $\log_e$. Logarithms to the base e are sometimes called Napierian logarithms, after the mathematician, Napier, while logarithms to the base 10 are sometimes called Briggsian logarithms, after their inventor, Briggs (A.D. 1617). Any number may be used for a base, although 10 and 2.71828

[1] This chapter may be omitted by those who understand how to use logarithms, including negative logarithms.

. . . are used more than any other numbers. In general, if $a^x = y$, $x = \log_a y$, where a is any number. For example, if $5^3 = 125$; $3 = \log_5 125$.

Change of Base.—It is often necessary to change from one base to another, and for this operation we have the following rules: $\text{Log}_{10}\ x$ may be changed to $\log_e x$ by multiplying by 2.303.

$$\mathbf{Log_e\ x = 2.303\ log_{10}\ x.}$$

The number 2.303 is the logarithm to the base 2.71828 of 10. As an example of this rule it is desired to find $\log_e 30.2$.

Natural logarithms are not readily available but from ordinary logarithm tables $\log_{10} 30.2$ is found to be 1.4801. Then $\log_e 30.2 = 2.303 \times 1.4801 = 3.410$. Log_e occurs in many formulas in physical chemistry, but it is written as $2.303\ \log_{10}$. Sometimes the constant 2.303 is absorbed with other constants in an equation to give a new constant.

$\text{Log}_e\ x$ may be changed to $\log_{10} x$ by multiplying by 0.4343.

$$\mathbf{Log_{10}\ x = 0.4343\ log_e\ x}$$

The number 0.4343 is $\log_{10} 2.71828$. To illustrate this rule, $\log_e 50$ is 3.915. Then $\log_{10} 50 = 0.4343 \times 3.915 = 1.699$.

In general $\log_a x$ may be changed to $\log_e x$ by multiplying by $\log_e a$.

$$\mathbf{Log_e\ x = log_e\ a(log_a\ x)\ and\ Log_a\ x = log_a\ e(log_e\ x).}$$

The reason for these rules is evident from the following operations.

b and g are numbers such that 10^b and e^g are both equal to the number c. That is, $10^b = c$ and $e^g = c$.

Then $b = \log_{10} c$ and $g = \log_e c$ by the definition of logarithms. It follows also that $e^g = 10^b$ since both are equal to c.

$$\log_e (e^g) = \log_e 10^b \text{ (taking } \log_e \text{ of both sides).}$$

$$g \log_e e = b \log_e 10 \text{ (rule for the logarithm of a power of a number).}$$

$$g = b \log_e 10 \text{ (since } \log_e e = 1).$$
$$\log_e c = \log_{10} c \times \log_e 10 \text{ (substituting for } b \text{ and } g).$$
$$\log_e c = 2.303 \log_{10} c \ (\log_e 10 = 2.303).$$

This is the first rule given above.

The second rule may be proved in a similar manner by taking $\log_{10}$ of both sides of the equation.

It is obvious that division by 0.4343 will convert $\log_{10} x$ to $\log_e x$ just as well as multiplication by 2.303, since $\frac{1}{2.303} = 0.4343$. Likewise division by 2.303 will convert $\log_e x$ to $\log_{10} x$.

The Use of Logarithms

The Characteristic and Mantissa.—Logarithms are divided into two parts, the characteristic which comes at the left of the decimal point and the mantissa which comes at the right. The latter determines the figures which make up the antilogarithm and the former gives the position of the decimal point. For example, in the logarithm 2.5933, 0.5933 is the mantissa and 2 is the characteristic. By reference to a logarithm table it is found that the mantissa 0.5933 corresponds to the sequence of figures 3920. The characteristic indicates that the number belongs in the hundreds, and that it should be pointed off to give 392.0.

The rule is that a characteristic of zero corresponds to units place and that for each increase of one in the characteristic the number has one more significant figure at the left of units place. Likewise for each decrease of one in the characteristic, the first significant figure is moved one place to the right of units place.

The following examples illustrate the proper use of the characteristic:

$$\log 5 = 0.6990 \qquad \log 50 = 1.6990 \qquad \log 5{,}000 = 3.6990$$
$$\log 0.5 = \bar{1}.6990 \qquad \log 0.05 = \bar{2}.6990 \qquad \log 0.0005 = \bar{4}.6990$$
$$\text{antilog } 0.6990 = 5 \qquad \text{antilog } 1.6990 = 50 \qquad \text{antilog } \bar{1}.6990 = 0.5$$

Multiplication and Division.—Since the addition of exponents is equivalent to the multiplication of the numbers (p. 266), and since logarithms are exponents, it is obvious that a multiplication process may be turned into the much easier process of addition by the use of logarithms. For example

$$100 \times 1{,}000 = 10^2 \times 10^3 = 10^{2+3} = 10^5 = 100{,}000.$$

Carrying out the same operation with logarithms,

$$\begin{aligned} \log 100 &= 2.0 \\ \log 1{,}000 &= 3.0 \\ & \overline{} \\ & \ 5.0 \text{ (adding)} \\ \text{antilog } 5.0 &= 100{,}000. \end{aligned}$$

Also in another example,

$$250 \times 30 \times 8 = 10^{2.3979} \times 10^{1.4771} \times 10^{0.9031} = 10^{4.7781} = 60{,}000;$$

or by logarithms,

log 250	2.3979
log 30	1.4771
log 8	0.9031
	4.7781 (adding)
antilog	4.7781 = 60,000.

In the same way division is turned into subtraction by expressing the numbers as logarithms. For example,

$$\frac{10{,}000}{100} = \frac{10^4}{10^2} = 10^{4-2} = 10^2 = 100$$

or by logarithms,

log 10,000 = 4.0
log 100 = 2.0

2.0 (subtracting)
antilog 2.0 = 100

Powers and Roots.—A number is easily raised to a power with the help of logarithms. The logarithm of the number is multiplied by the exponent and the antilogarithm of the product gives the desired answer. Likewise roots may be extracted by dividing the logarithm by the index of the required root. Several illustrations follow:

$$(100)^3 = (10^2)^3 = 10^{2\times 3} = 10^6 = 1{,}000{,}000$$

or by logarithms,

log 100 = 2
2 × 3 = 6 (multiplying the logarithm by the exponent)
antilog 6 = 1,000,000

$$3^3 = 3 \times 3 \times 3 = 27$$

or by logarithms,

log 3 = 0.4771
3 × 0.4771 = 1.4313
antilog 1.4313 = 27

$16^{2.3} = ?$
log 16 = 1.2041
× 2.3

2.7694
antilog 2.7694 = 588

$$\sqrt[3]{27} \text{ or } 27^{1/3} = ?$$
$$\log 27 = 1.4313$$
$$1/3 \times 1.4313 = 0.4771$$
$$\text{antilog } 0.4771 = 3.0$$

Other Properties of Logarithms.—In some operations it is convenient to remember that $\log_{10} 10 = 1$. This fact follows at once from the equation $10^1 = 10$ and the definition of logarithms. In general, the logarithm to any base of that base is 1. If a represents any constant,

$$\log_a a = 1.$$

In other problems it is important to remember that the logarithm to any base of 1 is zero. That is,

$$10^0 = 1 \text{ or } \log_{10} 1 = 0; \text{ also, } a^0 = 1, \text{ and } \log_a 1 = 0$$

Logarithm Tables.—A four-place logarithm table with proportional parts for interpolation is given on the back cover of this book. Such a table gives an accuracy up to one part in ten thousand or 0.01 of 1 per cent. For many purposes the slide rule described in the next chapter is as good and considerably more convenient. When greater accuracy is required, a five-place table is used. A graphical table to five places has been published recently, which is preferable to the older tables.[1] This table eliminates the numerical interpolation and is specially valuable for looking up antilogarithms.

NEGATIVE LOGARITHMS

Change of Sign of the Mantissa.—The use of logarithms for ordinary multiplication and division offers no particular difficulty, but in chemical calculations where the logarithm of a number must be used in algebraic equations, certain situations arise which prove to be puzzling for the student. Most of the difficulties are concerned with the negative sign. It is important to remember that logarithms are exponents, and as such they may be either positive or negative. The mantissae as given in ordinary logarithm tables, however, are always positive. The decimal point between the characteristic and the mantissa always signifies addition, as it does in ordinary decimals.

[1] LACROIX and RAGOT, "A Graphic Table Combining Logarithms and Antilogarithms," The Macmillan Company, New York, 1925.

When the minus sign is written above the characteristic, it applies only to the characteristic, but when it is written in front of the characteristic it applies to both mantissa and characteristic.

For example, the logarithm $\bar{2}.4$ = (minus two) + (four-tenths), but the logarithm -2.4 = (minus two) + (minus four-tenths).

By carrying out the addition suggested in the first case, the logarithm $(\bar{2}) + (0.4)$ becomes -1.6.

The form $\bar{2}.4$ is used for simple addition or subtraction of logarithms or looking up antilogarithms in tables, but the form -1.6 is used in operations involving the multiplication or division of logarithms. The first form $(\bar{2}.4)$ is not convenient for these operations because it is a peculiar number which is part positive and part negative. It is therefore changed over into a number (-1.6) which is all negative, by simply carrying out the addition signified by the decimal point.

On the other hand, the second form (-1.6) is not suitable for use with logarithm tables and it must be converted into the form which has a positive mantissa. This conversion may be effected by adding 1 to the mantissa and subtracting 1 from the characteristic, and in this way the value of the logarithm is unchanged. The operation is as follows,

$$-1.6 = (-1) \text{ plus } (-0.6) =$$
$$[(-1) - 1.00] \text{ plus } [(-0.6) + 1.00] =$$
$$-2 \text{ plus } +0.4 = \bar{2}.4$$

$(-2) + (.4)$ may then be written as $\bar{2}.4$, and the antilogarithm corresponding to the mantissa 0.4 is found in the logarithm tables.

When a mantissa is subtracted from 1, the new number which results is called the co-logarithm. Although tables of co-logarithms save this subtraction process they are not sufficiently common to warrant their discussion here.

In ordinary numerical calculations these difficulties of negative logarithms are conveniently eliminated by adding and subtracting 10 instead of 1, but in certain physical chemical calculations this artifice cannot be used. Experience has shown that students who are dependent entirely on the simplified system which adds and subtracts 10, are sure to have difficulty in physical chemistry, particularly with the calculation of hydrogen ion concentrations expressed in pH values. The logarithm of 0.5 may be written as $\bar{1}.6990$ or as $9.6990 - 10$. In either case it must be changed to -0.301 for algebraic operations, in the first case by adding 0.6990 to (-1) and in the second case by adding 9.6990 to (-10).

Illustrations.—The following calculations illustrate various ways in which negative logarithms may occur.

1. It is desired to find the value of y in the equation $10^{-0.3010} = y$. By the definition of logarithms, y is the antilogarithm of -0.3010. But logarithm tables do not contain negative mantissae and so -0.3010 is converted into a logarithm with a positive mantissa by adding and subtracting 1. $-0.3010 = (-0.3010) + (1.00) - (1) = (-1) + (0.6990) = \bar{1}.6990$. The antilogarithm of $\bar{1}.6990$ is 0.500. In other words $10.^{-0.3010} = 0.500$.

2. It is desired to find the value of x in the equation $10^x = 0.05$. $x = \log_{10} 0.05$ and logarithm tables show that $\bar{2}.6990$ is the logarithm of 0.05. The equation is not written $10.^{\bar{2}.6990} = 0.05$, but it is written $10^{-1.3010} = 0.05$ to avoid the mixed exponent which is part positive and part negative. It could be written also, $10^{-2} \times 10^{+0.6990}$. The expression $10^{-1.3010}$ is equivalent to the expression $1/10^{1.3010}$.

3. In extracting the cube root of 0.05, it is awkward to divide $\bar{2}.6990$ by 3, so $\bar{2}.6990$ is converted into -1.3010 by adding together 0.6990 and (-2). Then dividing -1.3010 by 3 gives -0.4337. But the number -0.4337 cannot be found in logarithm tables for such tables contain only positive mantissae. It is necessary to convert -0.4337 into a number with a positive mantissa by adding 1 and subtracting 1. Then $-0.4337 = \bar{1}.5663$ and the antilogarithm of $\bar{1}.5663$ is 0.3684. Therefore 0.3684 is the cube root of 0.05.

4. It is desired to find the value of y in the equation $e^{-5} = y$.

$$\begin{aligned}
\log_{10} e^{-5} &= \log_{10} y \text{ (taking } \log_{10} \text{ of both sides)} \\
-5 \log_{10} e &= \log_{10} y \\
\log_{10} y &= (-5)(0.4343) = -2.1715 = \bar{3}.8285 \\
y &= \text{antilog } \bar{3}.8285 = 0.006737
\end{aligned}$$

5. If $e^{-x} = 0.2982$, $x = ?$

$$\begin{aligned}
\log_{10} e^{-x} &= \log_{10} 0.2982 \\
-x \log_{10} e &= \bar{1}.4745 \\
-x(0.4343) &= -0.5255 \\
x &= \frac{0.5255}{0.4343} = 1.21
\end{aligned}$$

6. $\sqrt[5]{\dfrac{0.025}{8.50}} = ?$

$$\begin{aligned} \log 0.025 &= \bar{2}.3979 \\ \log 8.50 &= 0.9294 \\ &\overline{\bar{3}.4685 \text{ (subtracting)}} \end{aligned}$$

$$\bar{3}.4685 = (-3) + (0.4685) = -2.5315$$

$$\frac{-2.5315}{5} = -0.5063 = \bar{1}.4937$$

$$\text{antilog } \bar{1}.4937 = 0.3117$$

7. Logarithms of negative numbers are imaginary since no real exponent of 10 can yield a negative number. Nevertheless negative numbers may be used in logarithmic calculations by assuming them to be positive and applying the proper sign to the resulting answer. For example, if it is desired to extract the fifth root of −0.025/8.50 the problem is solved as shown for 0.025/8.50 in the preceding problem. The answer is 0.3117 and it is given the sign −0.3117 since the odd root of a negative number must be negative.

8. The following calculation occurs in the determination of the solubility of a salt by electromotive force measurements. It seems to be difficult for students.

$$0.0592 \log \frac{0.0093}{x} = 0.3984. \quad x = ?$$

$$\log 0.0093 - \log x = \frac{0.3984}{0.0592} = 6.73$$

$$\log x = \bar{3}.9685 - 6.73 = -2.0315 - 6.73 = -8.7615 = \bar{9}.2385$$

$$x = \text{antilog } \bar{9}.2385 = 1.732 \times 10^{-9}$$

The negative logarithm may be avoided entirely by carrying out the division after the operations on the logarithm are completed.

$$0.0592 \log \frac{0.0093}{x} = 0.3984. \quad x = ?$$

$$\text{if } z = \frac{0.0093}{x}, \text{ then } 0.0592 \log z = 0.3984$$

$$\log z = \frac{0.3984}{0.0592} = 6.73$$

$$z = \text{antilog } 6.73 = 5.370 \times 10^{6}$$

$$z = \frac{0.0093}{x} = 5.370 \times 10^{6}$$

$$x = \frac{9.3 \times 10^{-3}}{5.370 \times 10^{6}} = 1.732 \times 10^{-9}$$

Exercises

1. (*a*) $\log_{10} 1{,}234 = ?$ *Ans.:* 3.09132. (*b*) $\log 0.001234 = ?$
2. (*a*) $\text{antilog}_{10} 3.2468 = ?$ *Ans.:* 1765.2. (*b*) $\text{antilog } 4.2468 = ?$
3. What are the logarithms of the following numbers?
   (*a*) 250. *Ans.:* 2.39794. (*b*) 250,000.
   (*c*) 0.25. *Ans.:* $\bar{1}.39794$. (*d*) 0.00025.
4. What are the antilogarithms of the following logarithms?
   (*a*) 0.4567 *Ans.:* 2.8622 (*b*) $\underline{4}.567$
   (*c*) 456.700 *Ans.:* 5.01×10^{456} (*d*) $\overline{45}.670$
5. (*a*) $10^x = 932$. $x = ?$ *Ans.:* 2.96942. (*b*) $10^x = 52$. $x = ?$
6. (*a*) $10^{4.70} = x$. $x = ?$ *Ans.:* 5.012×10^4. (*b*) $10^{0.120} = x$. $x = ?$
7. (*a*) If $\log_{10} 1.234 = 0.09132$, $\log_e 1.234 = ?$ *Ans.:* 0.2103.
   (*b*) $\log_e 3.247 = ?$ (*c*) $\log_e 22.4 = ?$ (*d*) $\log_e 360 = ?$
8. (*a*) If $\log_e 25.66 = 3.2450$, $\log_{10} 25.66 = ?$ *Ans.:* 1.409.
   (*b*) If $\log_e 6.31 = 1.844$, $\log_{10} 6.31 = ?$
9. $2468 \times 0.002345 \div 49.23 = ?$ *Ans.:* 0.1176.
10. $14.92 \times 0.1927 \div 7.31 = ?$
11. (*a*) $9.3^3 = ?$ *Ans.:* 804.34. (*b*) $3{,}625^6 = ?$
12. (*a*) $3.198^5 = ?$ *Ans.:* 334.5. (*b*) $3.198^{52} = ?$
13. (*a*) $\sqrt[5]{11.7} = ?$ *Ans.:* 1.63. (*b*) $\sqrt[3]{(2.831)^2} = ?$
14. (*a*) $e^{5.2} = ?$ *Ans.:* 181.1. (*b*) $e^{52} = ?$
15. (*a*) $e^x = 250$. $x = ?$ *Ans.:* 5.5215. (*b*) $e^x = 1.70$. $x = ?$
16. (*a*) $10^x = 0.052$. $x = ?$ *Ans.:* −1.284. (*b*) $10^x = 0.52$. $x = ?$
17. (*a*) $10^x = 0.035$. $x = ?$ *Ans.:* −1.4559. (*b*) $10^x = 0.00932$. $x = ?$
18. (*a*) $10^{-3.42} = x$. $x = ?$ *Ans.:* 0.00038. (*b*) $10^{-4.7} = x$. $x = ?$
19. (*a*) $10^{-0.75} = x$. $x = ?$ *Ans.:* 0.1778. (*b*) $10^{-0.083} = x$. $x = ?$
20. (*a*) $10^x = -5.2$. $x = ?$ *Ans.:* Impossible. (*b*) $10^x = -0.039$. $x = ?$
21. (*a*) $\log_e 0.05 = ?$ *Ans.:* $\bar{3}.0038$. (*b*) $\log_e 0.007 = ?$
22. (*a*) $\log_e 0.63 = ?$ *Ans.:* $\bar{1}.5380$. (*b*) $\log_e 0.0051 = ?$
23. (*a*) $e^{-2.3} = ?$ *Ans.:* 0.1002. (*b*) $e^{-0.04} = ?$
24. (*a*) $e^{-0.013} = ?$ *Ans.:* 0.9871. (*b*) $e^{-2.01} = ?$
25. (*a*) $3.198^{-0.52} = ?$ *Ans.:* 0.5464. (*b*) $3.198^{-5.0} = ?$
26. (*a*) $5^{-12} = ?$ *Ans.:* 4.09×10^{-9}. (*b*) $12^{-5} = ?$
27. $\sqrt[5]{\dfrac{-0.0123}{5.678}} = ?$ *Ans.:* −0.2933.
28. $\sqrt[3]{\dfrac{-0.019}{5.62}} = ?$
29. $0.700 = 0.02 \log \dfrac{0.01}{x}$. $x = ?$ *Ans.:* 1×10^{-37}
30. $0.450 = 0.012 \log \dfrac{0.009}{x}$. $x = ?$

CHAPTER III

THE SLIDE RULE

The slide rule is a mechanical logarithm table. It is perhaps the greatest time saver ever invented for the calculations of the physical scientist. The student is strongly urged to purchase one ($8 to $10) for he will find his efficiency greatly increased not only in connection with this book but in later courses and throughout his professional career. A cheap slide rule for rough work may be purchased for $1.

A 10-in. slide rule gives an accuracy of about 1 part in 1,000, or $\frac{1}{10}$ of 1 per cent. This accuracy is frequently sufficient, and in fact many of the physical and chemical constants are not known with a greater accuracy. If an accuracy up to 1 part in 10,000 is required the four-place logarithm table on the back cover of the book is convenient. When an accuracy of 1 part in 100,000 is needed, a five-place table or the graphical table mentioned on page 11 is recommended.

A great many different mathematical operations can be carried out with the slide rule, and there are so many scales that the beginner is bewildered. The chemist, however, uses the instrument chiefly for multiplication and division, for logarithms and for the extraction of square roots; and these operations are very simple. The discussions in this book will be limited to these operations, involving only the *C* and *D* scales, the *L* scale, and the *A* scale. These scales are shown in Fig. 1*a*. The proper use of the other scales may be learned from the manufacturers' manual which is sold with the slide rule.

In purchasing a slide rule it is essential to choose one which has the *L* scale, with its uniform divisions, such for example, as the "Polyphase Duplex" slide rule.

MULTIPLICATION AND DIVISION

The *C* and *D* scales are used for multiplication and division. It will be observed in Fig. 1 that they are logarithmic scales,

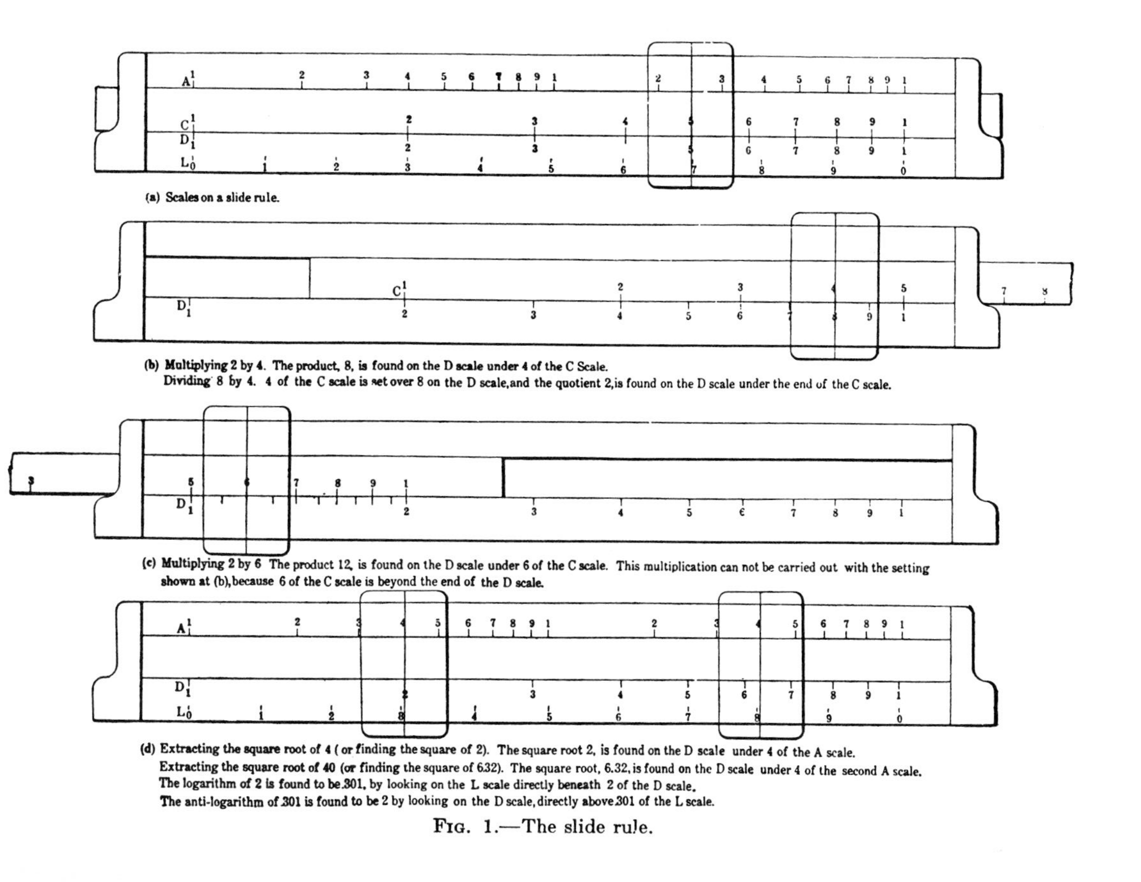

(a) Scales on a slide rule.

(b) Multiplying 2 by 4. The product, 8, is found on the D scale under 4 of the C Scale.
Dividing 8 by 4. 4 of the C scale is set over 8 on the D scale, and the quotient 2, is found on the D scale under the end of the C scale.

(c) Multiplying 2 by 6 The product 12, is found on the D scale under 6 of the C scale. This multiplication can not be carried out with the setting shown at (b), because 6 of the C scale is beyond the end of the D scale.

(d) Extracting the square root of 4 (or finding the square of 2). The square root 2, is found on the D scale under 4 of the A scale.
Extracting the square root of 40 (or finding the square of 6.32). The square root, 6.32, is found on the D scale under 4 of the second A scale.
The logarithm of 2 is found to be .301, by looking on the L scale directly beneath 2 of the D scale.
The anti-logarithm of .301 is found to be 2 by looking on the D scale, directly above .301 of the L scale.

FIG. 1.—The slide rule.

with the divisions becoming smaller and smaller as the numbers increase. The divisions correspond to the logarithms of the numbers from 1 to 10, but the numbers themselves rather than the logarithms of the numbers are engraved on the scale.

The *C* scale slides back and forth in grooves, so that a section of the *C* scale can be added to a section of the *D* scale just as 3 may be added to 5 by measuring off 5 cm. on one ruler and 3 cm. on a second ruler. If the divisions of the scales represented numbers, the operation would correspond to the addition of the numbers, but since they represent the logarithms of the numbers the addition of the scales corresponds to the multiplication of the numbers.

The advantage of the slide rule is brought out clearly in the following example where the operation is compared with logarithmic calculation and illustrated in Fig. 1*b*.

MULTIPLYING 2×4

Log Tables		Slide Rule
Log 2 is found in tables	= 0.30103	1 at the left end of the *C* scale is set over 2 on the *D* scale.
Log 4 is found in tables	= 0.60206	
Log 2 and log 4 are added	= 0.90309	The vertical hair line of the runner is set over 4 of the *C* scale.
Antilog 0.90309 is found in tables	= 8.000	On the *D* scale under the hair line the product is found to be 8.

The second operation is much shorter than the first. Finding the numbers on the scales is equivalent to looking up the logarithms in tables for the scale is marked to give the logarithms of the numbers. The setting of the *C* scale and the sliding of the runner are equivalent to the more tedious operation of arithmetical addition. Thanks to the logarithmic scale the reading of the number on the *D* scale under the runner is equivalent to looking up the antilogarithm in tables.

Division is just as simple as multiplication, the only difference being that the logarithms of the numbers are subtracted rather than added. The divisor is found on the *C* scale and set over the number on the *D* scale which is to be divided. The quotient is then found under the 1 at the end of the scale. In Fig. 1*b* the hair line gives the setting for dividing 8 by 4.

There is a little technique involved in the efficient setting of a slide rule. The greatest loss of time in multiplication occurs when one guesses wrongly as to which end of the C scale to use. If the product gives a number which is off the edge of the D scale, it is necessary to reset the scale with the other end of the C scale over the multiplier. For example, if it is desired to multiply 2×6, and the left end of the C scale is set over 2 on the D scale; the number under 6 on the C scale cannot be found, since 6 projects beyond the edge of the scale as shown in Fig. 1*b*. It is necessary then to shift the C scale to the other end so that the 1 at the right of the C scale lies above the 2 as shown in Fig. 1*c*. Then the answer 12 is found under the 6. If the first figures of the numbers give a product less than 10 the left end of the C scale is used, but if more than 10, the right end of the C scale is used. After a little practice one becomes proficient in deciding which end of the scale to use.

It is always advantageous to carry out a calculation with the smallest number of settings for then the mechanical errors are reduced to a minimum. With this end in view an operation involving several multiplications and divisions should be carried out in such a way that the division and multiplication follow each other alternately. If all the numbers of the numerator are multiplied together first, more settings are required.

The first part of the C and D scales is divided into hundredths, while the last part of the scale is divided into fiftieths or twentieths. In reading the slide rule it is absolutely necessary to note, always, which divisions are being read. This confusing difference is made necessary by the logarithmic scale which continuously gives narrower divisions as the numbers increase on the scale. On the 10-in. slide rule each division represents 0.01 between 1 and 2, 0.02 between 2 and 4, and 0.05 between 4 and 10. Probably most of the mistakes made in slide-rule calculations are due to a failure to distinguish between these different divisions.

The slide rule does not point off decimals and it is best not to bother with the decimal point until the calculation is finished. Usually the numbers may be rounded off so that a rough mental calculation will give the position of the decimal point.

For example in the calculation $\dfrac{3.21 \times 4.5 \times 90.1}{195}$ the slide rule gives 6,685. For fixing the decimal point, this may be written $\frac{3 \times 5 \times 100}{200} = \frac{1,500}{200} = 7.5$. The correct answer then is 6.685.

It is a great help in keeping track of the decimal point to adopt the system suggested on page 1, whereby all the numbers are written down with one figure at the left of the decimal point, and the true decimal point is indicated by multiplying by 10 with the proper exponent.

Logarithms

A slide rule can be used directly as a three-place logarithm table by simply sliding the runner along the *D* and *L* scales. A number on the *L* scale is the logarithm of the corresponding number on the *D* scale. In the same way antilogarithms are read off on the *D* scale. In Fig. 1*d* it is shown that 0.301 on the *L* scale is the logarithm of 2.00 on the *D* scale.

Although the slide rule renders the use of logarithms unnecessary, there are many cases in physical chemistry where the logarithms themselves must be multiplied or divided. In such cases the logarithms are found on the *L* scale and then transferred to the *D* scale.

In the example $3.05 \times \log 5.2$, log 5.2 is found by placing the runner over 5.2 on the *D* scale and reading 0.716 on the *L* scale. 0.716 is then found on the *D* scale and multiplied by 3.05 with the help of the *C* scale, giving 2.184. In the example $3.05 \times \log 520$, the characteristic, 2, must be added to the logarithm 0.7160 as found on the slide rule. The *D* scale is then set on 2.7160 instead of on 0.716. In case the characteristic is negative the operation is more complicated and the rules discussed on page 12 must be observed.

Any of the problems in roots and powers suggested in the preceding chapter may be solved with the slide rule, using the *L* and *D* scales for logarithms and antilogarithms and the *C* and *D* scales for multiplication and division.

For example $(3.19)^{5.23} = x$. $x = ?$

$$\log x = 5.23 \log 3.19$$

$$x = \text{antilog } 5.23 \log 3.19$$

Under 3.19 on the *D* scale is found 0.504 on the *L* scale. Then $\log 3.19 = 0.504$.

0.504 (on the D scale) $\times$ 5.23 (on the C scale) = 2.638 (on the D scale).

Above 0.638 on the L scale is found 435 on the D scale, and the characteristic 2 shows that the number is in the hundreds.

435 then is the antilog of 2.638. This is the number x which is required.

Roots and Powers

Squares and square roots are easily determined with the help of the A scale at the top. This scale is divided into two equal logarithmic scales, each complete from 1 to 10. These scales then are equivalent to the logarithmic scale (the D scale) multiplied by 2 (p. 10) and the numbers engraved on the A scales correspond then to the squares of the numbers on the D scale. On the slide rule, the vertical line of the slider is placed over the number on the D scale and the square is read off directly on the A scale.

The extraction of square roots is accomplished just as readily by passing from the A scale to the D scale with the help of the vertical line of the runner. This operation is equivalent to dividing the logarithm by 2, which in turn is equivalent to extracting the square root (p. 10).

There is a complication in extracting square root, because there are two scales to choose from. A rough mental calculation will show which one to use. The number under 4 of the first scale is 2 and the number under 4 of the second scale is 6.322 as shown in Fig 1*d*. Obviously 2 is the square root of 4 and 6.322 is the square root of 40. Likewise 20 is the square root of 400 and 63.22 is the square root of 4,000. Also 0.2 is the square root of 0.04 and 0.6322 is the square root of 0.4000.

The rule is that the left-hand scale is used for numbers greater than 1 which contain an odd number of figures and the right hand scale is used for numbers which contain an even number of figures at the left of the decimal point. In case the number is less than 1, an odd number of zeros between the decimal point and the first significant figure puts the number on the left-hand scale and an even number of zeros puts it on the right-hand scale.

Exercises

The following operations should be repeated two or three times with the slide rule. A comparison of the answers then gives an idea of the accuracy obtainable.

1. $2 \times 4 \times 3 = ?$ *Ans.:* 24.
2. $9.1 \times 3.6 \times 4.5 \times 7.1 = ?$
3. $8.23 \times 0.017 \times 365 = ?$ *Ans.:* 51.05.
4. $24 \div 6 = ?$
5. $\dfrac{6.41 \times 3.27}{9.18 \times 7.98} = ?$ *Ans.:* 0.2861.
6. $\dfrac{123 \times 161 \times 290 \times 981 \times 418}{260 \times 396 \times 656 \times 610} = ?$
7. $\log 3.215 = ?$ *Ans.:* 0.5072.
8. $\log 9.62 = ?$
9. $\text{antilog } 0.3270 = ?$ *Ans.:* 2.123.
10. $\text{antilog } 0.907 = ?$
11. $4.26 \times \log 7.96 = ?$ *Ans.:* 3.838.
12. $\dfrac{9.18 \times 3.26 \times \log 1.234}{1.375 \times 8.8} = ?$
13. What are the reciprocals of the following numbers?
    (*a*) 30. *Ans.:* 0.0333. (*b*) 0.025. (*c*) 317. (*d*) 0.04032.
14. What are the reciprocals of the following numbers?
    (*a*) 40. *Ans.:* 0.025. (*b*) 0.0360. (*c*) 9,025. (*d*) 41.90.
15. What are the squares of the following numbers?
    (*a*) 9. *Ans.:* 81. (*b*) 30. (*c*) 725. (*d*) 0.0412.
16. What are the squares of the following numbers?
    (*a*) 80. *Ans.:* 6,400. (*b*) 3.15. (*c*) 692. (*d*) 0.0013.
17. What are the square roots of the following numbers?
    (*a*) 81. *Ans.:* 9. (*b*) 8.1. (*c*) 340. (*d*) 0.034.
18. What are the square roots of the following numbers?
    (*a*) 49. *Ans.:* 7. (*b*) 4.9. (*c*) 721. (*d*) 0.0571.
19. (*a*) $4.2^{4.2} = ?$ *Ans.:* 415. (*b*) $306^{6.1} = ?$
20. (*a*) $96^{4.8} = ?$ *Ans.:* 3.27×10^9. (*b*) $71^{6.3} = ?$
21. $3.017:9.63::5.28:x.$ $x = ?$ *Ans.:* 16.85.
22. $41.7:0.027::x:543.$ $x = ?$
23. $\dfrac{273}{393} \times \dfrac{741}{760} \times \dfrac{35.7}{22{,}400} \times 6.06 \times 10^{23} = ?$ *Ans.:* 6.505×10^{20}.
24. $\dfrac{273}{298} \times \dfrac{750}{760} \times \dfrac{103}{22{,}400} \times 6.06 \times 10^{23} = ?$

CHAPTER IV

GRAPHICAL REPRESENTATION OF EQUATIONS

For every curve arising in physical chemistry there is an equation, and conversely for every equation in x and y for which there are real roots there is a corresponding curve. The curve gives at once the relation between variables and it is a great aid to clear thinking, but the equation usually gives the greater accuracy. It is the function of analytical geometry to show the relation between graphs and equations; and it is the purpose of the next three chapters to familiarize the student with the mathematical significance of simple graphs. Further examples will be found throughout the book. The plotting of curves or the graphing of equations is comparatively simple. The reverse process of finding the equation which corresponds to a given curve is more difficult, except in the case of straight lines. Discussions of this technique will be reserved for a later chapter (p. 227).

Rectangular Coordinates

The quantitative study of the influence of one variable on another constitutes one of the commonest operations in science. The influence of pressure on the volume of a gas, or of temperature on the solubility of a salt, or of time on the course of a chemical reaction are all examples. The simplest way to represent the relation between such variables is to draw a picture. Certain conventions have been adopted, which render this graphical representation extremely simple. The magnitude of one variable is plotted along a horizontal scale and the other variable is measured by a vertical scale.

This plotting on rectangular coordinates is familiar to everyone. The horizontal axis is called the X axis, and values along it are known as abscissas. The vertical axis is called the Y axis, and values along it are known as ordinates. Of course, any other letters may be used instead of X and Y; P and V, or S and T for example. The intersection of the two axes is called the origin. The positive and negative values of x and y are shown in Fig. 2. At the right of the Y axis, x has positive values and at the left it has negative values. Above the X axis, y is positive and below the X axis, it is negative.

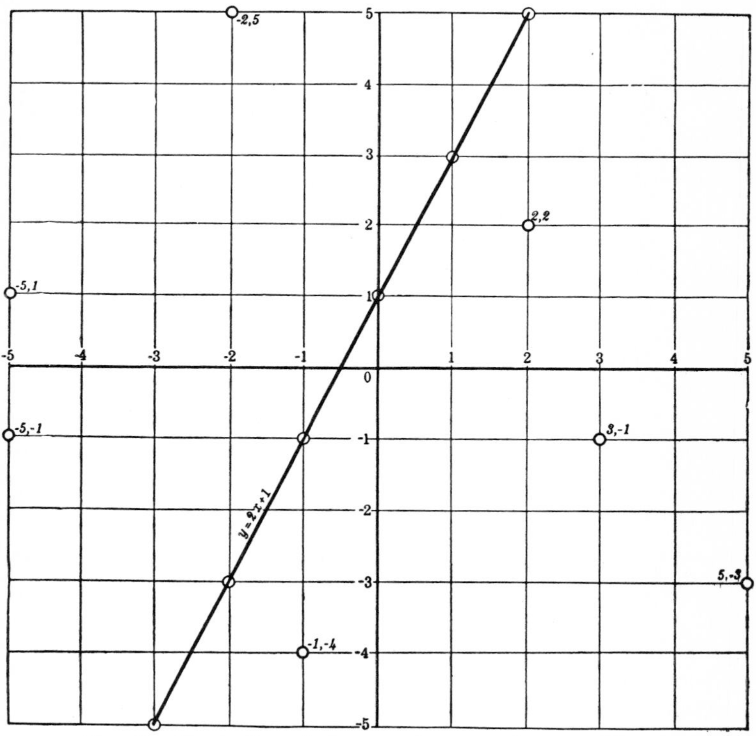

Fig. 2.—Illustration of the use of rectangular co-ordinates.

A point is defined by its two coordinates. The x value, or abscissa, is the distance out from the Y axis and it always comes first. The y value, or ordinate, is the distance above or below the X axis and it always comes second. Thus the point 5, -3, is the point which has 5 for its abscissa and -3 for its ordinate. Various other points are labeled in Fig. 2.

A line or curve is made up of an infinite number of points, each of which has coordinates which satisfy a given equation. For example, every point on the line $y = 2x + 1$ is such that the value of y is obtained by multiplying the x value by 2 and adding 1, and every point therefore satisfies the equation $y = 2x + 1$.

Straight Lines

Straight lines are the simplest graphs. They represent equations of the first degree, that is, algebraic equations containing no exponents other than 1. Conversely, an algebraic equation of the first degree gives a straight line. The most general form of the equation which gives a straight line is

$$Kx + K'y + K'' = 0$$

where K, K', and K'' are constants. This equation can be changed by transposing, and by dividing through by K',

$$y = -\frac{K}{K'}x - \frac{K''}{K'}$$

The equation can be changed into a simpler form as follows,

$$y = mx + b$$

where m and b are constants ($m = -K/K'$ and $b = -K''/K'$). Any equation of the first degree can be changed into this form, where y, the dependent variable, stands alone. This is the fundamental equation of the straight line.

m is defined as the slope of the line, or the ratio of the vertical increase to the corresponding horizontal increase. When the line is extended so that the abscissa increases from x to $x + \Delta x$, the ordinate increases also from y to $y + \Delta y$ as shown in Fig. 3*a*. Then

$$m = \frac{\Delta y}{\Delta x}$$

and the slope may be determined by measuring the vertical and horizontal increases.

The ratio $\Delta y/\Delta x$ also gives the tangent of the angle θ, which the line makes with the X axis, and so

$$m = \tan \theta$$

The slope m may then be determined by measuring the angle between the line and the X axis with a protractor as shown in Fig. 3b and looking up the tangent of the angle in tables (p. 279).

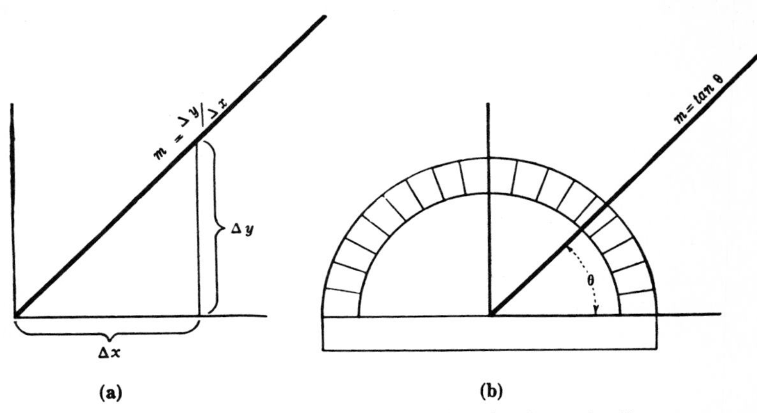

FIG. 3.—Two methods for measuring the slope of a line.

(a) The slope of the line = m = $\Delta y/\Delta x$
(b) The slope of the line = m = $\tan\theta$. The angle θ is measured with a protractor and $\tan\theta$ is obtained from tables of tangents.

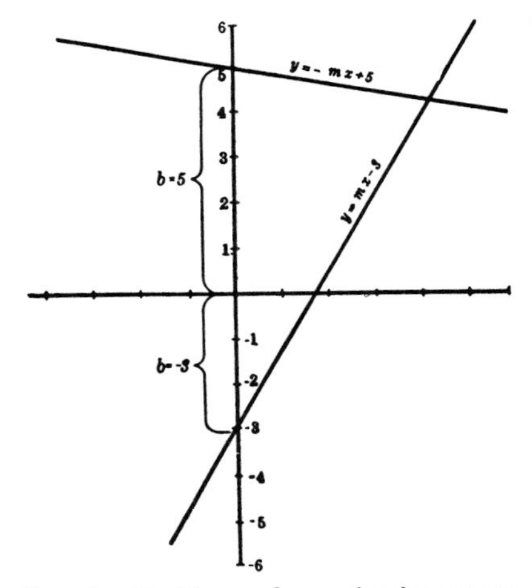

FIG. 4.—Positive and negative intercepts.

The intercept on the Y axis (usually called b) is equal to the distance along the Y axis between the origin and the intersection of the line with the Y axis. Graphically speaking, x is zero along the Y axis and algebraically speaking the term mx drops out when x is zero. The equation $y = mx + b$ then reduces to $y = b$. In other words b is the value which y assumes when x becomes zero. The significance of b is illustrated in the two lines of Fig. 4, where both positive and negative values are shown.

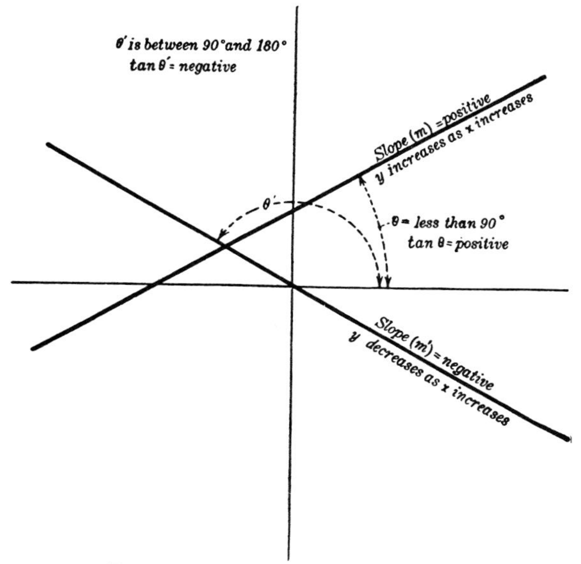

FIG. 5.—Positive and negative slopes.

When the slope of the line (m) is positive y increases as x increases but when it is negative y decreases as x increases as shown in Fig. 5. A decrease is an increase in a negative direction. It will be remembered that an acute angle has a positive tangent and that an obtuse angle (90° to 180°) has a negative tangent. The sign of the slope agrees then with the sign of the tangent as shown in Fig. 5.

THE SLOPE AND INTERCEPT EQUATION

The slope and intercept equation for the straight line, $y = mx + b$, has been discussed in the preceding section. Some special cases of this equation are shown in Fig. 6. When b is zero ($y = -2x$, for example) the line passes through the origin, for at the origin $y = 0$.

When the coefficient of x is one, ($y = x - 2$) the line makes an angle of 45° with the X axis, since the tangent of 45° is 1. When the x term is missing ($y = 4$) the line is horizontal, for the slope must be zero to make the x term drop out and only a horizontal line can give a tangent of zero. Also when the y

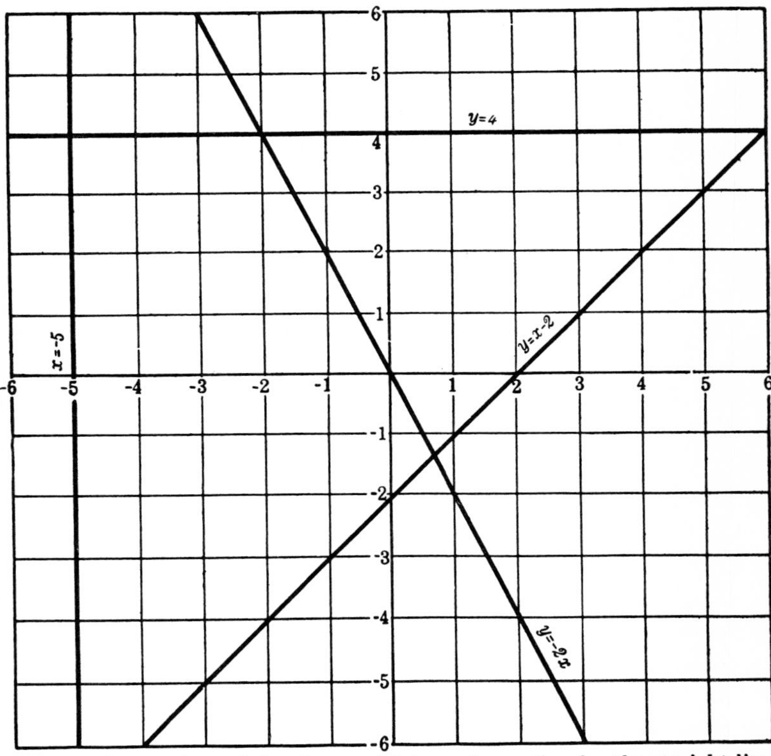

FIG. 6.—Illustrations of the slope and intercept equation for the straight line. $y = mx + b$.

term is missing ($x = -5$), x can have but one value (it must be a constant) and only a vertical line can give a line such that the x value is always the same, no matter what the value of y.

Through the application of the intercept equation many of the properties of the line may be defined at a glance. For example, the equation

$$y = 2x + 3$$

represents a straight line because the equation is of the first degree. Since b is 3 it crosses the Y axis at 3. Since m is 2, $\Delta y/\Delta x$ is 2, and any increase upward along the Y axis is twice as great as the increase outward along the X axis. The line starts in such a direction that the angle between it and the X axis is the angle whose tangent is 2, *i.e.*, the angle is approximately $63\frac{1}{2}°$.

These facts deduced from the equation are sufficient to determine the line. The graph is shown in Fig. 7. Like all graphs, it can be plotted also by substituting various numbers for x and solving for the corresponding values of y, as in the accompanying figure.

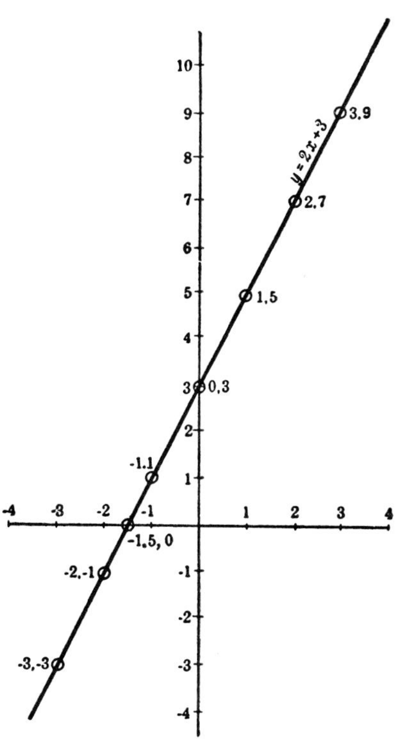

FIG. 7.—Determination of a line by plotting points. All points which correspond to solutions of the equation, $y = 2x + 3$, fall on the straight line.

The equation $2y + 8x - 10 = 0$, illustrates the important point that any equation must be transformed into the standard form with y, alone, on the left-hand side of the equation. In this equation 8 is *not* the slope and -10 is *not* the intercept. The slope is -4 and the intercept is $+5$. It is necessary to divide through the original equation by the coefficient of y (2 in this case), and to transfer the x term and the constant to the other side of the equation, thus

$$y = \frac{-8x}{2} + \frac{10}{2} = -4x + 5.$$

The intercept formula is convenient also in determining the equation of a straight line when the angle is known.

For example, what is the equation of the line which crosses the y axis at -2 and makes an angle of 30° with the X axis? The tangent of 30° is 0.577. The equation for the line is $y = 0.577x - 2$.

Again, what is the equation of the line which passes through the origin and makes an angle of 120° with the X axis? The angle 120° puts the line in the second quadrant (between 90° and 180°) where the ordinate is positive but the abscissa is negative and the tangent is therefore negative. The line must slope so that y increases as x decreases.

Tan 120° = $-$ tan (180° $-$ 120°) = $-$ tan 60 = -1.732. (p. 280). The equation is $y = -1.732x$.

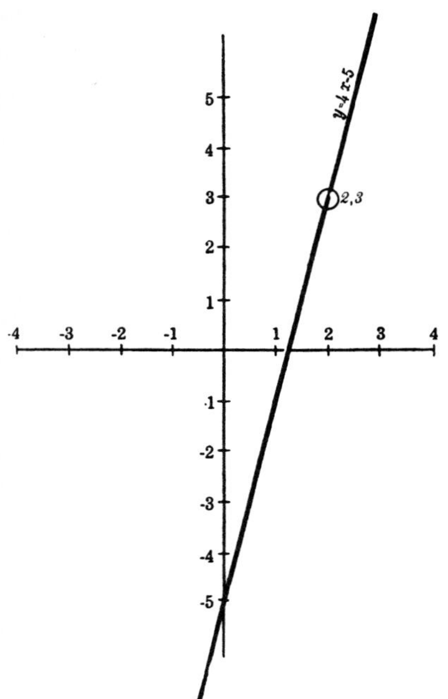

FIG. 8.—Graph of the equation which passes through the point 2, 3 and has a slope of 4.

The Slope and One-point Equation

The slope and one-point equation of the straight line is

$$(y - y_1) = m(x - x_1)$$

where x_1 and y_1 are the coordinates of a specified point.

The general equation given in the preceding section is $y = mx + b$, and since it is true for all points it must be true also for the special case at the point x_1, y_1, giving

$$y_1 = mx_1 + b$$

Subtraction of the special equation from the general equation gives the desired equation, as follows

$$\begin{array}{rcl} y &=& mx + b \\ -(y_1 &=& mx_1 + b) \\ \hline (y - y_1) &=& m(x - x_1) \end{array}$$

The use of this equation may be illustrated by finding the equation for the straight line which passes through the point 2, 3, and has a slope of 4. Since the specified point is 2, 3, it follows that $x_1 = 2$ and $y_1 = 3$. Substituting into the equation

$$y - 3 = 4(x - 2) = 4x - 8$$
$$y = 4x - 8 + 3 = 4x - 5$$

The line is shown in Fig. 8. It passes through the point 2, 3 and has a slope of 4.

In another example, a line making an angle of 135° with the X axis passes through the point 5, − 2. What is the equation of the line?

The tangent of the angle is -1, since $\tan 135° = -\tan(180° - 135°) = -\tan 45° = -1$

$$y - (-2) = -1(x - 5)$$

The equation of the line is

$$y = -x + 3.$$

The Two-point Equation

The two-point equation of the straight line is given by the equation

$$y - y_1 = \frac{y_2 - y_1}{x_2 - x_1}(x - x_1)$$

where x_1 and y_1 are the coordinates for one point, and x_2 and y_2 are the coordinates for a second point.

At the second point, $y_2 = mx_2 + b$
At the first point, $y_1 = mx_1 + b$

$$y_2 - y_1 = m(x_2 - x_1) \quad \text{(subtracting)}$$
$$m = \frac{y_2 - y_1}{x_2 - x_1}. \quad \text{(dividing by } x_2 - x_1\text{)}$$

Substituting the value of m into the slope and one-point equation, $y - y_1 = m(x - x_1)$,

$$y - y_1 = \frac{y_2 - y_1}{x_2 - x_1}(x - x_1)$$

This equation is the most convenient for finding the equation for the straight line which passes through two given points

For example, it is desired to find the equation of the straight line which passes through the two points 2, 3 and 5, 15.

$$x_1 = 2 \text{ and } y_1 = 3; \quad x_2 = 5 \text{ and } y_2 = 15$$

$$y - 3 = \left(\frac{+15 - 3}{5 - 2}\right)(x - 2) = \frac{12}{3}(x - 2) = 4(x - 2)$$

$$y - 3 = 4(x - 2)$$

$$y = 4x - 5$$

It is immaterial which point is considered to be x_1y_1 and which is x_2y_2. For example, the preceding equation could have been written

$$y - 15 = \frac{3 - 15}{2 - 5}(x - 5) = \frac{-12}{-3}(x - 5) = 4x - 20$$

$$y = 4x - 5, \text{ as before.}$$

In another illustration of the two-point equation, it is desired to find the equation for the line determined by the two points, -2, -300 and 5,000, -10. These points are so far apart that they cannot be graphed accurately on coordinate paper. The equation may be determined with perfect accuracy, however, as follows:

$$y - (-300) = \left[\frac{-10 - (-300)}{5{,}000 - (-2)}\right](x + 2) = \frac{290}{5{,}002}(x + 2)$$

$$y + 300 = 0.058(x + 2)$$

$$y = 0.058x - 299.884$$

Parallel Lines

Two straight lines are parallel if they have the same slope. If the slopes or tangents m and m' are equal, the corresponding angles θ and θ' must be equal, and if the lines make the same angles with the X axis they must be parallel to each other.

The lines $y = 2x$, and $y = 2x + 10$, and $2y = 4x - 500$ are all parallel since they have the same slope, or the same coefficient of x.

Likewise, $y = \dfrac{-x}{5} + 10$, $y = \dfrac{-x}{5} - 20$, and $y = \dfrac{-x}{5} + c$ are all parallel.

Points of Intersection

The point of intersection of two or more lines has coordinates which satisfy the equations of each of the intersecting lines.

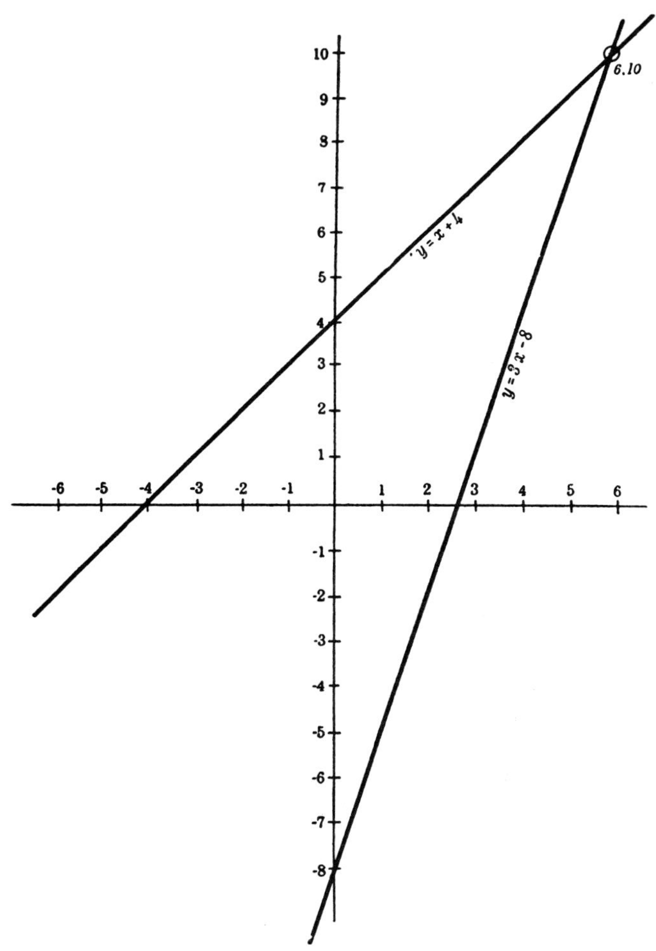

FIG. 9.—Point of intersection of the equations $y = x + 4$ and $y = 3x - 8$.

The point of intersection may be determined, then, by solving the equations of the lines simultaneously.

For example, at what point do the two lines $y = x + 4$ and $y = 3x - 8$ intersect? Solving the equations simultaneously:

$$\begin{array}{r} y = x + 4 \\ -(y = 3x - 8) \\ \hline 0 = -2x + 12 \\ x = 6 \end{array}$$

When

$$x = 6,\ y = 6 + 4 = 10$$

The two lines intersect where $x = 6$ and $y = 10$. When $x = 6$, $y = x + 4 = 10$, and $y = 3x - 8 = 10$. These coordinates satisfy both equations, and they are the only values which satisfy both. The lines are shown in Fig. 9.

In a second example it is desired to find which of the four lines given below intersect at one point, and what the coordinates of this point are:

$$\begin{aligned} y &= 2x \\ y &= -3x + 10 \\ 3y &= -2x - 3 \\ 2y - x - 6 &= 0 \end{aligned}$$

The first, second, and fourth equations all give $y = 4$ when $x = 2$. This fact is determined by solving the equations simultaneously in pairs. When x is given the value of 2, the third equation does not become 4. It may be concluded then that all the lines except the one corresponding to the third equation intersect at the point 2, 4. In every problem of this type, it is a good plan to sketch the graphs, as a check.

The Distance between Two Points

The distance, S, between two points, x_1, y_1 and x_2, y_2 is given by the equation

$$S = \sqrt{(x_2 - x_1)^2 + (y_2 - y_1)^2}$$

This relation follows at once from the familiar rule of geometry that the square of the hypotenuse of a right-angled triangle is equal to the sum of the squares of the other two sides. Figure 10 makes the reason clear.

To illustrate this rule, it is desired to find the shortest distance, S, between the point 12, 10 and the point 6, 2, as shown in Fig. 10.

$$S = \sqrt{(12 - 6)^2 + (10 - 2)^2} =$$
$$\sqrt{6^2 + 8^2} = \sqrt{36 + 64} = \sqrt{100} = 10$$

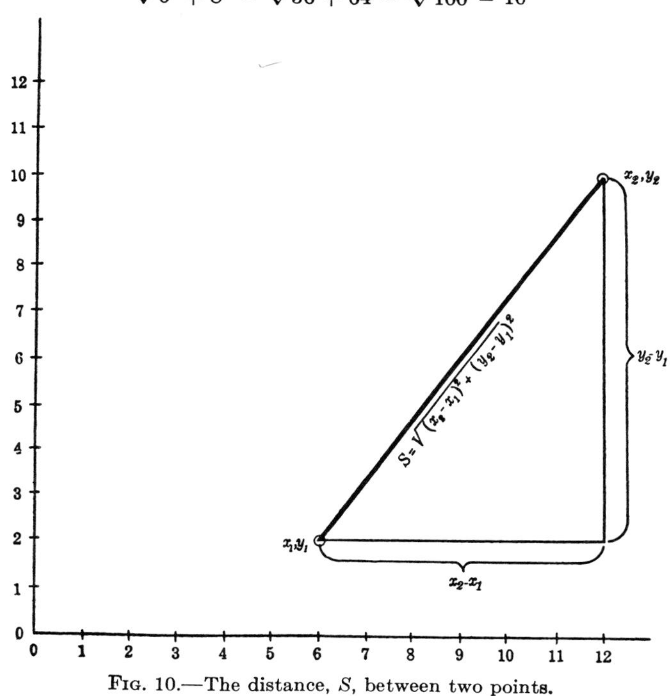

FIG. 10.—The distance, S, between two points.

It is immaterial which point is considered x_1, y_1 and which is considered x_2, y_2, for the answer is the same if the points are interchanged in the formula.

CHANGE OF ORIGIN

The reference point on any graph with rectangular coordinates is the origin, the point of intersection of the X and Y axes. If the axes are moved to new positions, each point has different coordinates, when referred to the new axes, than it had when referred to the old axes. The difference between the two sets of coordinates for any point is equal to the difference between the two origins. The point shown in Fig. 11, has the coordinates 8, 7. If the X axis is moved upward 5 units along the Y axis, the point then has an ordinate of 2. The new ordinate is obtained by subtracting from the old ordinate the distance upward through which the X axis was moved. If the Y axis is

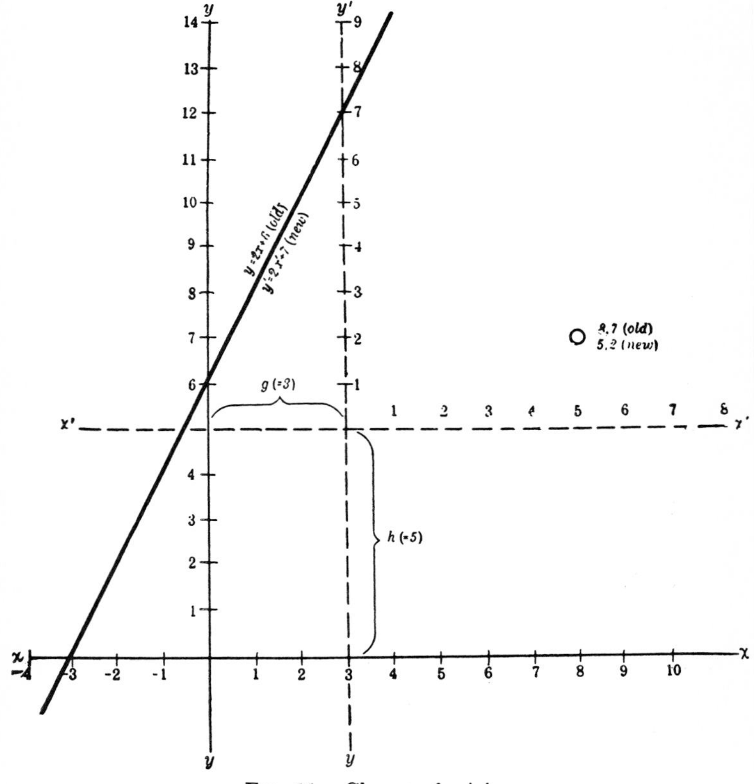

FIG. 11.—Change of origin.
The new axes $x'x'$ and $y'y'$ are represented by dotted lines. The new origin is located at the point g, h, when referred to the old axes. In this illustration $g = 3$ and $h = 5$.

moved +3 units along the X axis, the new abscissa of the point is then $8 - 3$ or 5. If both axes are moved, the point has the position 5, 2 on the new axes (the dotted lines).

In general, if g and h are the coordinates of the new origin with reference to the old axes, the new coordinates, x', y', of any point are related to the old coordinates, x and y by the equations

$$x' = x - g \text{ and } y' = y - h$$

or

$$x = x' + g \text{ and } y = y' + h.$$

A line has a different equation after the axes have been changed. Since the new axes are parallel to the old, the line must have the same slope, but the intercept on the Y axis will be different. The old equation may be transformed into the new by substituting into the old equation the equivalent of the old coordinates in terms of the new coordinates, as follows:

$$y = mx + b \text{ (old axes)}$$
$$(y' + h) = m(x' + g) + b \text{ (new axes)}$$

The line shown in Fig. 11 has the equation

$$y = 2x + 6, \text{ referred to the old axes.}$$

Referred to the new axes, where $h = 5$ and $g = 3$, the equation becomes

$$(y' + 5) = 2(x' + 3) + 6$$
$$y' = 2x' + 6 + 6 - 5 = 2x' + 7.$$

The correctness of the new equation can be checked readily by reference to the graph.

In another example, the axes are changed so that the new origin has the coordinates 5, -10 when referred to the old axes. What is the equation of the line, $y = -\frac{x}{5} + 2$ when referred to the new axes?

The new equation of the line is,

$$(y' + h) = -\frac{1}{5}(x' + g) + 2$$
$$y' - 10 = -\frac{1}{5}(x' + 5) + 2 = -\frac{x'}{5} - 1 + 2$$
$$y' = -\frac{x'}{5} + 11$$

Exercises

1. In Fig. 12, what are the coordinates of the points i, j, k, l, m, n?

   Ans.: $i = 4, 2.$

2. On coordinate paper the following points are to be located:

   (a) 5, 0. (b) 2, -5. (c) -5, 8. (d) -3, -1.

3. The lines corresponding to the following equations are to be sketched: (*a*) $y = x$. (*b*) $y = \frac{1}{2}x + 2$. (*c*) $y = -2x + 2$. (*d*) $y = 10x - 2$.

4. Given the following equations:
(*a*) $y = 2x + 8$. (*b*) $3y = -9x + 15$. (*c*) $2u + 4v + 6 = 0$. (*d*) $y = 5$.
Where does the line of each equation cut the Y axis, according to numerical calculations? What is the slope of each line? What angle does each line make with the X axis? (A table of tangents is given on p. 279.) The numerical answers should be checked with rough graphs.

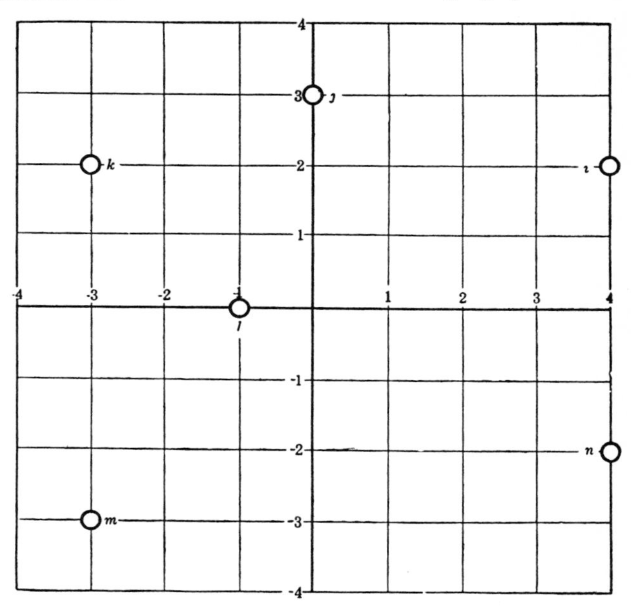

FIG. 12.—Points to be defined by x and y.

5. (*a*) What is the equation of the line which has a slope of 3 and intersects the ordinate at 5? *Ans.:* $y = 3x + 5$.
(*b*) What is the equation of the line which has a slope of $-\frac{1}{2}$ and intersects the ordinate at -10?

6. (*a*) What is the equation of the line which passes through the origin and makes an angle of 60° with the X axis? *Ans.:* $y = 1.732x$.
(*b*) What is the equation of the line which intersects the X axis at 5 and makes an angle of 150° with it? (It is well to check these equations with graphs.)

7. (*a*) What is the equation of the line which passes through the point 5, 10 and has a slope of 2? *Ans.:* $y = 2x$.
(*b*) What is the equation of the line which passes through the point -2, 6 and has a slope of $-\frac{1}{3}$?

8. (*a*) What is the equation of the line which passes through the point 3, −6 and makes an angle of 45° with the X axis?

   Ans.: $y = x - 9$.

   (*b*) What is the equation of the line which passes through the point −4, −4 and makes an angle of 120° with the X axis?

9. (*a*) What is the equation of the line which passes through the two points 1, 2 and −3, 4? *Ans.:* $x + 2y = 5$.

   (*b*) What is the equation of the line which passes through the two points 2, −5 and 10, −1?

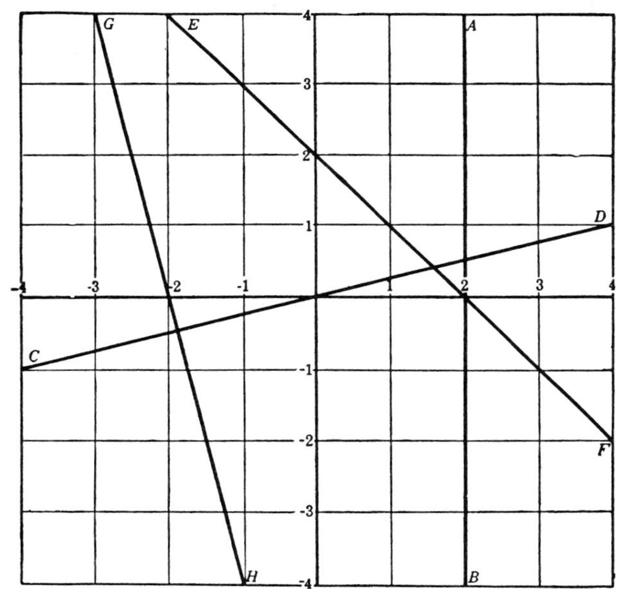

FIG. 13.—Straight lines to be defined with equations. **Prob. No. 11.**

10. (*a*) What is the equation of the line which passes through the two points 5, −250 and 1, 750? *Ans.:* $y = -250x + 1{,}000$.

    (*b*) What is the equation of the line which passes through the two points 3, 600 and 400, 1?

11. In Fig. 13, what are the equations of the lines AB, CD, and EF? What is the equation of the line that joins G and H?

12. Straight lines are drawn at random on rectangular coordinates and their equations are to be determined.

13. A straight line is drawn on rectangular coordinates and its equation is determined by three different equations.

14. (*a*) Given two lines $y - x = 0$, and $2y - 4x = 0$. What angle does each line make with the abscissa? By subtraction, what angle do the lines make with each other?

    Ans.: 45° and 63° 26′. 18° 26′ with each other.

(b) At what angle do the two lines $y = x/2 + 5$ and $y = 3x - 5$ intersect?

15. Two intersecting straight lines are drawn at random and the angle of intersection is measured and compared with the angle of intersection as calculated from the equations of the two lines.

16. It is to be shown by graphs that the following lines are parallel:

$$5y = 15x$$
$$y = 3x + 5$$
$$3y = 9x - 12$$

17. What is the equation of the line which passes through the point $-2, 3$ and is parallel to the line $4x + 2y = 535$? *Ans.:* $y = -2x - 1$.

18. What is the equation of the line which passes through the point 3, 5 and is parallel to the line $4x + y = 1$?

19. What is the equation of the line which passes through the point $-5, -5$, and is parallel to the line $y = 5{,}000$? *Ans.:* $y = -5$.

20. What are the co-ordinates of the point of intersection of the lines $x + y = 1$ and $y = x + 2$?

21. (a) Where do the equations $y = 5x - 1{,}700$ and $y = 2x + 1{,}400$ intersect?

(b) What angle does a line joining this point of intersection with the origin make with the x axis?

Ans.: (a) 1,033.3, 3,466.5. (b) 73° 24′ 44″.

22. It is to be shown that the three lines $y = 3x + 5$, $2y = -4x + 20$, and $y = x + 7$ intersect in a point (1, 8). Does the line $y = 2x + 5$ pass through this same point also?

23. Two intersecting straight lines are drawn at random and the point of intersection is determined graphically. The equation of each line is determined and the point of intersection calculated. What is the equation of the horizontal line which passes through this point of intersection?

24. What is the equation of the line which is steeper by 45° than the line $y = \frac{x}{3} - 5$ and intersects it at the point where $x = 9$.

25. What is the shortest distance between the points $-3, 2$ and $4, 5$? The calculations should be checked by plotting on coordinate paper and measuring. For extracting square roots, a slide rule should be used or the table on page 275. *Ans.:* 7.61.

26. What is the distance along a straight line between the the points $-2, 3$ and $3, -5$?

27. What is the equation of the line $y = 3x - 5$, after the reference axes have been moved so that the new origin corresponds to the point $-2, 5$ on the old axes?

28. What is the equation of the line formed by plotting the Fahrenheit temperature scale against the Centigrade scale (100° C. = 212° F. and 0° C. = 32° F.). What is the Centigrade temperature corresponding to 70° F., as obtained by interpolating on the graph and by substituting into the equation for the line? At what temperature do the Centigrade and the Fahrenheit scales register the same?

29. What is the equation which shows the relation between the absolute scale and the Fahrenheit scale? (212° F. = 373 K, 32° F. = 273 K.) What is the temperature of a furnace on the absolute scale if a pyrometer calibrated in terms of the Fahrenheit scale reads 1,500°? *Ans.:* 1,089° K.

30. Charles' Law states that the volume of a gas is proportional to the absolute temperature. At 273° K the volume of a gram molecule is 22.4 liters, and it increases $\frac{1}{273}$ of its volume for each degree rise in temperature. These facts are to be stated by a formula and represented by a graph. What is the volume of a gram molecule of a gas at 373° K? At 298° K?

31. The specific heat of liquid chloroform has the following values:

t° C.	20	30	40	50
Sp. Ht.	0.2311	0.2341	0.2371	0.2401

The data are plotted and the best straight line is drawn through the points. The equation of the line gives the specific heat of chloroform as a function of temperature (C.). What is its specific heat at 0°; at 36.52°?
Ans.: Sp. Ht. = $0.000304t + 0.225$. Sp. Ht.$_{0°}$ = 0.2250. Sp. Ht$_{36.52}$ = 0.2361.

32. The following temperatures are recorded while water is being heated in an open dish in the laboratory:

Time (minutes)	0	10	20	30
Temp. (C.)	35°	55°	75°	95°

What equation expresses the temperature as a function of time? What is the temperature after 15 min.? After 21.5 min.? After 35 min.?

(The extrapolation to 35 min. is incorrect because the water boils at 100° and does not rise beyond this temperature. Even short extrapolations are apt to be unsafe in physical or chemical phenomena. The extrapolated value has a significance. It is the temperature which the water would have, except for a superimposed effect—the boiling away of the liquid. The temperature could be realized by carrying out the experiment under increased pressure.)

33. The length, l, of a rod is given by the expression, $l = l_0 + l_0at$, where l_0 is the length at 0° C. and t is the temperature in centigrade degrees. a is a constant depending on the material of the rod. For copper a is 1.41×10^{-5} and for glass it is 8.33×10^{-6}. What is the length of a glass rod and a copper rod at 50°, if they were both 125 cm. long at 0° C? a is known as the coefficient of linear thermal expansion.

Ans.: Copper, 125.088 cm. Glass, 125.052 cm.

34. The coefficient of cubical expansion of a solid is three times as large as the coefficient of linear expansion. If the rods of Prob. 33 are 1 sq. cm. in cross sections at 0°, what is the volume of each at 50° C.?

CHAPTER V

GRAPHS OF EQUATIONS OF THE SECOND DEGREE

In the preceding chapter on straight lines y and x were never squared or cubed. In this chapter, the equations involving x^2 and y^2 and xy are discussed briefly. The graphs of these equations are not straight lines and the slope of the line changes as x changes. The most general equation of the second order is

$$Ax^2 + By^2 + Cx + Dy + Exy + F = 0$$

Frequently some of these terms become zero and it is rarely necessary to work with an equation containing all of them. Graphs corresponding to equations of this kind are called conic sections because they can all be produced by passing a plane through a cone in various ways.

The Circle

A circle is a curve, all points of which are equidistant from a point. It is drawn with a compass, which keeps the distance between the pencil point and the center of the circle always the same.

The simple equation for the circle is

$$x^2 + y^2 = r^2$$

A graph of the circle $x^2 + y^2 = 25$ is shown in Fig. 14.

Taking any point, A, on the circle and dropping a perpendicular to the X axis, it is evident that a right-angle triangle is formed, and therefore, $OA^2 = OB^2 + BA^2$. But OB and BA are the coordinates x and y of the point A, and OA is the radius, r, of the circle. Substituting $x^2 + y^2 = r^2$, the same relation is found to hold no matter where the point A is located on the circle.

This equation represents a circle whose center is on the origin. Usually the circle is placed in this position but if the center of the circle is not on the origin, a more general equation is necessary,

If the center is h units above the origin and g units to the right of the origin the equation becomes

$$(x - g)^2 + (y - h)^2 = r^2$$

as shown in Fig. 15.

$BA = y - h;\ OB = x - g,$ and $OA = r$

$(OB)^2 + (BA)^2 = (OA)^2$ (definition of the circle)

$(x - g)^2 + (y - h)^2 = r^2$ (substituting)

$(x^2 - 2gx + g^2) + (y^2 - 2hy + h^2) - r^2 = 0$ (expanding)

$x^2 + y^2 - 2gx - 2hy + (g^2 + h^2 - r^2) = 0$ (combining)

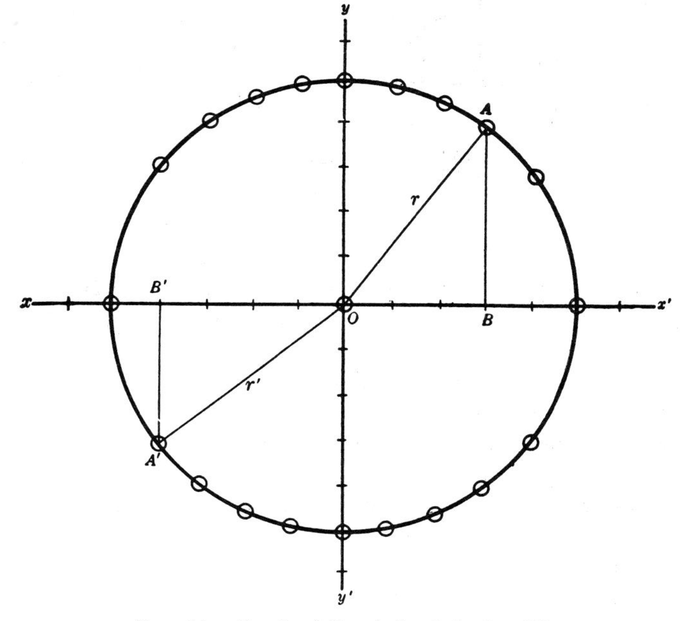

FIG. 14.—Graph of the circle $x^2 + y^2 = 25$.

x	±0	1	2	3	4	5	6
y	±5	4.9	4.6	4.0	3.0	0	imaginary

Since g, h, and r are constants the equation can be reduced to the form

$$x^2 + y^2 + Cx + Dy + F = 0$$

where

$$-2g = C,\ -2h = D, \text{ and } (g^2 + h^2 - r^2) = F$$

This is the general form of the circle, and whenever an equation has the same coefficient for x^2 and y^2 and contains no term in xy, the equation may be reduced to this form. If the x and y terms are missing the center of the circle is on the origin.

It is usually best in graphing an equation of a circle to determine the center and the radius. If the equation is solved for various values of x many of the values will be imaginary and consequently will not fall on the circle at all. Many trial calculations are necessary in this method of graphing; but the center and radius method enables one to draw the complete circle from a single calculation.

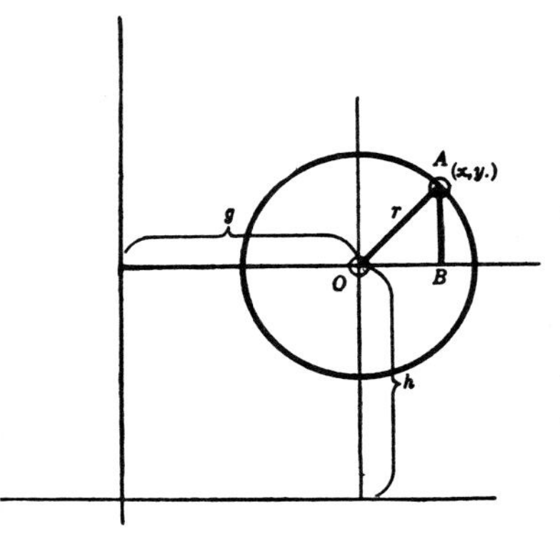

FIG. 15.—The equation of a circle which does not have its center on the origin is $(x - g)^2 + (y - h)^2 = r^2$.

The following problems illustrate the use of the equation of the circle:

What is the graph of the equation $x^2 + y^2 = 64$? The radius is the square root of 64, which is 8. Since there are no terms in x or y the center of the circle is at the origin. These two facts are sufficient to construct the graph with the aid of a compass.

Various points on the circumference may be located by solving the equation. Whenever a graph is to be sketched by solving the equation for several values of x and y, it is advisable to choose values which will render the calculation particularly simple. For example, the first values to be chosen are $x = 0$ and $y = 0$, for when $x = 0$, $y = \sqrt{64} = 8$ and -8. When $y = 0$, $x = \sqrt{64} = +8$ and -8. The intersections on the axes are given at once and four points are established. When x is greater than 8 or less than -8, y becomes equal to the square root of a negative number—in other words there is no real point on the circle beyond 8 or -8. Likewise there is no real point above 8 or below -8.

What is the equation of the circle which has its center at the point -5, $+10$, and has a radius of 4? A point on the circle has the abscissa x and the ordinate y, but when it is referred to the center of the circle the ordinates of the center must be subtracted and the abscissa becomes $x - (-5)$ and the ordinate becomes $y - 10$. The equation then is not $x^2 + y^2 = 16$ but

$$(x + 5)^2 + (y - 10)^2 = 16; \text{ or } x^2 + 10x + 25 + y^2 - 20y + 100 = 16,$$
$$x^2 + y^2 + 10x - 20y + 109 = 0.$$

Reversing the process it is possible to locate the center of the circle from the general equation, containing terms in x and y. Numbers are added to both sides of the equation so as to complete the squares. For example, 25 and 100 may be added to the equation $x^2 + y^2 + 10x - 20y + 109 = 0$, giving $(x + 5)^2 + (y - 10)^2 = (125 - 109)$ and the center is located at -5, $+10$ since $(x - g)^2 + (y - h)^2 = r^2$.

The Ellipse

An ellipse is a closed figure somewhat similar to a circle. An illustration of the ellipse is given in Fig. 16.

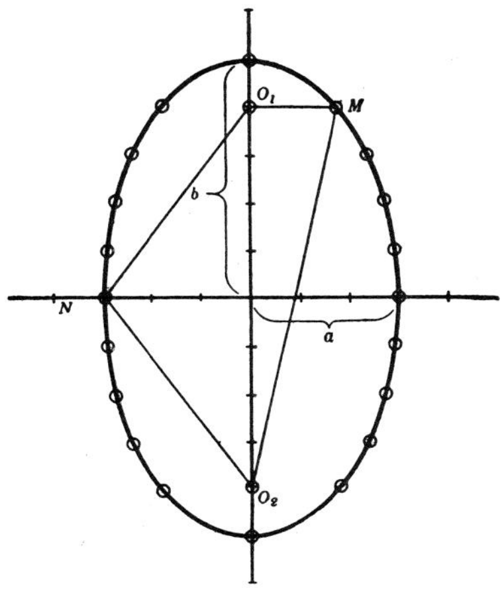

Fig. 16.—Graph of the ellipse $x^2/9 + y^2/25 = 1$.

y	± 0	1	2	3	4	5	6
x	± 3	2.94	2.75	2.40	1.80	0.00	imaginary

The planets revolve around the sun in elliptical orbits, and it is probable that the atoms of chemistry consist of similar systems with negative electrons revolving in elliptical orbits around positive nuclei. With the help of this assumption, it has been possible to calculate with extraordinary accuracy the position of the lines in the spectrum of hydrogen and helium.

Every point on an ellipse is situated so that the sum of the distances from the point to two given points is a constant. For example, $O_1M + O_2M = O_1N + O_2N$.

The equation for the simple ellipse is $\frac{x^2}{a^2} + \frac{y^2}{b^2} = 1$.

The horizontal axis is equal to $2a$ and the vertical is equal to $2b$. When a and b are equal, the axes become equal and the ellipse becomes a circle. The circle is in fact a special case of the ellipse.

The values of y and x may be determined by extracting square roots of the equation just given.

$$y = \pm \frac{b}{a}\sqrt{a^2 - x^2},$$

and

$$x = \pm \frac{a}{b}\sqrt{b^2 - y^2}$$

These equations show that for every value of x between a and $-a$ there is a positive and a negative value of y, and that there are two values of x for every value of y between b and $-b$.

If a term in y is included in the equation, the horizontal axis is not on the X axis. If a term in x is present the vertical axis of the ellipse does not coincide with the Y axis. If there is a term in xy, the axes of the ellipse are not parallel to the X and Y axes.

The Parabola

In a parabola the square of one variable is proportional to the first power of another variable; that is, $y^2 = kx$. When a projectile is thrown out horizontally from the top of a tower, a parabolic curve is described by the projectile because the horizontal distance traveled is proportional to the time, but the distance traveled downward is proportional to the square of the time. The best reflectors for searchlights are parabolic in shape, for all the light coming from a point of light at the "focus" is reflected in parallel beams. A liquid gives a parabolic depression when it is rotated rapidly.

A typical parabola is shown in Fig. 17. It represents the equation $y^2 = 8x$.

Each point on a parabola is equidistant from a point and a straight line. $FM = CM$, and $FN = ND$. The point, F, is called the focus and the line CD is called the directrix. This line is situated at

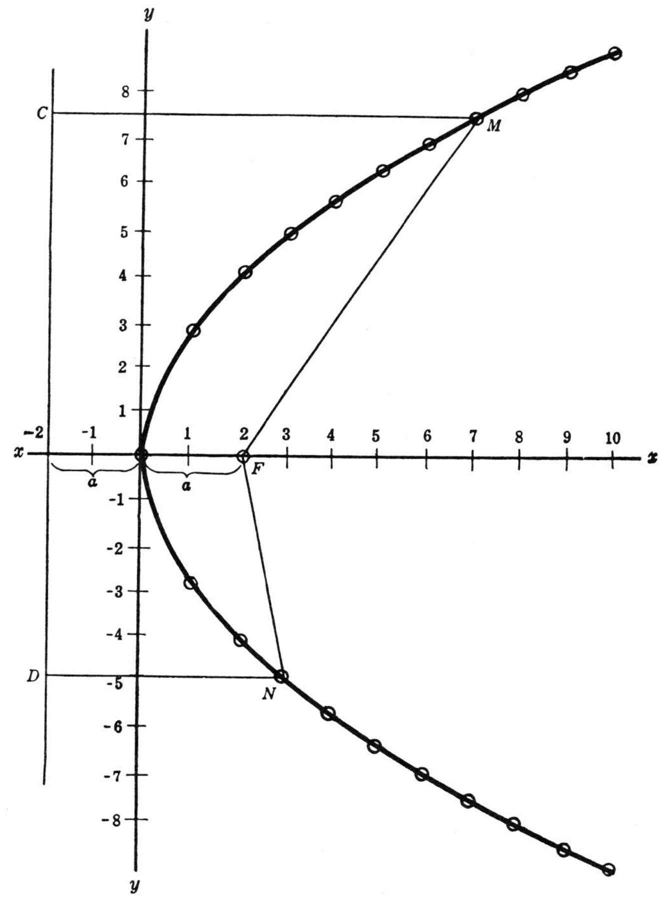

FIG. 17.—Graph of the parabola $y^2 = 8x$.

x	0	1	2	3	4	5	6	7	8	9	10
y	±0	2.8	4.0	4.9	5.7	6.3	6.9	7.5	8.0	8.5	8.9

the distance a from the origin, and the focus is situated at an equal distance on the other side of the origin. The distance a is one-fourth of the constant, k, defined by the equation $y^2 = kx$, and accordingly the equation of the parabola may be written $y^2 = 4ax$. The equation which is graphed in Fig. 17 may then be written $y^2 = 4 \times 2x$, and it is known that a is equal to 2 and that the directrix is situated two units at the left of the origin. It follows also that the focus is 2 units at the right of the origin.

Certain properties of the parabola should be pointed out. Since $y^2 = 4ax$ it follows that $y = \pm 2\sqrt{ax}$. The parabola, $y^2 = 4ax$, passes through the origin since y becomes zero when x is zero. The value of y is equal to $\sqrt{4a}$, when x is equal to 1. For every positive value of x there is one positive and one negative value of y. The part of the parabola which lies below the X axis is a mirror image of the part which lies above it.

The parabola $y = \pm 2\sqrt{ax}$ lies wholly at the right of the ordinate, since square roots of negative values of x are imaginary.

The equation $x^2 = 4ay$; or $x = \pm 2\sqrt{ay}$ gives a parabola similar to the one already discussed, except that it lies along the Y axis rather than along the X axis. It lies wholly above the X axis and the part at the left of the Y axis is a mirror image of the part at the right.

The term parabola is reserved for equations of the general type $y = kx^2$; but curves of the general type $y = kx^n$ where n is a positive number are called parabolic curves. The function $y = kx^3$ is known as the cubic parabola and the function $y = k\sqrt[2]{x^3}$ the semicubic parabola.

Nearly Straight Lines

In Fig. 17 it can be seen that the curve of the parabola becomes more and more like a straight line as the parabola is extended out from the origin. As x increases indefinitely y also increases indefinitely, and taking out different sections along the parabola one can find a considerable variety of curvature. Now there are many relations among physical-chemical phenomena which can be expressed by lines which are nearly straight lines but which have some curvature. The parabola is valuable in cases of this kind, for, by appropriate choice of numerical values it is often possible to find a section of the parabola which passes through the experimentally determined points. It is usually best in reporting the influence of one variable on another to express the relation in the form of an equation. This is more accurate and less clumsy than tables or graphs.

The general equation of a curve of this type which is nearly a straight line is

$$y = a + bx + cx^2$$

At first sight it may not be recognized that this equation is a parabola. It represents a parabola, the axes of which are not on the X and Y axes. The following derivation shows the relation between this equation and the simple parabola:

$y = kx^2$ represents a parabola on the origin.

When the axes of the parabola are removed from the origin by g and h the equation becomes

$$(y - h) = k(x - g)^2$$
$$(y - h) = k(x^2 - 2gx + g^2) = kx^2 - 2kgx + kg^2$$
$$y = (kg^2 + h) - 2kgx + kx^2$$
$$y = a + bx + cx^2$$

Where $a = (kg^2 + h)$, $b = (-2kg)$ and $c = k$.
a, b, c, g, h, and k are all constants.

A few examples of these parabolic equations, taken from physical chemical data are given below. The graph of the first one is given in Fig. 18.

1. Specific heat (s) of ethyl alcohol as a function of temperature (t).

$$s = 0.5068 + 0.00286t + 5.4 \times 10^{-6}t^2$$

2. Specific heat, s, of mercury as a function of temperature, t.

$$s = 0.003458 - 0.000{,}01074t + 0.000{,}000{,}0385t^2$$

3. The length l, of copper as a function of temperature, t,

$$l_t = l_0(1.67 \times 10^{-5}t + 4.03 \times 10^{-9}t^2)$$

l_0 is the length measured at 0°

When these equations are given it is easy to find the value of the dependent variable corresponding to any specified value of the independent variable. For example, when the temperature is 50° C. the specific heat of alcohol is

$$s = 0.5068 + 0.00286 \times 50 + 5.4 \times 10^{-6} \times 2{,}500$$
$$= 0.5068 + 0.1430 + 0.0135 = 0.6633$$

Likewise it is possible to find the independent variable corresponding to any specified value of the dependent variable.

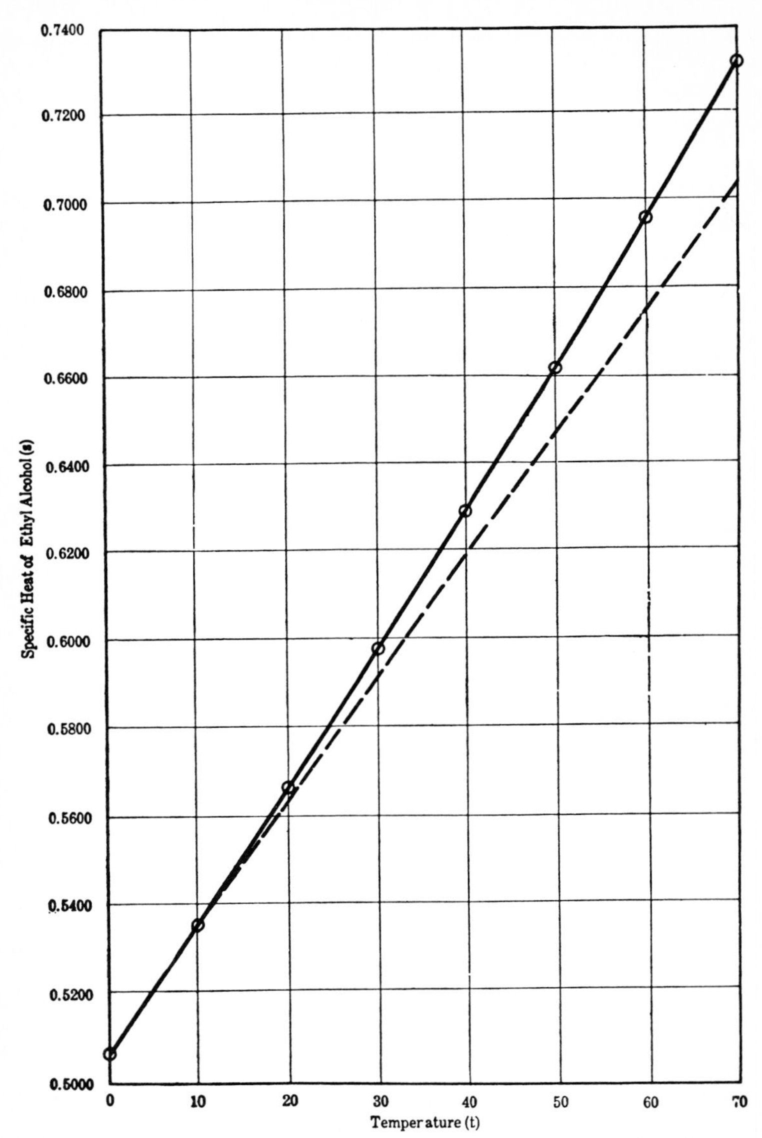

FIG. 18.—The specific heat, s, of ethyl alcohol plotted as a function of temperature. An example of a nearly straight line fitted by a parabolic equation. $s = 0.5068 + 0.00286t + 5.4 \times 10^{-6}t^2$. The straight line, $s = 0.5068 + .00286t$, is shown with the broken line.

This operation involves the solving of a quadratic equation or it may be carried out approximately by graphical interpolation.

Taking the same example of ethyl alcohol, it is desired to find the temperature, t, at which the specific heat, s, is 0.5100.

$$0.5100 = 0.5068 + 0.00286t + 5.4 \times 10^{-6}t^2.$$
$$5.4 \times 10^{-6}t^2 + 0.00286t - 0.0032 = 0$$

Solving by the formula for a quadratic equation given on page 266

$$t = \frac{-0.00286 \pm \sqrt{0.00286^2 - 4 \times 5.4 \times 10^{-6} \times -(0.0032)}}{2 \times 5.4 \times 10^{-6}}$$

$$t = \frac{-0.00286 + \sqrt{8.18 \times 10^{-6} + 6.9 \times 10^{-8}}}{1.08 \times 10^{-5}} = \frac{-0.002860 + 0.002872}{1.08 \times 10^{-5}}$$

$t = 1.11°$. (The negative root has no physical significance.)

The finding of the constants a, b, and c for a set of experimental data is somewhat tedious. Suggestions for evaluating these constants will be given later on page 229.

In case it is found impossible to express experimental values with sufficient accuracy by an equation of the type $y = a + bx + cx^2$, an equation of the type $y = a + bx + cx^2 + dx^3$ may be tried. The value of y can be determined for any corresponding value of x, as before, but it is not easy to solve the equation for the value of x corresponding to a given value of y, except by graphical interpolation. It is rarely necessary to carry the equation out farther to a term in x^4.

The volume of mercury, v, is given as a function of temperature, t, in the following cubic equation:

$$v_t = v_0(1 + 1.8055 \times 10^{-4}t + 1.244 \times 10^{-8}t^2 + 2.54 \times 10^{-11}t^3)$$

v_0 is the volume measured at 0°.

To illustrate the use of the equation, 10.000 cc. of mercury is measured out at 0° and it is desired to find the volume occupied at 25°.

$$v_{25} = v_0(1 + 1.8055 \times 10^{-4} \times 25 + 1.244 \times 10^{-8} \times 25^2 + 2.54 \times 10^{-11} \times 25^3)$$

$$v_{25} = 10.000(1 + 0.0045138 + 7.7 \times 10^{-6} + 4 \times 10^{-7} = 10.04522 \text{ cc.}$$

The Hyperbola

Equations of the type $y = x^n$ are parabolic equations if n is a positive number, and they are *hyperbolic* equations if n is a

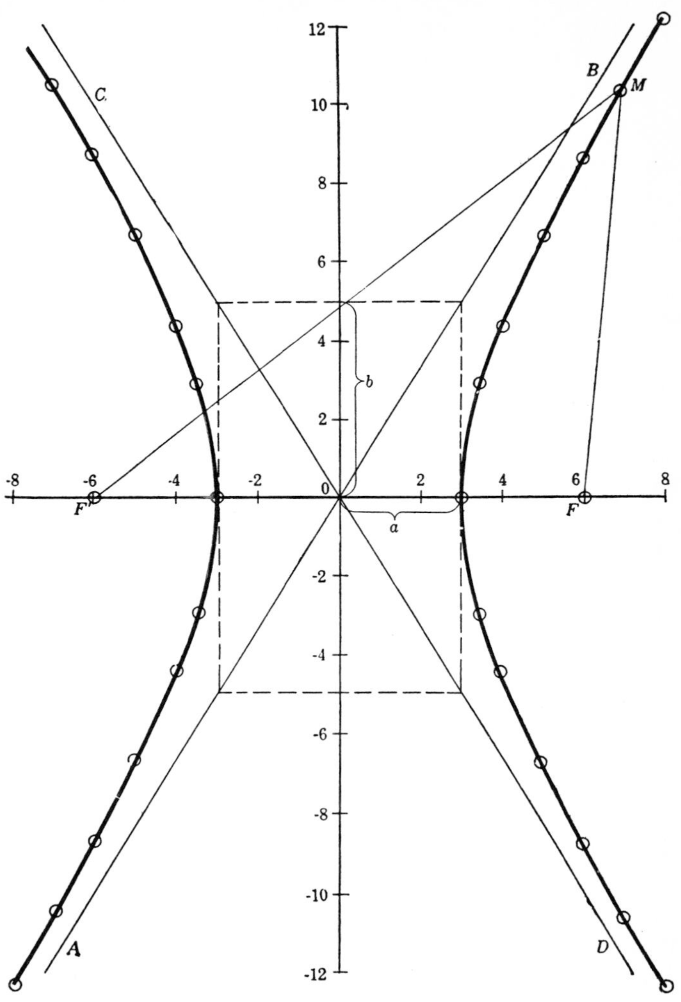

FIG. 19.—Graph of the hyperbola. $\frac{x^2}{9} - \frac{y^2}{25} = 1.$

$$y = \pm \frac{b}{a} \sqrt{x^2 - a^2}$$

x..............	±	1	2	3	3.5	4	5	6	7	8
y..............	±	imaginary	imaginary	0.0	2.9	4.4	6.7	8.7	10.5	12.3

negative number. The simple equation of the hyperbola with its axes on the X and Y axes is

$$\frac{x^2}{a^2} - \frac{y^2}{b^2} = 1$$

The hyperbola is generated by a point moving in such a way that the difference of its distance from two fixed points is always constant. An example of the hyperbola is given in Fig. 19, where the equation $\frac{x^2}{9} - \frac{y^2}{25} = 1$ is plotted.

The equation may be transposed into a form which can be solved readily for y, as follows

$$b^2x^2 - a^2y^2 = a^2b^2$$
$$a^2y^2 = b^2x^2 - a^2b^2$$
$$= b^2(x^2 - a^2)$$
$$y = \pm\frac{b}{a}\sqrt{x^2 - a^2}$$

and

$$x = \pm\frac{a}{b}\sqrt{y^2 + b^2}$$

The hyperbola is symmetrical with respect to both axes. For each value of x there is a positive and negative value of y and in the same way, for each value of y there are two values of x which are equal but opposite in sign.

The two foci of the hyperbola are shown at F and F'. M is any point on the hyperbola and the difference between $F'M$ and FM has a certain fixed value. This difference is a constant no matter where the point M is placed on the hyperbola.

As the hyperbola extends out from the origin it approaches a straight line called an asymptote, but it never coincides exactly with this straight line. If carried far enough from the origin it can be made to approach the straight line within any desired degree of accuracy. The asymptotes are shown at AB and CD in Fig. 19.

Reciprocals

There are several phenomena in physical chemistry in which one factor is inversely proportional to another. Expressed in another way, one factor is directly proportional to the reciprocal of the other. The mathematical relation is given by the equations

$$y \propto \left(\frac{1}{x}\right); \text{ or } y = k\left(\frac{1}{x}\right); \text{ or } yx = k.$$

Among the familiar illustrations is the law of Boyle, according to which the volume of a gas is inversely proportional to the

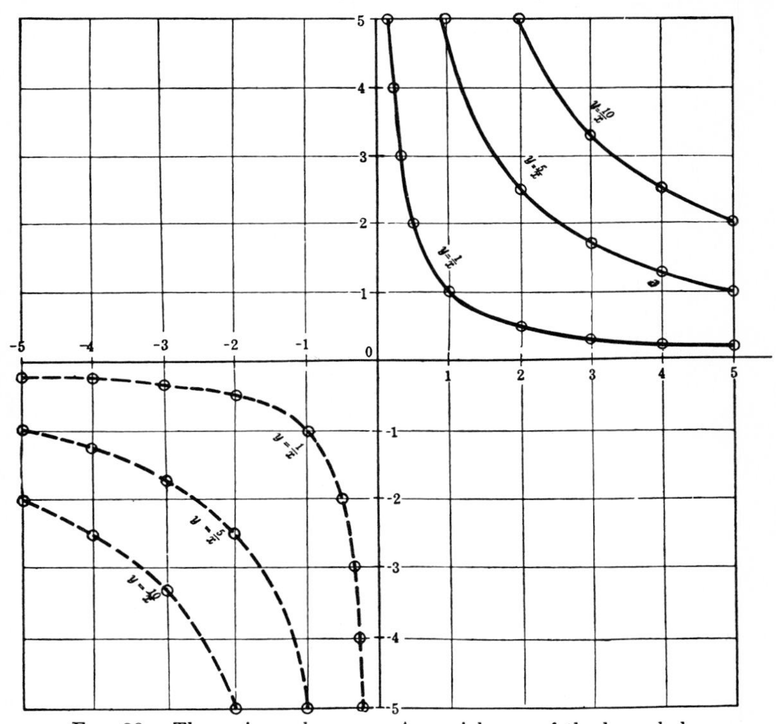

Fig. 20.—The reciprocal curve. A special case of the hyperbola.

x	0.0	0.2	0.25	0.33	0.5	1	2	3	4	5
$10/x$	∞						5.0	3.3	2.5	2.0
$5/x$	∞					5	2.5	1.7	1.3	1.0
$1/x$	∞	5.0	4.0	3.0	2.0	1	0.5	0.33	0.25	0.20

pressure. Also, the depth of solution in a colorimeter is inversely proportional to the concentration of the colored material. The conductance of a solution is equal to the reciprocal of its resistance. The product of the concentration of the hydrogen ions and the hydroxyl ions in an aqueous solution is a constant. Representative equations of this type are graphed in Fig. 20.

It will be noticed that in all cases y approaches infinity as x approaches zero, and that x approaches infinity as y approaches zero.

If $xy = k$, it is also true that $(-x)(-y) = k$. For every graph shown in Fig. 20 then, there is a corresponding graph in the third quadrant, shown by dotted lines. Although the dotted curves are just as important mathematically as those in the first quadrant they often lose significance when applied to physical problems, for a gas cannot have negative volume nor a solution negative concentration.

It will be observed that the curves of Fig. 20 bear a striking resemblance to the curve of Fig. 19. As a matter of fact the curves of Fig. 20 are actually hyperbolas. The reciprocal curve, $xy = k$ is a special case of the hyperbola, in which $a = b$, and the hyperbola is referred to its asymptotes as axes.

When $a = b$, the equation of the hyperbola, $\frac{x^2}{a^2} - \frac{y^2}{b^2} = 1$, becomes $x^2 - y^2 = a^2$.

The tangent of the angle between the asymptote and the X axis is given by b/a, and when $b = a$ the tangent becomes 1. The tangent of 45° is 1. Likewise, the other asymptote which has a tangent of $b/-a$ makes an angle of 135° with the abscissa when $a = b$. The angle between the two asymptotes is therefore 90° when $a = b$, and such an hyperbola is called a right angle hyperbola.

If now the axes are rotated through an angle of $-45°$ they will coincide with the asymptotes, and the hyperbola will be referred to its own asymptotes as axes. It can be shown by a theorem not given here that when the axes are rotated through $-45°$ the equation $x^2 - y^2 = a^2$ becomes $xy = a^2/2$ or $xy = k$.

There are other curves of this general reciprocal type such as $x^2y = 1$, $xy^3 = 1$, and $x^3y = 1$, etc. They are classed as hyperbolic curves and they approach the X and Y axes asymptotically.

Axes with Unequal Scales

In mathematical studies, the scales of the X axis and the Y axis are almost always equal but very often in plotting physical or chemical relations the two factors are so different in magnitude that this plan cannot be followed. A unit of distance along X may represent different values of x, according to the arbitrary scale used. It must be borne in mind, always, that the relation between variables is given by the scales which are assigned to the abscissa and the ordinate, rather than by the number of squares which can be counted out from the origin.

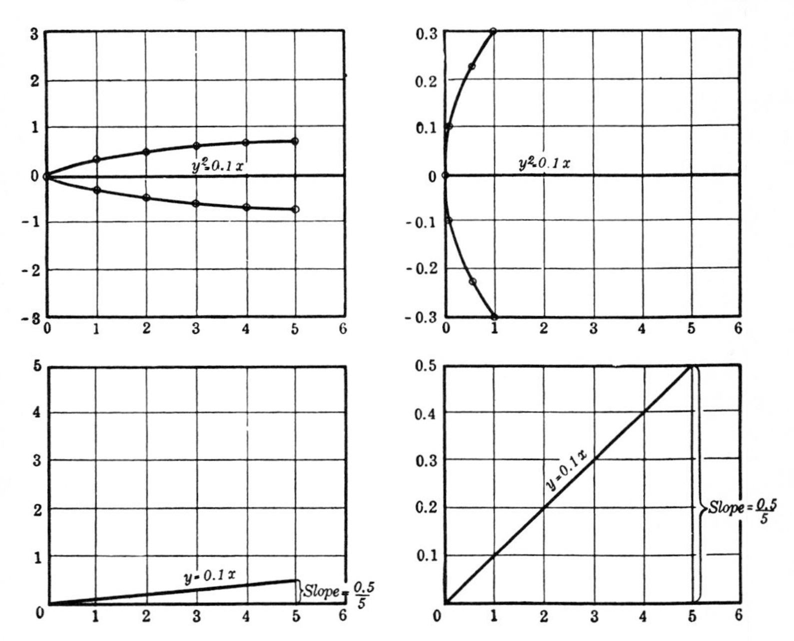

FIG. 21.—Axes with unequal scales. The two parabolas shown above represent the same equation. They appear different because the scale of plotting is different. The straight line at the left is plotted in the usual manner, while the one at the right represents the same equation plotted on axes of unequal scales. In finding equations for graphs of this kind, it is necessary to obtain Δy and Δx from the scales rather than from the number of squares.

For example, the two different parabolas shown in Fig. 21 both represent the same equation and the only difference is that the scales along the axes are different in the two cases.

The straight line $y = 0.1x$ lies too close to the X axis for convenience when plotted in the standard way, but if plotted on axes such that each division of the ordinate corresponds to 0.1 unit instead of 1 unit, a much better graph is obtained. The two graphs are shown in Fig. 21. It is obvious that the second one is more useful and that it permits a more accurate interpolation.

In working with axes of unequal scales care must be taken in drawing conclusions. The line of the second graph appears to make an angle with the X axis of 45°, giving a tangent of 1 and the equation $y = x$. Of course this is incorrect. The slope of the line AB/OA is not $5/5$ but $0.5/5$. The equation then is not $y = x$, but $y = 0.1x$.

Very frequently it is not convenient to have the origin of the graph coincide with the lower left-hand corner of the coordinate paper. The full utilization of the coordinate paper is the first criterion in deciding how to plot a curve from experimental

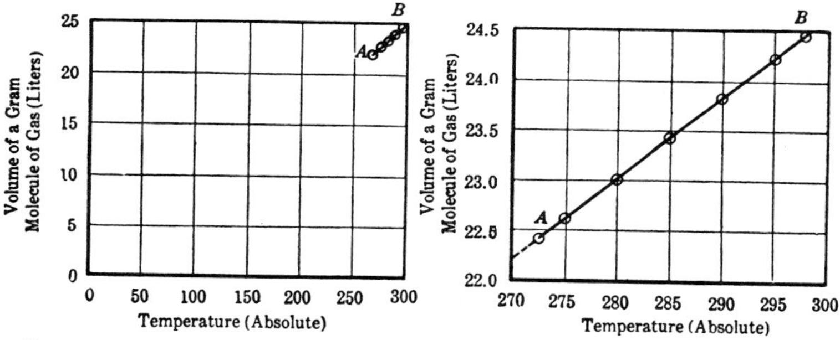

FIG. 22.—Two ways of plotting a graph. The first graph is poorly planned, because much of the paper is not utilized. In plotting experimental data, it is not necessary to start the scales at zero.

data. For example, if the volume of a gram molecule of gas is measured over a range of temperature from 273° to 298° (absolute temperature), the graph shown at the right of Fig. 22 is a much better way of representing the facts than the graph shown at the left.

When the scales do not start at zero, care must be used in the interpretation of the curves. For example in the second graph the line cuts the ordinate at the extreme left at the point 22.2. This is not the intercept on the Y axis, however. The Y axis is the ordinate at which $x = 0$, and it would be way off the page at the left. In finding the equation for the line it is best not to try to find the intercept on the Y axis but to use the two-point formula (p. 31) in the region where the experiments have been carried out. For example, the equation calculated from the two points A and B (Fig. 22) is $y = 0.082x$. The points A and B have the coordinates 273, 22.38 and 298, 24.43.

Exercises

1. The circle $x^2 + y^2 = 25$ is to be graphed.

Ans.: Center on origin. Radius = 5.

2. The circle $3x^2 + 3y^2 - 75 = 0$ is to be graphed.

3. Given the circle $(x - 2)^2 + (y - 3)^2 = 1$. Where is the center of the circle and what is the radius? The circle is to be graphed with a compass. *Ans.:* Center on 2, 3. Radius = 1.

4. Given the circle $x^2 + y^2 - 4x - 6y + 12 = 0$. The circle is to be graphed by solving the equation for various values of x, such as 0, 1, 2, 3, and 4. The equation may also be arranged in the form of Prob. 3, and then the circle may be graphed more easily with a compass.

5. The circle $x^2 + y^2 + 8x + 4y - 5 = 0$ is to be graphed.

6. A circle is drawn at random and its equation is determined.

7. What is the equation for the circumference of a nickel with its center on the origin?

The numerical constants will change with the values of the coordinates. For example, millimeter paper will give one equation and paper ruled in inches will give another.

8. What is the equation for the circumference of a half-dollar which lies in the third quadrant and is tangent to the Y axis and the X axis?

9. A straight line is drawn so that it intersects a circle. The points of intersection are to be located by finding the equations of the line and the circle, and solving the two equations simultaneously. These points are to be checked with the points located graphically.

10. Where does the line $y = 2x + 1$ intersect the circle $x^2 + y^2 = 64$?

11. What is the equation of the circle if its center is located at 2, 4, and its circumference touches the X axis? *Ans.:* $x^2 + y^2 - 4x - 8y + 4 = 0$.

12. What is the equation of the circle whose diameter is the line joining the two points 3, 6 and −2, 1?

13. Where is the center and what is the radius of the circle represented by the equation $x^2 + y^2 + 3x - 9y = 0$?

Ans.: Center −1.5, 4.5, radius 4.74.

14. Where is the center and what is the radius of the circle represented by the equation, $3x^2 + 3y^2 - 15x + 12y = 0$?

15. Given the ellipse $\frac{x^2}{16} + \frac{y^2}{36} = 1$, or $9x^2 + 4y^2 = 144$. The points corresponding to $x = 0, \pm 1, \pm 2, \pm 3, \pm 4$, and ± 5 are to be calculated. The ellipse is then drawn through these points.

16. The ellipse $\frac{x^2}{25} + \frac{y^2}{9} = 1$ is to be graphed.

17. The parabola $y^2 = 36x$ is to be graphed.

18. The parabola $x^2 = 36y$ is to be graphed.

19. At what points does the line $y = x + 3$ intersect the parabola $y^2 = 36x$?
Ans.: 0.3, 3.3 and 29.7, 32.7.

20. A projectile is thrown off horizontally from the top of a tower 500 ft. (152.4 m.) high, with a velocity of 50 ft. (15.24 m.) per second. Neglecting the resistance of the air, the horizontal distance traveled is $50t$ and the vertical distance traveled is $16t^2$, where t is the time in seconds. The position of the projectile is calculated at each second, and the graph is drawn showing the path of the projectile, referred to the top of the tower as the X axis.

21. What is the equation for the path of the projectile described in Prob. 20? The curve must be parabolic because the ordinate is proportional to the square of the time and the time is proportional to the abscissa. The constant of the parabola may be found by substituting into the general equation of the parabola, the specific value of the vertical and the horizontal distance at any time.

How high above the ground will the projectile be when it is 100 ft. (30.48 m.) out from the tower?
Ans.: 436 ft.

At what distance from the tower will the projectile hit the ground?

Where will the projectile be with reference to the ground after it has travelled 1,000 ft. (304.8 m.) out from the tower?

22. The following equations are to be graphed:

(*a*) $y = 1 + 2x + 1x^2$.
(*b*) $y = 1 + 2x + 0.1x^2$.
(*c*) $y = 1 + 2x + 0.01x^2$.

23. The equation $l = 1 + 1.67 \times 10^{-5}t + 4.03 \times 10^{-9}t^2$ is to be graphed.

24. The hyperbola $\frac{x^2}{16} - \frac{y^2}{4} = 1$ is to be graphed.

25. The hyperbola $\frac{x^2}{75} - \frac{y^2}{27} - \frac{1}{3} = 0$ is to be graphed.

26. The equation $xy = 36$ is to be graphed.

27. The equation $\frac{x^2}{9} - \frac{y^2}{9} = 1$ is to be graphed. The curves are traced on thin paper and rotated through 45°. What is the equation of the curve after rotation, the axes remaining unchanged?
Ans.: $xy = 4.5$.

28. The gas law is $PV = RT$. When the temperature, T, is 273° and the pressure, P, is 1 atmosphere, the volume, V, is 22.4 liters. These measurements show that the value of $R = \frac{1 \times 22.4}{273} = 0.082$ liter-atmospheres. The equation $PV = 0.082 \times 273$ is to be graphed. What are the values of V when $P = 1$; 10; 0.3; 0.0001?

29. The product of the concentration of hydrogen ions and hydroxyl ions is a constant in water, or in an aqueous solution.

$$C_{H^+} \times C_{OH^-} = 10^{-14}.$$

This equation is to be graphed. At what concentration of H^+ ions is the number of hydrogen ions and hydroxyl ions exactly the same?

What is the value of C_{OH}^- when $C_H^+ = 0.01$; 0.0001; 10^{-13}; 10^{-15}; 10?
Ans.: when $C_H^+ = 0.01$, $C_{OH}^- = 10^{-12}$.

CHAPTER VI

GRAPHS OF LOGARITHMIC AND TRIGONOMETRICAL FUNCTIONS

In the preceding chapter variables were raised to constant powers, as $y = x^2$; and in this section constants are raised to variable powers, as $y = 2^x$. Equations of this kind in which

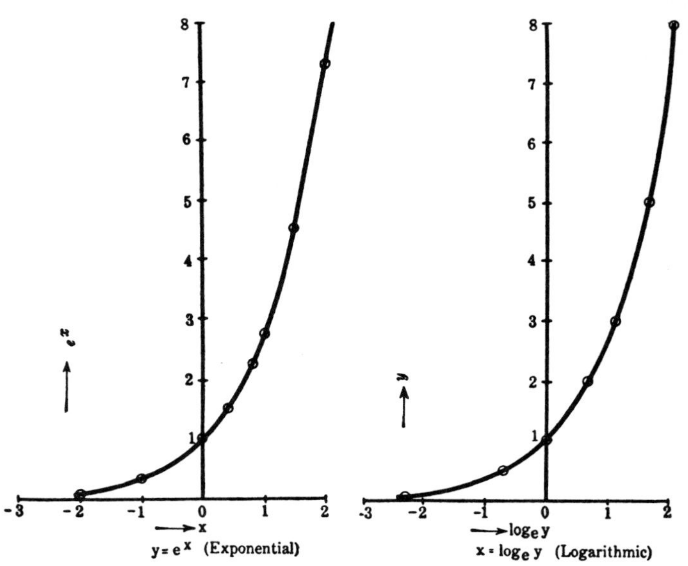

Fig. 23.—A comparison of the exponential and logarithmic equations. The two curves are identical.

Exponential........										
	x	x	-2	-1	0	0.4	0.8	1	1.5	2
	y	e^x	0.14	0.37	1.00	1.49	2.23	2.72	4.48	7.39

Logarithmic........									
	y	y	0.1	0.5	1	2	3	5	8
	x	$\log_e y$	-2.30	-0.69	0	0.69	1.10	1.61	2.08

the exponent is a variable are called exponential equations. The trigonometrical functions are usually grouped with the exponential or logarithmic functions and called transcendental functions as distinguished from algebraic functions.

Graphs of Exponential Equations

The most important exponential equation is shown in Fig. 23, where e^x is plotted against x (e is defined on pages 7 and 126). The fact that the logarithmic and exponential equations are equivalent, is well illustrated in this figure where both $y = e^x$, and $x = \log_e y$ are plotted. It is evident that the curves are identical.

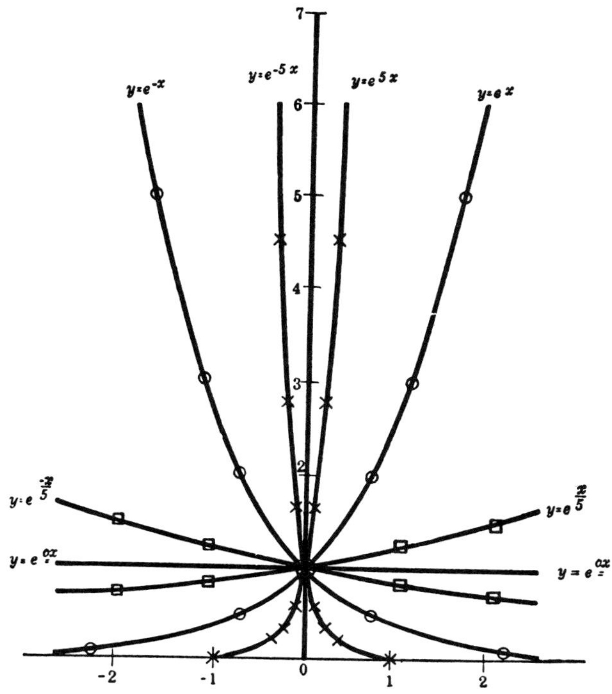

Fig. 24.—Graphs of $y = e^{ax}$ for various values of a.

A more general form of the exponential equation is $y = e^{ax}$. The change in the curve caused by giving various values to a is illustrated in Fig. 24. Since a number raised to the zero power is 1, all the curves must intersect the Y axis at $y = 1$.

Semilogarithmic Coordinates

Exponential or logarithmic equations are very common in physical chemical phenomena. They will be discussed more fully in Chapter XII. One of the best ways of determining whether or not a given set of phenomena can be expressed by a exponential or logarithmic equation is to plot the logarithm of one property against another property. Frequently a straight line is obtained and its equation may be found readily by the two-point formula (p. 31).

In the following table the concentration of hydrogen peroxide, determined experimentally, is given as a function of time:

Time (minutes)	0	5	10	15	20
Concentration	22.8	17.6	13.8	10.5	8.25
Log concentration	1.358	1.245	1.140	1.021	0.916

These facts are represented in three different ways in Fig. 25. In the first graph the experimental measurements are plotted directly. In the second, the logarithms of the concentrations are plotted as ordinates on rectangular coordinate paper. In the third, the concentrations are plotted on semilogarithm paper.

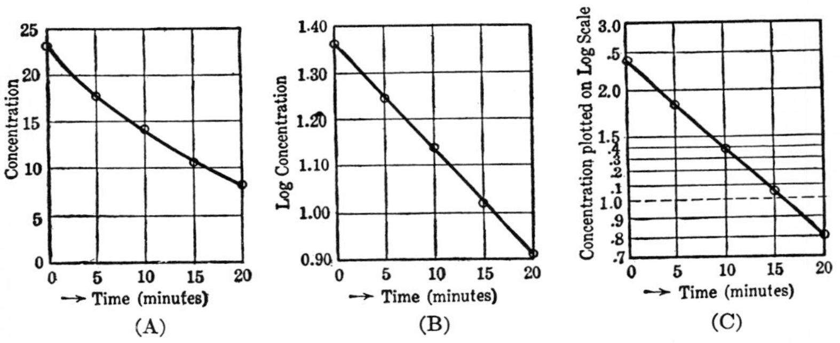

FIG. 25.—Illustration of the semilogarithmic graph. Three different ways of plotting the concentration of hydrogen peroxide, y, as a function of time, x.

(A) y plotted against x on rectangular coordinates.
(B) log y plotted against x on rectangular coordinates.
(C) y plotted against x on semilogarithmic paper.

The second and third graphs are equivalent. In the second graph, it is necessary to look up in tables the logarithms of the concentrations, while in the third the concentrations are plotted directly on the logarithmic scale of the ordinates. Graphing is much easier in the third case because the graph paper itself takes the place of a logarithm table.

The second method, however, possesses certain advantages of elasticity over the third. If the property which is plotted along the ordinate changes by tenfold, or somewhat less, the semilog paper is entirely satisfactory. If it changes by twofold, however, only a part of the graph paper can be utilized, while if the

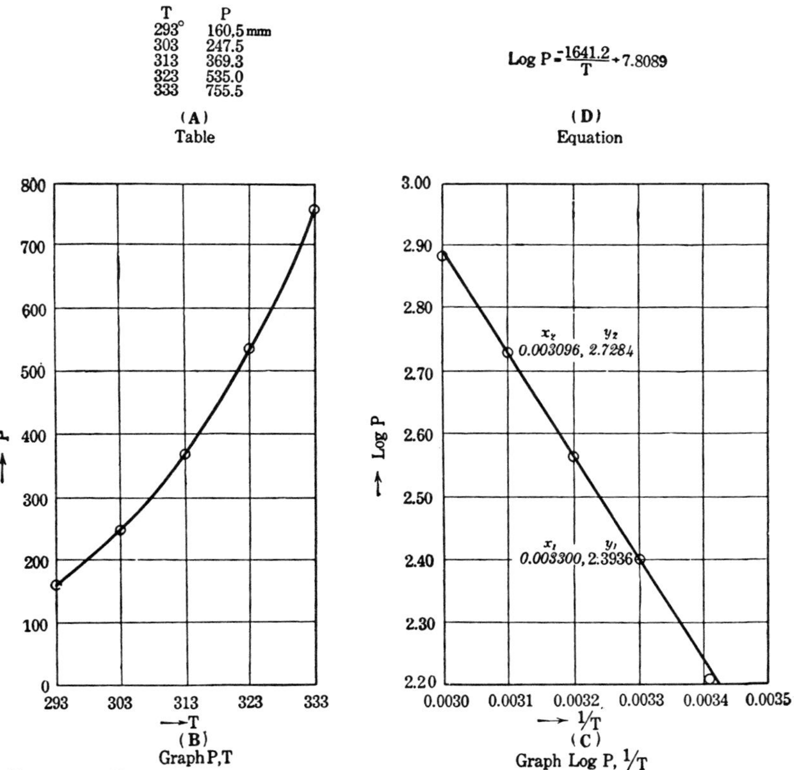

FIG. 26.—Four different methods for representing the vapor pressure, P, of chloroform as a function of temperature, T.

characteristic changes it is necessary to attach a second sheet of paper above the first, as shown by the dotted line in Fig. 25C. With the rectangular coordinates of the second method, the ordinates may always be adjusted to take advantage of the whole paper.

It is difficult to find the equation for an exponential curve plotted as in the first graph, but it is usually quite simple when plotted as in the second or third graphs. If the line is straight the equation may be found by the two-point formula (p. 31). If the line is nearly straight its equation may be usually found by adding one or more extra constants to the equation, as for example

$$\log y = a + bx + cx^2 + dx^3 \text{ (p. 229)}$$

Another example of the use of the semilogarithm graph is given in Fig. 26 where the vapor pressure of chloroform is recorded as a function of temperature in four different ways. It is obvious that the representation of the facts by a mathematical equation is much the most compact and accurate, and in general the most convenient. The vapor pressure can be calculated at any temperature by simply substituting into the formula the desired value of T. The other methods involve inaccurate interpolations. The equation is obtained from the two-point formula as follows:

$$y - y_1 = \frac{y_2 - y_1}{x_2 - x_1}(x - x_1)$$

$$y - 2.3936 = \frac{2.7284 - 2.3936}{0.003096 - 0.003300}(x - 0.00330)$$

$$y - 2.3936 = -\frac{0.3348}{0.000204}(x - 0.00330)$$

$$y = -1641.2x + 7.8089$$

$$\log P = \frac{-1641.2}{T} + 7.8089.$$

Log-log Coordinates

It was shown in the preceding section that functions of the type $y = a^x$ and $y = \log x + C$ give straight lines when plotted with logarithms along one axis. Functions of the type $y = bx^a$ give straight lines when plotted with logarithms along both axes. When logarithms of both sides of the equation are taken the equation becomes

$$\log y = \log b + \log x^a = a \log x + \log b$$

If values of $\log y$ are plotted as ordinates and values of $\log x$ are plotted as abscissas, a straight line results, because a and $\log b$ are both constants.

In the case of the simpler formula $y = x^a$, the logarithmic equation is $\log y = a \log x$, and the straight line representing the equation passes through the origin.

As in the case of the semilog equation the graph may be plotted on rectangular coordinates by looking up the logarithms; or it may be plotted on non-uniform coordinates which are ruled in such a way that the numbers rather than their logarithms, are plotted directly.

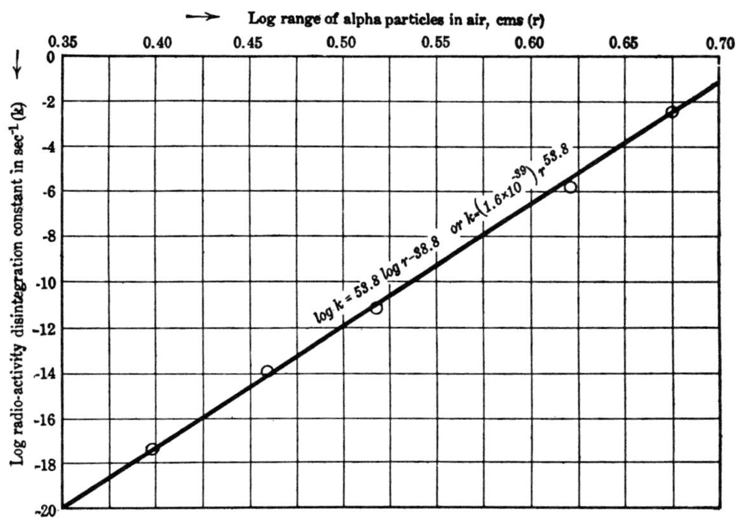

FIG. 27.—Log-log graph used in determining constants for the radio-active disintegration of elements in the radium series. The equation is $k = ar^b$.

Element	Ur_1	Ur_2	Ra	Nt	RaA
k (sec^{-1})	4.6×10^{-18}	1.1×10^{-14}	1.1×10^{-11}	2.1×10^{-6}	3.8×10^{-3}
$\log k$	-17.337	-13.959	-10.959	-5.681	-2.415
r (cms)	2.50	2.90	3.30	4.16	4.75
$\log r$	0.398	0.462	0.518	0.619	0.677

The following problem taken from the field of radio-activity serves as an example of this type of graph. The radio-activity constant, k, of an element in the Uranium family is related to the range, r, of the alpha particles by the equation $k = ar^b$. k represents the fraction of radio-atoms decomposing per minute and a and b are constants determined from experimental data. When the equation is graphed on log-log paper as shown in Fig. 27, it is easy to determine the value of k for any experimentally determined value of r. In this way it is possible through a measurement of the range of the alpha particle to evaluate k for elements which have too long or too short a life to measure directly.

The log-log graph is useful also in determining the constants of a parabolic or hyperbolic equation for these equations give straight lines when plotted on log-log paper. The slope of the straight line and the intercept on the Y axis give respectively a and log b, in the equation $y = bx^a$.

Trigonometrical Graphs

The sine curve is important because it is the simplest case of periodic or harmonic motion. Alternating electrical currents, vibrations of stretched strings, and motions of electrons within atoms are examples of periodic motion. By various combinations of sine curves it is possible also to build up any kind of a complex curve as in the application of Fourier's series p. 205.

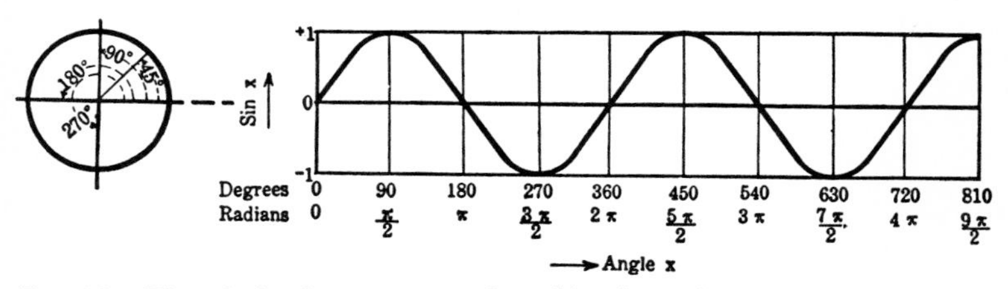

FIG. 28.—The relation between an angle and its sine. An example of a periodic function.

When a point moves around the circumference of a circle the corresponding sine goes through a series of values between $+1$ and -1. At every half-revolution around the circle the sine reverts to the same value. In Fig. 28 the sine of the angle is plotted against the angle. A collection of trigonometrical definitions and theorems may be found on page 269.

It should be noted that as α increases indefinitely sin α increases and decreases, giving a series of recurring values. For each complete angle, α, there is only one value of sin α, but for each value of sin α there is an indefinite number of different values of α. The distance between peaks or between troughs is the *wave length.* In the case of the sine curve it is numerically equal to 2π radians (or 360°).

The *period* is the length of time required for the wave to move one wave length, or the time taken for a trough to coincide with the position of the next trough. The *frequency* is the number of periods per unit of time. The *velocity* of the wave is equal to the frequency multiplied by the wave length. For example the frequency of blue light is 7.5×10^{14} vibrations per second and its wave length is 4×10^{-5} cm. The velocity of light then is $(7.5 \times 10^{14}) \times (4 \times 10^{-5})$ or 3×10^{10} cm. per second.

The graph of $\cos x$ plotted against x gives a curve very similar to that of $\sin x$, but it lags behind by 90°, or $\frac{\pi}{2}$ radians. For example, when x is 0, $\sin x = 0$, and $\cos x = 1$, and when x is $\frac{\pi}{2}$ radians or 90°, $\sin x = 1$ while $\cos x = 0$.

The graph of $\tan x$ plotted against x gives a discontinuous curve, passing from 0 to $+\infty$ and then abruptly to $-\infty$ and so on. The tangent is a periodic function having the period π.

Exercises

The following equations are to be graphed on rectangular coordinates, plotting values of y against x.

1. $y = 2^x$.
2. $y = 10^x$; and $x = \log y$.
3. $y = e^{-x}$; and $-x = \log_e y$.
4. $y = 10^{2x}$; and $y = 10^{-2x}$.
5. $y = 10^{x/4}$; and $y = 10^{-x/4}$.

The following equations are to be graphed on semilogarithm paper or on rectangular coordinates with logarithms plotted along one axis.

6. $x = \log y$.
7. $x = \frac{1}{2} \log y$.
8. $\log y = x/2$.
9. $\log y = x^3 + 5$.
10. A straight line is drawn at random on semilogarithm paper and its equation is determined.

The following equations are to be graphed on log-log paper:

11. $\log y = \log x$.
12. $\log y = \log x^2$.
13. $y = x^{1.7}$.
14. $y = 0.31\ x^{0.76}$.
15. A straight line is drawn at random on log-log paper and its equation is determined.
16. The two following equations are to be graphed.
    $y = 2^x$.
    $y = x^2$.
17. Where do the two curves $y = 10^x$ and $y = e^x$ intersect? *Ans.:* 0,1.
18. Where do the lines $\log y = x$ and $\log_e y = x$ intersect?

19. The vapor pressure of carbon tetrachloride has the following values:

Temperature, °C	0.	20.	40.	60.
Vapor pressure, mm	33.1	89.5	210.9	439.0

What is the equation of the line obtained by plotting the logarithm of the vapor pressure against the reciprocal of the *absolute* temperature (Centigrade temperature + 273 = absolute temperature)? At what temperature does carbon tetrachloride boil; *i.e.*, at what temperature does the pressure become 760 mm.?

Ans.: $\log p = \frac{-1706.4}{T} + 7.7760.$

20. The vapor pressure of mercury has the following values:

Temperature, °C	100.	200.	300.	400
Vapor pressure, mm	0.745	19.90	242.1	1,588

What is the equation for the logarithm of these vapor pressures as a function of the absolute temperature? What is the vapor pressure of mercury at 25° C.?

21. The solubility of calcium acetate in water is given approximately by the following data:

Temperature °C	0.	20.	40.	60.
Solubility, (moles per liter)	37.4	35.0	33.7	32.7

What is the equation of the line produced by plotting the logarithm of the solubility against the reciprocal of the absolute temperature? What is the solubility at 27° C.? *Ans.*: $\log S = \frac{66.66}{T} + 1.318.$

22. The specific reaction rate, k, for the decomposition of nitrogen pentoxide is as follows:

Temperature, °C	0°	25°	35°	45°	55°	65°
k (minutes)	0.000,0472	0.00203	0.00808	0.0299	0.0900	0.292

What numerical equation connects the logarithms of these specific reaction rates with the reciprocals of the absolute temperature. What is the value of k at 100° C.?

23. The vapor pressure, p in mm., of liquid arsenic trioxide is given by the equation

$$\log p = \frac{-2722}{T} + 6.513.$$

What is the vapor pressure at 300° C.? At what temperature is the vapor pressure 100 mm.? Ans.: 57.89 mm. 330.1° C.

24. According to Lambert's law the intensity of light I, after passing through the thickness, l, of an absorbing liquid is given by the equation

$I = I_0e^{-kl}$, where I_0 and k are constants. What graph is best suited to the application of this equation to experimental data?

25. Draw a line on log-log paper and find its equation expressed as an exponential equation.

CHAPTER VII

DIFFERENTIAL CALCULUS

Theory

Calculus is the mathematics of continually changing quantities, and its applications in science are many. The volume of a gas decreases when the pressure is increased, or the absorption of heat varies with the temperature, or the concentration of a chemical substance changes with the time. In general it may be said that the dependent variable, y, changes when the independent variable x changes. In mathematical language, y is said to be a function of x, and this relation is expressed by the equation

$$y = f(x)$$

The object of differential calculus is to find what change in y is produced by a small change in x. The symbol Δ signifies a change and Δy stands for a change in y, and Δx for a change in x. The expression dx (pronounced dee-eks) is called a differential and it signifies a vanishingly small change in x as distinguished from a finite change, Δx. It does not mean that x is multiplied by d, but that a very small amount of x is taken. Likewise dy and du stand for very small quantities of y and u. The expression dy/dx is called the derivative of y with respect to x. It is the definite value to which the ratio $\Delta y / \Delta x$ approaches as Δy and Δx become smaller and smaller. It is the goal of differential calculus to find and use this derivative. dy/dx may be subjected to algebraic operations and otherwise treated as an ordinary fraction, but it is in no sense equivalent to the ratio 0/0 which, of course, would be indeterminate.

When the relation between y and x is known, the relation between dy and dx can be determined by the process of differentiation. The next few pages show how this relation may be determined by finding the derivative, dy/dx. If there is no relation between y and x, there can be no relation between dy and dx and differentiation is impossible. Usually the relation may be found from laboratory experiments. For example, it has been found experimentally that the distance, y, traveled by a falling body in time, x, is given by the equation $y = kx^2$ where k is a constant. It will be shown that in this case $dy/dx = 2kx$. If x had stood for temperature, however, dy/dx could not have been calculated for there is no relation between the distance traveled and the temperature.

In general the first letters of the alphabet are used to represent constants and the last letters to represent variables. y and x are used most often for variables, but any other letters or symbols may be used, $z = f(u)$ or $t = f(s)$, etc. Although x is usually taken as the independent variable and its values plotted as abscissas, y or any other variable may be considered the independent variable as the occasion demands.

The specific calculations in Table I illustrate the meaning of dy/dx and its relation to $\Delta y/\Delta x$. A square of metal having x for its sides and y for its area is made to expand by heating until the sides increase by Δx. The new area is larger by an amount Δy. Since the area is equal to the square of the side the original area, y, is equal to x^2, and the enlarged area, $y + \Delta y$, is equal to $(x + \Delta x)^2$. Starting with a side of 1 cm. and an area of 1 sq. cm. and using different values of Δx, the following values of Table I are obtained.

TABLE I.—CALCULATIONS SHOWING THAT $\Delta y/\Delta x$ APPROACHES A LIMITING VALUE AS Δx APPROACHES ZERO. THE LIMITING VALUE IS DEFINED AS THE DERIVATIVE dy/dx

Original side x	Original area y or x^2	Increase of side Δx	New side $x + \Delta x$	New area $y + \Delta y = (x+\Delta x)^2$	Increase of area $\Delta y = (y+\Delta y) - y$	Ratio of increase $\Delta y/\Delta x$
1	1	1.0	2.0	4.00	$4-1=3$	$3/1=3.0$
1	1	0.8	1.8	3.24	$3.24-1=2.24$	$2.24/0.8=2.8$
1	1	0.6	1.6	2.56	$2.56-1=1.56$	$1.56/0.6=2.6$
1	1	0.4	1.4	1.96	$1.96-1=0.96$	$0.96/0.4=2.4$
1	1	0.2	1.2	1.44	$1.44-1=0.44$	$0.44/0.2=2.2$
1	1	0.1	1.1	1.21	$1.21-1=0.21$	$0.21/0.1=2.1$
1	1	0.01	1.01	1.0201	$1.0201-1=0.0201$	$0.0201/0.01=2.01$
1	1	0.001	1.001	1.002001	$1.002001-1=0.002001$	$0.002001/0.001=2.001$

It is apparent that the value of $\Delta y/\Delta x$ becomes smaller as Δx is taken smaller, but it does not approach zero. The limiting value to which $\Delta y/\Delta x$ approaches as Δx approaches zero may be obtained by extrapolating to the point where $\Delta x = 0$. The values of $\Delta y/\Delta x$ in Table I are plotted against Δx in Fig. 29 and an extrapolation of the line shows that $\Delta y/\Delta x$ approaches the value 2 when Δx approaches zero. The smaller Δx becomes, the more nearly does $\Delta y/\Delta x$ approach the value 2 but no value of Δx, however small, can make $\Delta y/\Delta x$ less than 2. As already

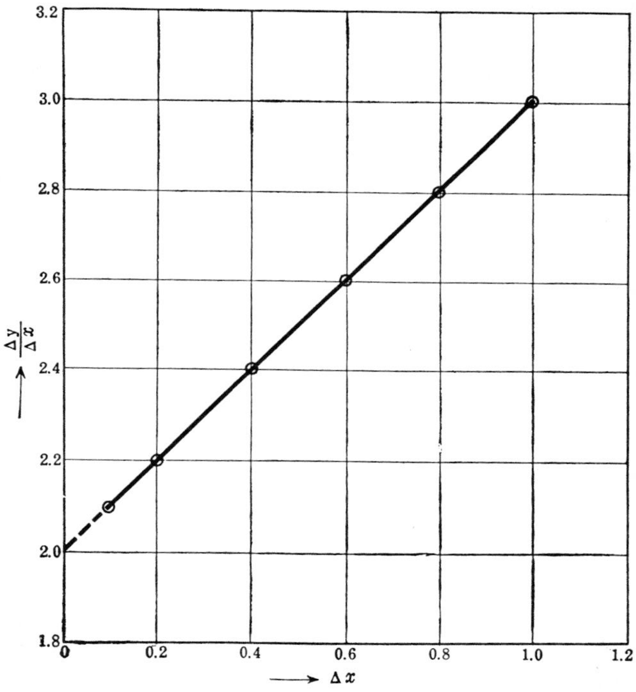

FIG. 29.—$\Delta y/\Delta x$ plotted against Δx from the data of Table I. The derivative dy/dx is obtained by extrapolating $\Delta x/\Delta y$ to the point where $x = 0$. In this example, $dy/dx = 2$.

shown, the limit to which $\Delta y/\Delta x$ approaches as Δx approaches zero is called the derivative and written dy/dx. It is a more useful function than $\Delta y/\Delta x$ because it has a definite value whereas $\Delta y/\Delta x$ changes with the size of Δx.

In Table I, x was always equal to 1 and in Table II similar calculations are made for the case where $x = 2$. Both calculations apply only for the equation $y = x^2$.

TABLE II.—$\Delta y/\Delta x$ FOR VARIOUS VALUES OF Δx WHEN $x = 2$ IN THE EQUATION $y = x^2$

Δx	1	0.8	0.6	0.4	0.2	0.1	0.01	0.001
$\Delta y/\Delta x$	5	4.8	4.6	4.4	4.2	4.1	4.01	4.001

As in the first table, it is observed that extrapolation to very small values of x gives a definite numerical value for $\Delta y/\Delta x$. In this case the number is 4. If still other values of x were treated in a similar way it would be found that the limiting value of $\Delta y/\Delta x$ when Δx is taken very small, is always equal to twice the value of x. In mathematical language when $y = x^2$. $dy/dx = 2x$.

RULES FOR DIFFERENTIATION

Instead of finding dy/dx by evaluating $\Delta y/\Delta x$ for various values of Δx and extrapolating to very small values of Δx, it is far easier and much more accurate to apply general rules for differentiation. The former method is clumsy and must be gone through with for every specific example. It is possible, however, to carry out the calculation once and for all in such a way that it may be applied to any number of specific examples. There are about a dozen simple rules for differentiation and it is well to memorize these few as one would memorize the multiplication table. It is necessary to become familiar with these rules by solving many problems, but the solutions themselves are not without interest. In fact the solution may be considered as a game with a few well-defined rules, similar to a chess game in which the different men can be moved according to certain arbitrary rules. The proof of these rules is accomplished in a simple manner, as shown by the following general procedure:

y is a function of x, or $y = f(x)$

x is increased a very little by adding Δx, and at the same time y is automatically increased by Δy. Then $y + \Delta y = f(x + \Delta x)$.

By subtracting y from $(y + \Delta y)$ it is found that $\Delta y = f(x + \Delta x) - f(x)$.

$\Delta y/\Delta x$ is then obtained by dividing the equation by Δx.

dy/dx is obtained by making Δx smaller and smaller and finding the limit to which $\Delta y/\Delta x$ approaches as Δx approaches zero.

Powers

If $y = x^2$, the general procedure just outlined may be followed and

$$
\begin{aligned}
y &= x^2 \\
(y + \Delta y) &= (x + \Delta x)^2 \\
y + \Delta y &= x^2 + 2x\Delta x + (\Delta x)^2 \text{ (expanding by the binomial theorem)} \\
y &= x^2 \\
\hline
\Delta y &= 2x\Delta x + (\Delta x)^2 \text{ (subtracting)}
\end{aligned}
$$

$$\frac{\Delta y}{\Delta x} = \frac{2x\Delta x + (\Delta x)^2}{\Delta x} = 2x + \Delta x \text{ (dividing by } \Delta x)$$

Now as Δx becomes smaller and smaller and approaches zero, the last term, Δx, of the equation drops out; and $\Delta y/\Delta x$ becomes dy/dx by definition of the derivative.

$$\frac{dy}{dx} = 2x + 0 = 2x.$$

This value, $2x$, checks with the value found before by a clumsy graphical method.

In a similar way it can be shown that if $y = x^3$, then $\dfrac{dy}{dx} = 3x^2$.

$$
\begin{aligned}
y + \Delta y = (x + \Delta x)^3 &= x^3 + 3x^2\Delta x + 3x(\Delta x)^2 + (\Delta x)^3 \\
y &= x^3 \\
\hline
\Delta y &= 3x^2\Delta x + 3x(\Delta x)^2 + (\Delta x)^3 \text{ (subtracting)}
\end{aligned}
$$

$$\frac{\Delta y}{\Delta x} = \frac{3x^2\Delta x}{\Delta x} + \frac{3x(\Delta x)^2}{\Delta x} + \frac{(\Delta x)^3}{\Delta x} =$$

$$3x^2 + 3x(\Delta x) + (\Delta x)^2 \text{ (dividing by } \Delta x)$$

$$\frac{dy}{dx} = 3x^2 \left(\text{the limit of } \frac{\Delta y}{\Delta x} \text{ as } \Delta x \text{ approaches zero}\right)$$

In the limiting value of $\Delta y/\Delta x$ all the terms containing Δx, or Δx raised to any power drop out when Δx approaches zero as a limit.

In a similar way it can be shown that if $y = x^n$, $dy/dx = nx^{n-1}$. It can be shown also by the same methods that this equation holds true if n is a fraction or a negative number.

The first rule for differentiation is as follows: If y is equal to a power of x, the derivative is equal to the product of the exponent and the variable raised to a power which is one less than the original exponent. It is better expressed in mathematical terms as follows:

$$\textbf{If } \mathbf{y = x^n, \frac{dy}{dx} = nx^{n-1}.} \qquad \text{(Rule 1)*}$$

* All the rules for differential calculus are collected on page 288.

Several examples are offered to show the application of the rule:

$$y = x^2. \quad \frac{dy}{dx} = 2x^{2-1} = 2x$$

$$y = x^5. \quad \frac{dy}{dx} = 5x^{5-1} = 5x^4$$

$$y = x. \quad \frac{dy}{dx} = x^{1-1} = x^0 = 1$$

$$y = \sqrt{x} = x^{1/2}. \quad \frac{dy}{dx} = \frac{1}{2}x^{1/2-1} = \frac{1}{2}x^{-1/2} = \frac{1}{2\sqrt{x}} \; *$$

$$y = \sqrt{x^3} = x^{3/2}. \quad \frac{dy}{dx} = \frac{3}{2}x^{3/2-1} = \frac{3}{2}x^{1/2} = \frac{3}{2}\sqrt{x}$$

$$y = \frac{1}{x^2} = x^{-2}. \quad \frac{dy}{dx} = -2x^{-2-1} = -2x^{-3} = \frac{-2}{x^3}$$

$$u = \sqrt[6]{v^5} = v^{5/6}. \quad \frac{du}{dv} = \frac{5}{6}v^{5/6-6/6} = \frac{5}{6}v^{-1/6} = \frac{5}{6\sqrt[6]{v}}$$

Added and Subtracted Functions

If y is equal to the sum or difference of any number of functions, the derivative dy/dx, is equal to the sum or difference of the derivatives of all the functions.

For example, if

$$y = x^2 - x, \quad \frac{dy}{dx} = \frac{d(x^2)}{dx} - \frac{d(x)}{dx} = 2x - 1.$$

The proof of this special case is as follows:

$$\begin{array}{lll} y + \Delta y & = (x + \Delta x)^2 - (x + \Delta x) & = x^2 + 2x\Delta x + (\Delta x)^2 - x - \Delta x \\ y & = & x^2 \qquad\qquad\qquad\qquad - x \\ \hline \Delta y & = & 2x\Delta x - \Delta x + (\Delta x)^2 \text{ (subtracting)} \end{array}$$

$$\frac{\Delta y}{\Delta x} = \frac{2x\Delta x}{\Delta x} - \frac{\Delta x}{\Delta x} + \frac{(\Delta x)^2}{\Delta x} = 2x - 1 + \Delta x \text{ (dividing by } \Delta x)$$

$$\frac{dy}{dx} = (2x - 1), \text{ (the limit of } \frac{\Delta y}{\Delta x} \text{ as } \Delta x \text{ approaches zero).}$$

In a similar manner the rule could be proved for any other functions. For example, if

$$y = x^4 - x^2 + x, \frac{dy}{dx} = \frac{d(x^4)}{dx} - \frac{d(x^2)}{dx} + \frac{d(x)}{dx} = 4x^3 - 2x + 1$$

In general, if

$$\mathbf{y = f_1(x) + f_2(x), \quad \frac{dy}{dx} = \frac{df_1(x)}{dx} + \frac{df_2(x)}{dx}.} \qquad \text{(Rule 2)}$$

* It will be remembered that fractional exponents refer to roots, and that a minus sign before an exponent puts the number in the denominator with a positive sign (p. 265).

Constants

If a constant, k, had been added to the equation of the preceding section the derivative would have been unchanged, as shown in the following example:

$$
\begin{aligned}
y &= x^2 + k \\
y + \Delta y &= (x + \Delta x)^2 + k = x^2 + 2x\Delta x + (\Delta x)^2 + k \\
y &= x^2 \qquad\qquad\qquad\qquad + k \\
\hline
\Delta y &= 2x\Delta x + (\Delta x)^2 \text{ (subtracting)} \\
\frac{\Delta y}{\Delta x} &= 2x\frac{\Delta x}{\Delta x} + \frac{(\Delta x)^2}{\Delta x} = 2x + \Delta x \text{ (dividing by } \Delta x) \\
\frac{dy}{dx} &= 2x \text{ (letting } \Delta x \text{ approach zero)}
\end{aligned}
$$

This procedure shows that the added constant drops out on differentiation, and it could have been made to drop out in any other example. The third rule follows: The derivative of an added constant is zero.

$$\mathbf{If\ y = k,\ \frac{dy}{dx} = 0.} \qquad \text{(Rule 3)}$$

For example,

$$\text{if } y = x - b, \frac{dy}{dx} = 1$$

$$\text{if } y = x^2 - 2, \frac{dy}{dx} = 2x$$

$$\text{if } y = 6, \frac{dy}{dx} = 0.$$

If the function of x had been multiplied or divided by a constant, this constant would have been found in the derivative. If

$$y = kx^2, \quad y + \Delta y = k(x + \Delta x)^2 = kx^2 + 2kx\Delta x + k(\Delta x)^2$$

Subtracting

$$
\begin{aligned}
y &= kx^2 \\
\hline
\Delta y &= 2kx\Delta x + k(\Delta x)^2 \quad \text{(subtracting)} \\
\frac{\Delta y}{\Delta x} &= 2kx + k\Delta x \quad \text{(dividing by } \Delta x) \\
\frac{dy}{dx} &= 2kx \text{ (letting } \Delta x \text{ approach zero)}
\end{aligned}
$$

The constant would have followed through to the final derivative, no matter what function of x was involved.

The fourth rule for differentiation, then, is as follows: Multiplied or divided constants remain unchanged on differentiation.

$$\textbf{If } \mathbf{y} = \mathbf{kf(x)}, \ \frac{\mathbf{dy}}{\mathbf{dx}} = \mathbf{k}\frac{\mathbf{df(x)}}{\mathbf{dx}}. \qquad \text{(Rule 4)}$$

For example,

$$\text{if } y = 5x^2, \frac{dy}{dx} = 2 \times 5x = 10x$$

$$\text{if } y = \frac{x^3}{6}, \frac{dy}{dx} = \frac{1}{6}3x^2 = \frac{x^2}{2}$$

$$\text{if } s = \frac{2}{3}\sqrt{t} + 3, \frac{ds}{dt} = \frac{2}{3} \times \frac{1}{2}\, t^{\frac{1}{2}-1} = \frac{1}{3\sqrt{t}}$$

Products

If y is equal to the product of two or more variables, each of which is a function of x, another rule is necessary.

$$y = uv$$

$$y + \Delta y = (u + \Delta u)(v + \Delta v) = uv + u\Delta v + v\Delta u + \Delta u\Delta v$$

$$y = uv$$

$$\Delta y = u\Delta v + v\Delta u + \Delta u\Delta v \text{ (subtracting)}$$

$$\frac{\Delta y}{\Delta x} = u\frac{\Delta v}{\Delta x} + v\frac{\Delta u}{\Delta x} + \Delta u\frac{\Delta v}{\Delta x} \text{ (dividing by } \Delta x)$$

When Δx approaches zero, the last term $\Delta u\frac{\Delta v}{\Delta x}$ drops out since Δu and Δv are functions of x and approach zero when Δx approaches zero. Also when Δx approaches zero, $\Delta y/\Delta x$, $\Delta u/\Delta x$, and $\Delta v/\Delta x$ become derivatives and the preceding equation may be written $\frac{dy}{dx} = u\frac{dv}{dx} + v\frac{du}{dx}$. A similar proof could be given for any other product, and the general rule is deduced as follows: The derivative of a product of two variables is equal to the first times the derivative of the second plus the second times the derivative of the first.

$$\textbf{If } \mathbf{y} = \mathbf{uv}, \ \frac{\mathbf{dy}}{\mathbf{dx}} = \mathbf{u}\frac{\mathbf{dv}}{\mathbf{dx}} + \mathbf{v}\frac{\mathbf{du}}{\mathbf{dx}}. \qquad \text{(Rule 5)}$$

A graphical demonstration of this rule as given in Fig. 30 is helpful. The original area, y, is equal to uv. u is increased by Δu and v by Δv. The new area Δy is bounded by the dotted lines, and consists of $(u\Delta v) + (v\Delta u) + \Delta u\Delta v$. When Δu and Δv are made very small, their product represented by the little square in the corner, becomes negligibly small and $\Delta y = u\Delta v + v\Delta u$. Dividing by Δx the equation becomes

$$\frac{\Delta y}{\Delta x} = u\frac{\Delta v}{\Delta x} + v\frac{\Delta u}{\Delta x}.$$

The resemblance of this equation to Rule 5 is evident. It is approximately true, whereas Rule 5 is exactly true.

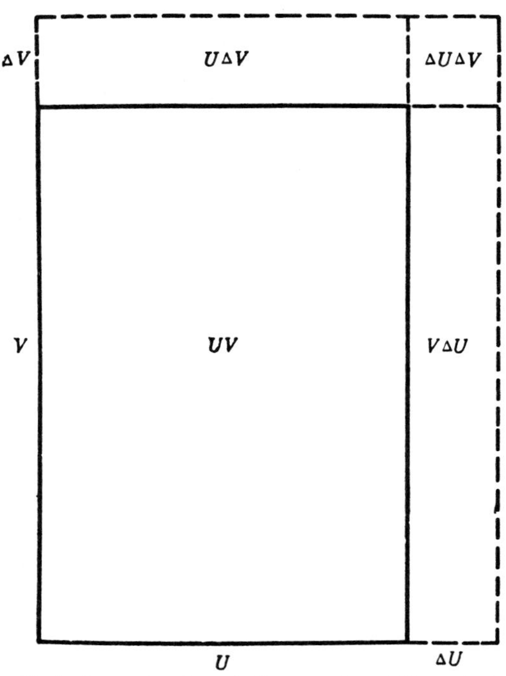

FIG. 30.—Graphical demonstration of the rule for differentiating a product.

If three or more variables are involved, the rule must be expanded.

$$\text{If } y = uvw, \frac{dy}{dx} = uv\frac{dw}{dx} + uw\frac{dv}{dx} + vw\frac{du}{dx}.$$

The following examples illustrate the application of the rule of products:

$$\begin{aligned} y = x^3(x^2 - 2), \frac{dy}{dx} &= x^3\frac{d(x^2 - 2)}{dx} + (x^2 - 2)\frac{d(x^3)}{dx} \\ &= x^3(2x) + (x^2 - 2)\ (3x^2) \\ &= 2x^4 + 3x^4 - 6x^2 \\ &= 5x^4 - 6x^2 \end{aligned}$$

This problem could have been solved by carrying out the multiplication before differentiating,

$$y = x^3(x^2 - 2) = x^5 - 2x^3, \frac{dy}{dx} = 5x^4 - 6x^2.$$

Sometimes one method is simpler and sometimes the other, depending on the particular problem.

Quotients

The rule for differentiating a quotient is a little more complicated than the others which have gone before.

$y = u/v$, where u and v are both functions of x.

When x increases by Δx, u increases by Δu, v increases by Δv, and y increases by Δy, and

$$y + \Delta y = \frac{u + \Delta u}{v + \Delta v} = \frac{u}{v} + \frac{\Delta u}{v} - \frac{u\Delta v}{v^2} - \frac{\Delta u \Delta v}{v^2}$$

Since the last step in algebraic long division is sometimes forgotten it is given here in detail:

$$\begin{array}{r|l|l}
v + \Delta v & u + \Delta u & \frac{u}{v} + \frac{\Delta u}{v} - \frac{u\Delta v}{v^2} - \frac{\Delta u\Delta v}{v^2} + \cdots \\
 & u + \frac{u\Delta v}{v} & \\
\hline
 & \Delta u - \frac{u\Delta v}{v} & \\
 & \Delta u + \frac{\Delta u\Delta v}{v} & \\
\hline
 & -\frac{u\Delta v}{v} - \frac{\Delta u\Delta v}{v} & \\
 & -\frac{u\Delta v}{v} - \frac{u(\Delta v)^2}{v^2} & \\
\hline
 & -\frac{\Delta u\Delta v}{v} + u\frac{(\Delta v)^2}{v^2} &
\end{array}$$

The division does not need to be carried farther for the further remainders are quantities of the second order, containing products of increments, and these terms always drop out later when Δx approaches zero.

The terms of the quotient may be rearranged giving

$$y + \Delta y = \frac{u}{v} + \frac{v\Delta u - u\Delta v - \Delta u\Delta v}{v^2}$$

and subtraction of $y = u/v$ gives

$$\Delta y = \frac{v\Delta u - u\Delta v}{v^2}$$

and dividing through by Δx,

$$\frac{\Delta y}{\Delta x} = \frac{v\frac{\Delta u}{\Delta x} - u\frac{\Delta v}{\Delta x} - \Delta u\frac{\Delta v}{\Delta x}}{v^2}$$

When Δx approaches zero Δu also approaches zero, $\Delta u \frac{\Delta v}{\Delta x}$ and later terms drop out, and $\Delta y/\Delta x$ becomes dy/dx.

The rule may be stated in words as follows:

The derivative of a quotient is equal to the denominator times the derivative of the numerator minus the numerator times the derivative of the denominator, all divided by the square of the denominator. (It must be remembered that the denominator comes first and that the numerator has the minus sign.) In mathematical language, it may be stated as follows·

$$\text{If } y = \frac{u}{v}, \quad \frac{dy}{dx} = \frac{v\frac{du}{dx} - u\frac{dv}{dx}}{v^2}. \qquad \text{(Rule 6)}$$

A few problems help to show how the rule is used.

$$y = \frac{3+x}{3-x}, \quad \frac{dy}{dx} = \frac{(3-x)\frac{d(3+x)}{dx} - (3+x)\frac{d(3-x)}{dx}}{(3-x)^2} =$$

$$\frac{(3-x)(1) - (3+x)(-1)}{(3-x)^2} = \frac{3 - x + 3 + x}{(3-x)^2} = \frac{6}{(3-x)^2}$$

In a second example,

$$y = \frac{2x+1}{x^2+2}.$$

$$\frac{dy}{dx} = \frac{(x^2+2)\frac{d(2x+1)}{dx} - (2x+1)\frac{d(x^2+2)}{dx}}{(x^2+2)^2} =$$

$$\frac{(x^2+2)2 - (2x+1)2x}{(x^2+2)^2} = \frac{2x^2 + 4 - 4x^2 - 2x}{(x^2+2)^2} = \frac{-2(x^2+x-2)}{(x^2+2)^2}$$

If possible, a quotient should be simplified first, as in the following example,

$$y = \frac{x^5 + x^3}{x^2} \qquad y = x^3 + x. \qquad \frac{dy}{dx} = 3x^2 + 1.$$

$$y = \frac{x^2 + 2x + 1}{x+1} \qquad y = x + 1. \qquad \frac{dy}{dx} = 1$$

Exercises

1. $y = x^3$. $\frac{dy}{dx} = ?$ *Ans.:* $3x^{3-1} = 3x^2$. 2. $y = x^4 + 3$. $\frac{dy}{dx} = ?$

3. $y = 3x^3 + 2x^2 + 5$. What is the ratio of a slight change in y to the corresponding change in x? *Ans.:* $9x^2 + 4x$.

4. $y = x^{-3} + 5x^{2/3}$. $\frac{dy}{dx} = ?$

5. $y = 2x^4 - \dfrac{3}{\sqrt[4]{x^7}} + 9x - 10.\quad \dfrac{dy}{dx} = ?$ *Ans.:* $8x^3 + \dfrac{21}{4\sqrt[4]{x^{11}}} + 9.$

6. $y = \sqrt[3]{x^5} - \dfrac{5}{x^2} + 2x.\quad \dfrac{dy}{dx} = ?$

7. $u = 2v^3 - \dfrac{1}{2}v^2 + 396.\quad \dfrac{du}{dv} = ?$ *Ans.:* $6v^2 - v.$

8. $z = \dfrac{3}{4} + t^{4/3} - \sqrt{t^4} - t + \dfrac{2}{t^3}.\quad \dfrac{dz}{dt} = ?$

9. $y = (x^3 + 3)(2x^2 + 5).\quad \dfrac{dy}{dx} = ?$

applying the rule for a product,

$$\frac{dy}{dx} = u\frac{dv}{dx} + v\frac{du}{dx} = (x^3 + 3)4x + (2x^2 + 5)3x^2 = 10x^4 + 15x^2 + 12x$$

multiplying out,

$$y = 2x^5 + 5x^3 + 6x^2 + 15$$

$$\frac{dy}{dx} = 10x^4 + 15x^2 + 12x.$$

10. $y = (x^2 + 4)2x^3.\quad \dfrac{dy}{dx} = ?$

11. $y = (3x^2 + 2x)5x.\quad \dfrac{dy}{dx} = ?$ *Ans.:* $45x^2 + 20x.$

12. $y = x^3 + 3x(x^2 + 2).\quad \dfrac{dy}{dx} = ?$

13. $y = (x^3 + 3x)(x^2 + 2).\quad \dfrac{dy}{dx} = ?$ *Ans.:* $5x^4 + 15x^2 + 6.$

14. $y = (3x^2 - 5x)(2x^4 + 3x).\quad \dfrac{dy}{dx} = ?$

15. $y = (bx^n)(ax^2 + c) = abx^{n+2} + bx^nc.\quad \dfrac{dy}{dx} = ?$ *Ans.:* $(n + 2)abx^{n+1} + bcn\, x^{n-1}.$

16. $y = ax^{3n}\left(\dfrac{b}{x^2} + c\right).\quad \dfrac{dy}{dx} = ?$

17. $y = (x^2 + 2)(x^3 - 2x)\dfrac{1}{x}.\quad \dfrac{dy}{dx} = ?$ *Ans.:* $4x^3.$

18. $y = 2\sqrt{x}\left(x^3 + 2x - \dfrac{1}{x^2}\right).\quad \dfrac{dy}{dx} = ?$

19. $y = \dfrac{2x^2}{3x^2 + 2}.\quad \dfrac{dy}{dx} = ?$ *Ans.:* $\dfrac{8x}{(3x^2 + 2)^2}.$

20. $y = \dfrac{x^3 + 3x}{x^2 + 2}.\quad \dfrac{dy}{dx} = ?$

21. $y = \dfrac{2x^3 + 5}{3x^2 + 2x + 1}.\quad \dfrac{dy}{dx} = ?$ *Ans.:* $\dfrac{6x^4 + 8x^3 + 6x^2 - 30x - 10}{(3x^2 + 2x + 1)^2}.$

22. $y = \dfrac{4x^5 + \dfrac{6}{x^2} - \sqrt[4]{x^5}}{2x}.\quad \dfrac{dy}{dx} = ?$

23. Water is dropping from a burette on a clean plate, so that a circular pool is gradually increasing in area and in radius.

What is the ratio of the increase in area to the increase in radius? At all times the area, y, $= \pi r^2$, and $dy/dr = 2\pi r$. This is the ratio at all times between the increase of the area and the increase of the radius, when the latter increase is very small.

What is the numerical value of this ratio when the radius is 2 cm.? $dy/dr = 2\pi r = 2\pi 2 = 12.56$. This equation means that the area is increasing 12.56 times as fast as the radius. When the radius is 10 cm., $dy/dr = 2\pi 10 = 62.8$ and the area is increasing 62.8 times as fast as the radius.

24. Air is blown into a soap bubble at a constant rate. What general expression will give the ratio of the rate of increase of volume to the rate of increase of radius? The volume of a sphere is $\frac{4}{3}\pi r^3$. What is the numerical value of this ratio when the radius is 5 cm.? 50 cm.?

25. The distance, S, traveled by a falling body is given by $S = \frac{1}{2}gt^2$. What expression will give the velocity of the body at any time? Velocity = distance/time. If g is 9.8 m. per second and t is expressed in seconds, how fast is the body falling at the end of the first second; at the end of the fourth second; the tenth; the sixtieth; the one-hundredth?

Ans.: Fourth second, 39.2 m. per second.

26. If the quantity of heat, Q, necessary to raise the temperature of a gram of a certain metal from 0° to t° is given by $Q = at + bt^2 + ct^3$, what is the specific heat at t°; at 100°; at 200°? Specific heat $= dQ/dt$.

27. The specific heat of liquid ethyl alcohol is given by the expression $s = 0.5068 + 0.00286t + 5.4 \times 10^{-6}t^2$, where t is any centigrade temperature between 0° and 60°. What is the ratio of the increase in heat capacity to the increase in temperature (ds/dt)? What is the numerical value of this ratio at 20° C. and at 50° C.? *Ans.:* 0.003076 at 20° C.

28. The frequency of vibration, v, of a stretched wire is given by the following equation: $v = (1/be)\sqrt{kf/\pi s}$ in which b = diameter of the wire and e = its length, s = its specific gravity and f = the force with which it is stretched. k and π are constants. What is the rate of change in vibration when b, e, s, and f are varied one at a time, all other factors being kept constant? Four separate problems are involved.

Ans.: $dv/db = (-1/eb^2)\sqrt{kf/\pi s}$.

CHAPTER VIII

DIFFERENTIATION

Functions of a Function

One of the first difficulties in differential calculus is the differentiation of an expression which is a function not of a simple variable but of a quantity which is already a function of the variable. It is perhaps the greatest stumbling block in elementary calculus. When y is some function not of x, but of x^2, a function of a function is involved. For example if $y = (x^2)^3$ it might be expected that dy/dx would be equal to $3(x^2)^2$. This differentiation is incorrect, however, as is evident from the fact that $y = (x^2)^3 = x^6$, and $dy/dx = 6x^5$. The difficulty lies in the fact that the latter operation differentiates with respect to x, and gives dy/dx, but the former operation differentiates with respect to x^2 and gives $dy/d(x^2)$. dy/dx is always desired.

A function of a function may be differentiated correctly by carrying out the operation in two stages, thus:

The first function of x is set equal to s, and differentiated with respect to x. Then the whole function is differentiated with respect to s. The two equations are then multiplied and the intermediate function, s, drops out. In this way dy/dx is obtained.

$$\frac{dy}{ds} \times \frac{ds}{dx} = \frac{dy}{dx}.$$

In general,

$$\textbf{If } \mathbf{y = f_2(f_1x), \frac{dy}{dx} = \frac{df_2(f_1x)}{d(f_1x)} \times \frac{d(f_1x)}{dx}}. \qquad \text{(Rule 7)}$$

This rule is illustrated by the following examples:

$$y = (x^2)^3 \quad \frac{dy}{dx} = ?$$

Let $s = x^2$, and then $y = s^3$

$$\frac{ds}{dx} = 2x, \quad \text{and} \quad \frac{dy}{ds} = 3s^2. \quad \text{(Differentiating)}$$

$$\frac{dy}{dx} = \frac{dy}{ds} \times \frac{ds}{dx} = 2x(3s^2). \quad \text{(Multiplying the two equations)}$$

$$3s^2(2x) = 3(x^2)^2 2x = 6x^5. \quad \text{(Substituting the value of } s\text{)}$$

In another example,

$$y = \frac{1}{\sqrt[3]{x^2 + k}}$$

Let $s = x^2 + k$ and then $y = s^{-1/3}$

$$\frac{ds}{dx} = 2x \text{ and } \frac{dy}{ds} = \frac{-1}{3}s^{-4/3}$$

$$\frac{dy}{dx} = \frac{dy}{ds} \times \frac{ds}{dx} = \left(-\frac{1}{3}s^{-4/3}\right)(2x) = \frac{-2x}{3\sqrt[3]{(x^2 + k)^4}}$$

Sometimes with complicated functions it is desirable to carry out the differentiation in three or more stages as in the following example:

$$y = x^3 + \sqrt{x^2 + 2}$$

Let

$$s = x^2 + 2, \quad \text{then } \frac{ds}{dx} = 2x$$

Let

$$r = \sqrt{x^2 + 2} = \sqrt{s}, \text{ then } \frac{dr}{ds} = \frac{1}{2}s^{-1/2} = \frac{1}{2\sqrt{x^2 + 2}}$$

and

$$\frac{dr}{dx} = \frac{dr}{ds} \times \frac{ds}{dx} = \left(\frac{1}{2\sqrt{x^2 + 2}}\right)(2x) = \frac{x}{\sqrt{x^2 + 2}}$$

Going back to the original equation,

$$y = x^3 + \sqrt{x^2 + 2} = x^3 + r, \text{ and}$$

$$\frac{dy}{dx} = 3x^2 + \frac{dr}{dx}.$$

Substituting the value of dr/dx obtained above,

$$\frac{dy}{dx} = 3x^2 + \frac{x}{\sqrt{x^2 + 2}}$$

Frequent examples of the differentiation of functions of functions are found in the trigonometrical and the logarithmic functions as well as in the algebraic expressions. The same methods apply in differentiating such equations as $y = \log x^3$; $y = \sin 2x$; $y = \log^2 x = (\log x)^2$ and others which will be met later.

After some practice it becomes possible to carry out these intermediate steps mentally without the necessity of writing down ds and dt but this longer, safer method is recommended at first.

LOGARITHMS

In a previous section a rule was given for differentiating a variable raised to a constant power, as for example, $y = x^2$, and in this section it will be found how to differentiate a constant raised to a variable power, as for example, $y = a^x$. All the derivations of this section depend on the remarkable property of a certain mathematical series—the property of remaining unchanged upon differentiation. This series is,

$$y = 1 + x + \frac{x^2}{1 \times 2} + \frac{x^3}{1 \times 2 \times 3} + \frac{x^4}{1 \times 2 \times 3 \times 4} + \frac{x^5}{1 \times 2 \times 3 \times 4 \times 5} + \cdots$$

and it may be differentiated to give

$$\frac{dy}{dx} = 0 + 1 + \frac{2x}{1 \times 2} + \frac{3x^2}{1 \times 2 \times 3} + \frac{4x^3}{1 \times 2 \times 3 \times 4} + \frac{5x^4}{1 \times 2 \times 3 \times 4 \times 5} + \cdots$$

$$\frac{dy}{dx} = 1 + x + \frac{x^2}{1 \times 2} + \frac{x^3}{1 \times 2 \times 3} + \frac{x^4}{1 \times 2 \times 3 \times 4} + \cdots$$

It is evident that the derivative, dy/dx, has the same value as the original function, y. Although the first equation has one more term than the differentiated expression, this extra term may be made negligible by carrying out the series to a sufficiently large number of terms (p. 127).

Now it is possible to obtain this series by raising the quantity e to the x power, and since

$$e^x = 1 + x + \frac{x^2}{1 \times 2} + \frac{x^3}{1 \times 2 \times 3} + \frac{x^4}{1 \times 2 \times 3 \times 4} + \cdots$$

it follows that

$$\textbf{If } \mathbf{y = e^x}, \qquad \frac{\mathbf{dy}}{\mathbf{dx}} = \mathbf{e^x}. \qquad \text{(Rule 8)}$$

In other words, the derivative of e^x is e^x.

The reason that e^x is equal to this particular series is now explained. e is defined as the limit of the series $\left(1 + \frac{1}{n}\right)^n$, when n becomes very large. It has the numerical value 2.71828 . . . , as stated already (p. 7). Since $e = \left(1 + \frac{1}{n}\right)^n$ when n is made very large, it follows that $e^x = \left(1 + \frac{1}{n}\right)^{nx}$ when n is made very large.

Expansion by the binominal theorem (p. 265) gives the following series:

$$e^x = \left(1 + \frac{1}{n}\right)^{nx} = 1^{nx} + \frac{nx1^{nx-1}(1/n)}{1} + nx(nx - 1)\frac{1^{nx-2}(1/n)^2}{1 \times 2} + nx(nx - 1)(nx - 2)\frac{1^{nx-3}(1/n)^3}{1 \times 2 \times 3} + \cdots$$

Rearranging terms,

$$e^x = \left(1 + \frac{1}{n}\right)^{nx} = 1 + \frac{nx}{n} + \frac{n^2x^2 - nx}{1 \times 2 \times n^2} + \frac{n^3x^3 - 3n^2x^2 + 2nx}{1 \times 2 \times 3 \times n^3} = 1 + x + \frac{x^2 - \frac{x}{n}}{1 \times 2} + \frac{x^3 - \frac{3x^2}{n} + \frac{2x}{n^2}}{1 \times 2 \times 3} + \cdots$$

When the denominators, n, become very large, the fractions become very small, and since n can be made as large as desired the fractions in which n occurs in the denominators can be made entirely negligible, and

$$e^x = \left(1 + \frac{1}{n}\right)^{nx} = 1 + x + \frac{x^2}{1 \times 2} + \frac{x^3}{1 \times 2 \times 3} + \frac{x^4}{1 \times 2 \times 3 \times 4} + \cdots$$

As explained before this is the series whch gives identically the same value after it is differentiated.

With this one fact that the derivative of e^x is e^x, it is possible to build up rules for differentiating any logarithmic or exponential function.

The derivative of $\log_e$ of a variable is the reciprocal of the variable.

If $y = \log_e x$, $\frac{dy}{dx} = \frac{1}{x}$. (Rule 9)

This rule follows directly from the definition of logarithms and Rule 8 for differentiating e^x.

If $y = \log_e x$, then $e^y = x$

$$\frac{dx}{dy} = e^y \quad \text{(interchanging variables in Rule 8)}$$

$$\frac{dx}{dy} = x \quad \text{(substituting } x \text{ for } e^y\text{)}$$

$$\frac{dy}{dx} = \frac{1}{x} \quad \text{(inverting)}$$

This is one of the most useful theorems in all calculus and it is one which finds wide application in physical chemistry.

The derivative of $\log_{10} x$ is the reciprocal of x multiplied by 0.4343.

$$\textbf{If } \mathbf{y = \log_{10} x,\ \frac{dy}{dx} = \frac{0.4343}{x}.} \qquad \text{(Rule 10)}$$

This rule is obtained by conversion into $\log_e$ and application of Rule 9 as follows:

$$y = \log_{10} x = (0.4343)\ (\log_e x)^*$$

$$\frac{dy}{dx} = 0.4343\frac{(1)}{x} \quad \text{(by Rule 4 for a multiplied constant and Rule 9)}$$

The derivative of $\log_a x$ is the reciprocal of x multiplied by $\log_a e$.

$$\textbf{If } \mathbf{y = \log_a x,\ \frac{dy}{dx} = \frac{\log_a e}{x}.} \qquad \text{(Rule 11)}$$

This is the general rule of which Rule 10 is a special case. It follows from the rule for the conversion of logarithms given on page 8 and the rule for multiplied constants given on page 77. a may be any constant. For example, if

$$y = \log_5 x, \quad \frac{dy}{dx} = \frac{\log_5 e}{x}$$

The derivative of a^x is the product of a^x and $\log_e a$.

$$\textbf{If } \mathbf{y = a^x,\ \frac{dy}{dx} = a^x \log_e a.} \qquad \text{(Rule 12)}$$

This rule may be proved by taking $\log_e$ of both sides.

$$y = a^x$$

$$\log_e y = \log_e a^x = x \log_e a$$

$$x = (\log_e y)\frac{(1)}{\log_e a}. \quad \text{(Dividing by } \log_e a)$$

$$\frac{dx}{dy} = \frac{1}{y}\frac{1}{\log_e a}. \quad \left(\text{Rules 4 and 9. } \frac{1}{\log_e a} \text{ is a multiplied constant}\right)$$

$$\frac{dy}{dx} = y \log_e a = a^x \log_e a. \quad \text{(Inverting and substituting } a^x \text{ for } y)$$

a is any constant; for example,

$$\text{If } y = 5^x, \quad \frac{dy}{dx} = 5^x \log_e 5.$$

* $\log_e x$ is converted into $\log_{10} x$ through multiplication by 0.4343, (p. 8).

It must be remembered that the rule for exponents of a contains the term $\log_e a$, while the rule for logarithms to the base a contains the term $\log_a e$. These two are easily confused.

ALGEBRAIC SIMPLIFICATION

It is obvious that an expression should be converted, by the ordinary operations of algebra, into that form which is most easily differentiated. No general rules are given because the transformation in each case must be handled as a separate problem. Usually it is better to carry out an operation of division or expansion or multiplication first, if it can be done easily.

The following examples serve as specific illustrations:

1. $$y = (x^2 + 9x)\left(2x^3 + 3x + \frac{1}{x}\right)$$

Since this multiplication is a little long, it is easier to differentiate as a product.

2. $$y = (ax + bx^2)^2$$

This equation can be expanded by the binomial theorem before differentiation, thus avoiding the differentiation of a function of a function.

3. $$y = (ax + bx^2)^{1/2}$$

Since the indicated operation is difficult, the expression should be differentiated as a function of a function, without attempting a simplification.

4. $$y = \frac{\log_{10}^2 (x + 5)}{\sqrt{\log_{10} (x + 5)}}$$

This expression should be changed to $y = \log_{10}^{3/2} (x + 5)$ and differentiated as a function of a function. The differentiation as a quotient would be much more tedious.

5. $$y = \frac{x - 4}{2x^2 - 2x - 24}$$

Whenever possible a fraction should be reduced to its lowest terms. In this case

$$y = \frac{x - 4}{2(x - 4)(x + 3)} = \frac{1}{2(x + 3)}$$

$$\frac{dy}{dx} = \frac{-1}{2(x + 3)^2}.$$

6. $$y = \frac{x^3 - 2x^2 - 2}{x^2}$$

In the case of improper fractions like this, it is better to carry out the division even if a remainder is left. In this example

$$y = x - 2 - \frac{2}{x^2}$$

$$\frac{dy}{dx} = 1 + \frac{4}{x^3}.$$

7.
$$y = \frac{x^3 + 2x + 1}{x + 1}$$

This improper fraction may be reduced by the process of division to the expression, $y = x^2 - x + 3 - \dfrac{2}{x + 1}$, which is easier to differentiate.

8.
$$y = \sqrt{5x + 2}$$

Sometimes it is advantageous for certain reasons to differentiate the inverse function of an expression. In this example

$$y^2 = 5x + 2 \quad \text{or} \quad x = \frac{y^2}{5} - \frac{2}{5}$$

and

$$\frac{dx}{dy} = \frac{2}{5}y \text{ and } \frac{dy}{dx} = \frac{1}{(2/5)y} = \frac{5}{2\sqrt{5x + 2}}.$$

Exercises

The following expressions are to be differentiated:

1. $y = (x^3 - 2)^2$. $\dfrac{dy}{dx} = ?$ *Ans.:* $6x^5 - 12x^2$.

2. $y = \sqrt{a - 2x^5}$. $\dfrac{dy}{dx} = ?$

3. $y = (ax^2 + bx)^n$. *Ans.:* $n(ax^2 + bx)^{n-1}(2ax + b)$. 4. $y = \dfrac{1}{\sqrt{2x^3 - b^2}}$.

5. $y = \sqrt[5]{\dfrac{2x^2 + x}{2x}}$. *Ans.:* $\dfrac{1}{5\sqrt[5]{\left(\dfrac{2x + 1}{2}\right)^4}}$. 6. $y = \sqrt[b]{2x^4 + 3x^2 + 5}$.

7. $y = \left(x^3 + \dfrac{2}{x^2} + \sqrt{x^5}\right)^2$.

Ans.: $\left(x^3 + \dfrac{2}{x^2} + \sqrt{x^5}\right)\left(6x^2 - \dfrac{8}{x^3} + 5\sqrt{x^3}\right)$.

8. $y = \sqrt[3]{(6x^2 - 3x^5)^2}$. 9. $y = x^3 + \sqrt{x^4 + 3}$. *Ans.:* $3x^2 + \dfrac{2x^3}{\sqrt{x^4 + 3}}$.

10. $y = \sqrt[3]{\sqrt{u^3} + \sqrt[3]{u^2} + \sqrt{\dfrac{1}{u^3}}}$.

11. $y = e^{ax}$.

$s = ax. \quad y = e^s.$

$\frac{ds}{dx} = a. \quad \frac{dy}{ds} = e^s.$

$\frac{dy}{dx} = \frac{dy}{ds} \times \frac{ds}{dx} = ae^{ax}.$

12. $y = e^{-2x}$.

13. $y = e^{3x^2}$. *Ans.:* $6xe^{3x^2}$.

14. $y = 2e^{\sqrt{x}}$.

15. $y = \sqrt{e^x + ae^{bx}}$. *Ans.:* $\frac{e^x + abe^{bx}}{2\sqrt{e^x + ae^{bx}}}$.

16. $y = 4e^{-2x^2}$.

17. $y = ae^{\frac{q}{rt}}$. $\frac{dy}{dt} = ?$ *Ans.:* $\frac{-aq}{rt^2}e^{\frac{q}{rt}}$.

18. $y = be^{3x+7}$.

19. $y = \log_e x^2$. *Ans.:* $\frac{2}{x}$.

$s = x^2. \quad y = \log_e s.$

$\frac{ds}{dx} = 2x. \quad \frac{dy}{ds} = \frac{1}{s}.$

$\frac{dy}{dx} = 2x\frac{(1)}{s} = \frac{2x}{x^2} = \frac{2}{x}.$

20. $y = a \log_e x^3$.

21. $y = \log_e^3 x$. *Ans.:* $\frac{3}{x} \log_e^2 x$.

22. $y = 3 \log_e^2 x^2$.

23. $y = 3e^x \log_e x$. *Ans.:* $\frac{3e^x}{x} + 3e^x \log_e x$.

24. $y = \frac{x^2}{\log_e x}$.

25. $y = \sqrt{\log_{10} x}$. *Ans.:* $\frac{0.2171}{x\sqrt{\log_{10} x}}$.

26. $y = \log_{10}^2 x$.

27. $y = \log_{10} (x^2 + 4)$. *Ans.:* $\frac{0.8686x}{x^2 + 4}$.

28. $y = \sqrt{\log_{10} x^4}$.

29. $y = \log_{10} (x^3 + a)^2$. *Ans.:* $\frac{2.6058x^2}{x^3 + a}$.

30. $y = \frac{1}{\log_a x}$.

31. $y = b \log_a x^c = bc \log_a x$. *Ans.:* $\frac{bc \log_a e}{x}$.

32. $y = \log_5 (x^2 + 5)^2$.

33. $y = 3^{x^2}$. *Ans.:* $(2x)3^{x^2} \log_e 3$.

34. $y = \frac{1}{2} a^{2x}$.

35. $y = 3a^{ax^2/2}$. *Ans.:* $3xa^{1 + \frac{ax^2}{2}} \log_e a$.

36. $y = \frac{3x^3}{3^x}$.

37. $y = e^x a^x$. *Ans.:* $e^x a^x \log_e a + a^x e^x$.

38. $y = 5^{2x}$.

39. $y = \frac{2(x + 3)}{\sqrt{x + 3}}$. *Ans.:* $\frac{1}{\sqrt{x + 3}}$.

40. $y = \frac{3x^4 - x^2 - 1}{x^2}$.

41. $y = \frac{(x^2 + 2)(x^2 - 2)}{x^8 - 6x^4 + 8}$. *Ans.:* $\frac{-4x^3}{(x^4 - 2)^2}$.

42. $y = \frac{3x^3 + 6x^2 + 3x}{x + 1}$.

43. $y = x^3\left(x + 2\sqrt{x} - \frac{a}{x}\right)$. *Ans.:* $x(4x^2 + 7\sqrt{x^3} - 2a)$.

44. $y = \frac{2}{\sqrt{3 - x}}$.

NOTE. $\log x^a = \log (x)^a$ and $\log^a x = (\log x)^a$.

CHAPTER IX

GRAPHS AND CALCULUS

Graphical Significance of Differentiation

It is easy to become so absorbed in the process of differentiation discussed in the preceding chapter that the significance of differentiation is lost. The real meaning of calculus may be clearly visualized with the help of graphs. It will be remembered that the slope of a line is the tangent of the angle which the line makes with the X axis, and that this tangent is given by the ratio of the increase of the ordinate to the increase of the abscissa. With straight lines, the ratio of finite increases, $\Delta y/\Delta x$, gives the slope, but with curved lines the slope is changing continuously, and *the slope of the curve at any point is given by* dy/dx. A straight line which is drawn tangent to the curve at a given point is the graphical representation of dy/dx as shown in Fig. 31 for the curve $y = x^2/5$. The numerical value of the tangents may be determined in this way at different points, and it is found by reference to Table I that every tangent to the curve has a value equal to two-fifths of the value of x at the point of tangency.

Since the slopes of the tangents are the dy/dx's, their values may be obtained directly by the rule for differentiation. If $y = x^2/5$, $dy/dx = \frac{2}{5}x$. The results by the graphical method and the calculus methods are identical, but the latter is more accurate and more convenient, while the former clumsy method helps to visualize the meaning of differentiation.

In ascending curves such as $y = x^2/5$, y increases when x increases. Both dy and dx are positive and so their ratio, dy/dx is positive. The fact that the curve of Fig. 31 becomes steeper as x increases, shows that dy/dx is becoming larger. In descending curves, such for example as $y = -x^2/5$ where $dy/dx = -\frac{2}{5}x$, y decreases when x increases and the slope is negative.

In connection with the plotting of equations, it should be noted that the added constant in the equation, *i.e.*, the intercept on the Y axis, does not affect the slope of the curve. In terms of calculus, the added constant drops out on differentiation and dy/dx

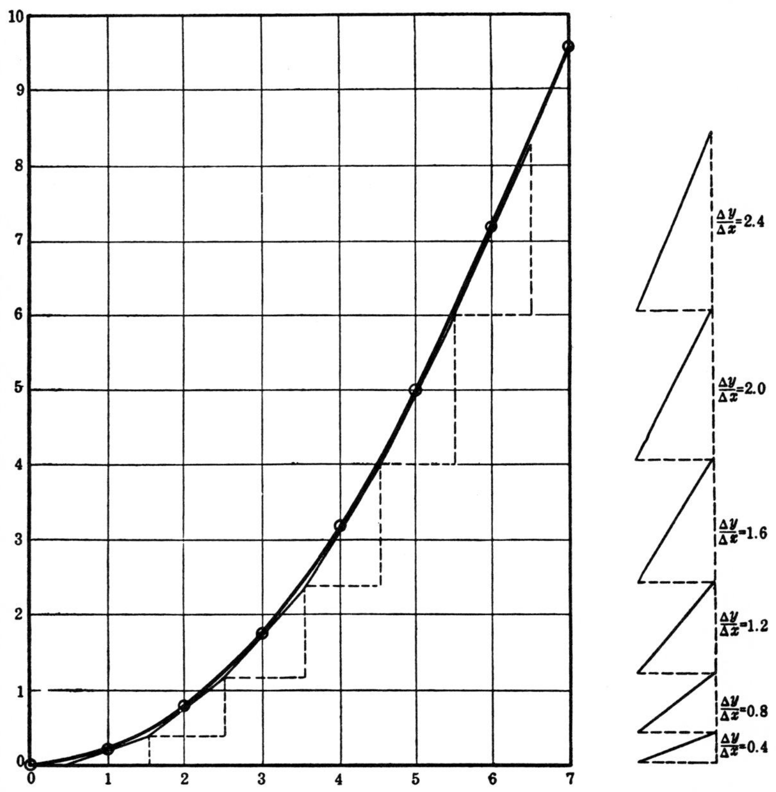

FIG. 31.—The slope of the curve $y = x^2/5$ at various values of x. The slope is determined at any point by drawing a tangent and evaluating $\Delta y/\Delta x$. The slope may be found more accurately by differentiating the equation of the line. For the curve shown, $dy/dx = 0.4x$.

has the same value no matter what added constant is contained in the original equation.

The equation $dy/dx = \frac{2}{5}x$ (Table I) means in words that y increases at a different rate than x and that the ratio of the rates is always numerically equal to two-fifths the value of x.

TABLE I.—THE SLOPE OF THE LINE $y = x^2/5$

x	y	$\Delta y/\Delta x$ (Determined graphically)	$dy/dx = \frac{2}{5}x$ (Determined by differentiation)
0	0	0	$\frac{2 \times 0}{5} = 0$
1	0.2	$\frac{0.4}{1} = 0.4$	$\frac{2 \times 1}{5} = 0.4$
2	0.8	$\frac{0.8}{1} = 0.8$	$\frac{2 \times 2}{5} = 0.8$
3	1.8	$\frac{1.2}{1} = 1.2$	$\frac{2 \times 3}{5} = 1.2$
4	3.2	$\frac{1.6}{1} = 1.6$	$\frac{2 \times 4}{5} = 1.6$
5	5.0	$\frac{2.0}{1} = 2.0$	$\frac{2 \times 5}{5} = 2.0$
6	7.2	$\frac{2.4}{1} = 2.4$	$\frac{2 \times 6}{5} = 2.4$

The real meaning of the derivative should be kept in mind even for more complicated expressions. For example, if $y = \sqrt[5]{(x^3 + 2x^2 + 5x)^3}$, dy/dx is found by the rules for differentiation to be $\frac{3(3x^2 + 4x + 5)}{5\sqrt[5]{(x^3 + 2x^2 + 5x)^2}}$. This means, in words, that when x increases a little, y increases $\frac{3(3x^2 + 4x + 5)}{5\sqrt[5]{(x^3 + 2x^2 + 5x)^2}}$ times as much. Taking a concrete case, when $x = 2$, and x increases a little, y increases $\frac{3(3 \times 2^2 + 4 \times 2 + 5)}{5\sqrt[5]{(2^3 + 2 \times 2^2 + 5 \times 2)^2}}$ or 4.075 times as much. Graphically, this differentiation means that when $\sqrt[5]{(x^3 + 2x + 5x)^3}$ is solved for various values of x and these solutions are plotted on the Y axis against x, the tangent to the curve at any point is given by the expression $\frac{3(3x^2 + 4x + 5)}{5\sqrt[5]{(x^3 + 2x^2 + 5x)^2}}$. At the point on the curve where $x = 2$, the tangent has the numerical value 4.075 and this is the slope of the curve at that point. Reference to a table of tangents shows that the tangent at this point makes an angle of 76° 12′ with the X axis.

Rate of Change of the Slope. Successive Differentiation

If the slopes of the curves $y = x^2/5 + 2$ and $y = -x^2/5 + 2$ are plotted against x, the straight lines of Fig. 32

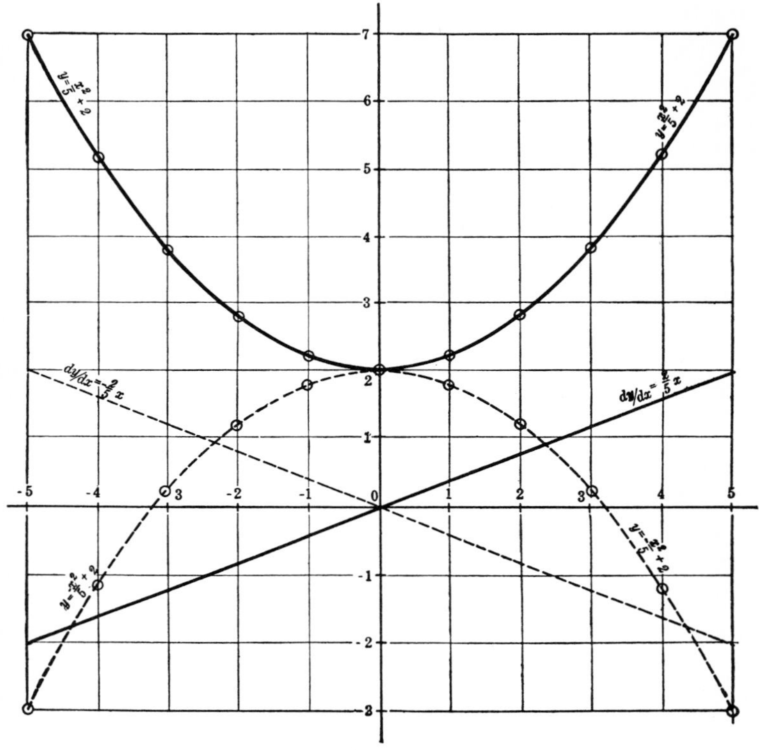

Fig. 32.

In full lines:

The graph of the equation $y = \frac{x^2}{5} + 2$. y is plotted against x.

The graph of the derivative of $y = \frac{x^2}{5} + 2$. $\frac{dy}{dx}$ is plotted against x.

In dotted lines:

The graph of the equation $y = \frac{-x^2}{5} + 2$. y is plotted against x.

The graph of the derivative of $y = \frac{-x^2}{5} + 2$. $\frac{dy}{dx}$ is plotted against x.

are obtained. Now the slope of the new line is given by the ratio of the ordinate increase to the abscissa increase, or $d(dy/dx)/dx$. This expression is equivalent to $d(dy)/(dx)^2$ or d^2y/dx^2. It is called the second derivative of y with respect to x. In words, it is the rate of change of the slope of the original curve, and it shows how an increase in x changes the slope of the original curve.

It is easier to visualize this second derivative by reference to a phenomenon, for which the derivatives have well-defined physical meanings. Experiments have shown that the distance traveled by a falling stone in time, t, is given by the equation $s = \frac{1}{2}gt^2$

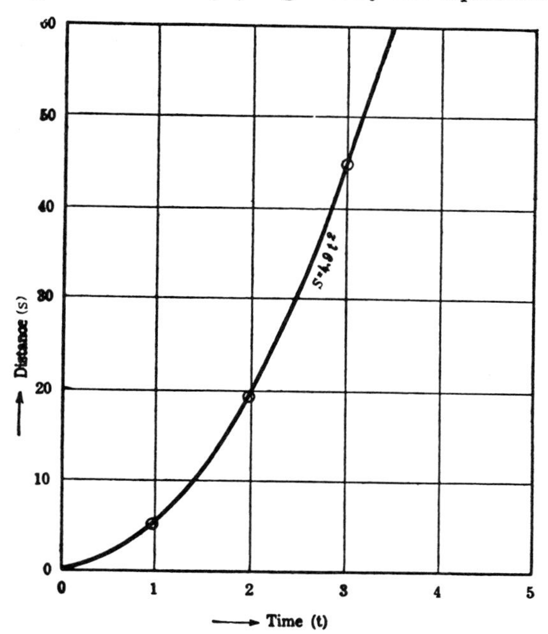

FIG. 33.—Graph of the equation. $s = 4.9t^2$.

$= 4.9t^2$. The first derivative, ds/dt, represents the rate at which the distance increases and is known as the velocity. Applying the rules for differentiation, $ds/dt = gt$. Graphically, the velocity at any time is the tangent to the curve shown in Fig. 33. At the end of the first second $t = 1$, and the velocity is $ds/dt = g \times 1 = 9.8 \times 1 = 9.8$ m. per second. At the end of the third second the velocity is 9.8×3 or 29.4 m. per second, and at the end of the tenth second it is 98 m. per second.

When the velocity is plotted against time, the straight line, shown in Fig. 34 is obtained. The straight line means that, although the velocity increases with time, it does so at a uniform rate. The derivative of velocity with respect to time, or the second derivative of distance with respect to time, is known as acceleration. It gives the increase in velocity for an increase in time. The value of g is obtained by differentiating gt. Graphically, the acceleration is given by the tangent of the curve of Fig. 34. Obviously, since the curve is a straight line, the

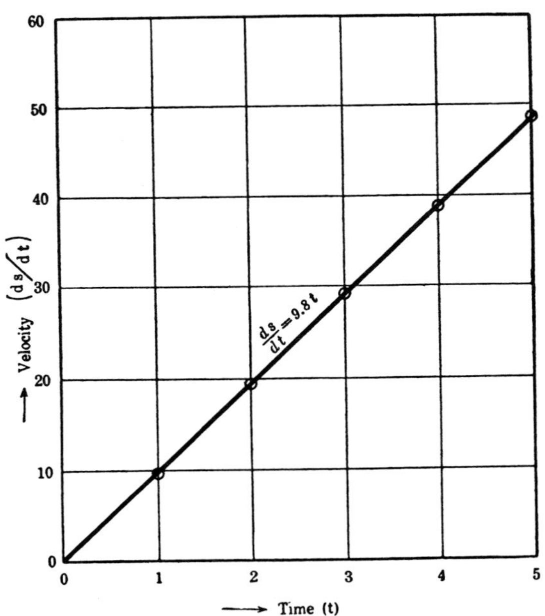

FIG. 34.—Graph of the first derivative of the equation $s = 4.9t^2$ (Fig. 33). $\frac{ds}{dt} = 9.8t$.

slope or tangent (d^2s/dt^2) is the same for all values of t. In other words, the acceleration is a constant. This fact is represented by Fig. 35 in which acceleration or d^2s/dt^2 is plotted against time, giving a horizontal line, which has the value of 9.8 for every value of t.

Going still farther, the slope of this horizontal line is zero, since the tangent of the angle zero is zero. By calculus, also, the derivative of the constant g is zero, *i.e.*, $d^3s/dt^3 = 0$. Since there is no change in acceleration with time, the acceleration must be constant.

This process for obtaining, by calculus, the rate of change of a slope of a curve is known as successive differentiation. Differ-

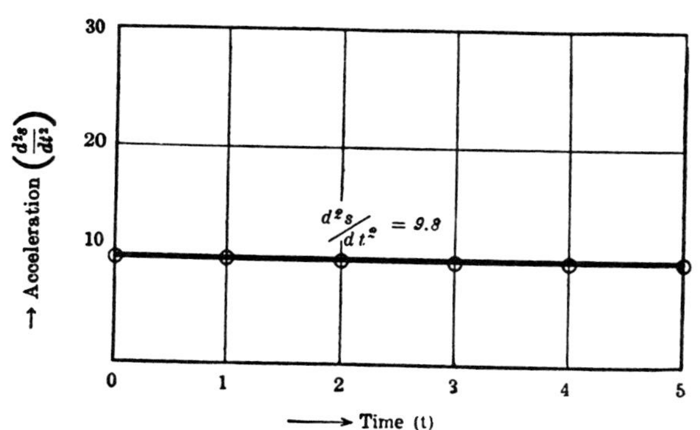

FIG. 35.—Graph of the second derivative of the equation $s = 4.9t^2$. (Fig. 33). $\frac{d^2s}{dt^2} = 9.8$.

entiation of a derivative gives the second derivative and differentiation of the second derivative gives the third. The process may be continued through many differentiations as indicated in the examples of Table II.

TABLE II.—SUCCESSIVE DIFFERENTIATIONS

Original equation....	......	$y = 5x^4$	$y = \log_e x$
First derivative......	dy/dx	$20x^3$	$+1/x$
Second derivative....	d^2y/dx^2	$60x^2$	$-1/x^2$
Third derivative.....	d^3y/dx^3	$120x$	$+2/x^3$
Fourth derivative....	d^4y/dx^4	120	$-6/x^4$
Fifth derivative......	d^5y/dx^5	0	$+24/x^5$
Sixth derivative......	d^6y/dx^6	....	$-120/x^6$
and etc.	......		

MAXIMA AND MINIMA

Problems of maxima and minima may be solved by the method of trial, as illustrated in the following problem. What are the dimensions of the rectangle of greatest area with a perimeter of 10 cm.? The area is the product of the two sides, and the sum of the two sides is half the perimeter, or 5 cm. Various divisions of this 5 cm. are given in tabular form and plotted in Fig. 36.

TABLE III.—AREAS OF VARIOUS RECTANGLES HAVING A PERIMETER OF 10 CM.

Length	Width	Area
4.9	0.1	0.49
4.0	1.0	4.00
3.0	2.0	6.00
2.0	3.0	6.00
1.0	4.0	4.00
0.1	4.9	0.49
2.25	2.75	6.19
2.50	2.50	6.25
2.75	2.25	6.19

The table and the graph both indicate that somewhere between 2 and 3 the area is largest. Since the graph indicates that the maximum value of the area comes when the side is about halfway between 2 and 3, more areas are calculated in this region as shown by the last three calculations. When the sides are each taken as 2.5 the area has its maximum value.

Now this is a natural, but a very clumsy method for determining what value of the independent variable will give to the whole function a maximum value. Differential calculus provides a clever and simple way of solving problems of this type. The idea may be understood best by reference to a graph such as that of Fig. 37. When the maximum point of the curve is being approached the curve is ascending, and in mathematical language dy/dx is positive. When the maximum has been passed the curve is descending, and dy/dx is negative. Just at the peak the slope changes from positive to negative and must go through zero. In other words, the tangent to the curve at the maximum is horizontal. Now in solving this problem it is only necessary to determine what value of x will make dy/dx become zero, and

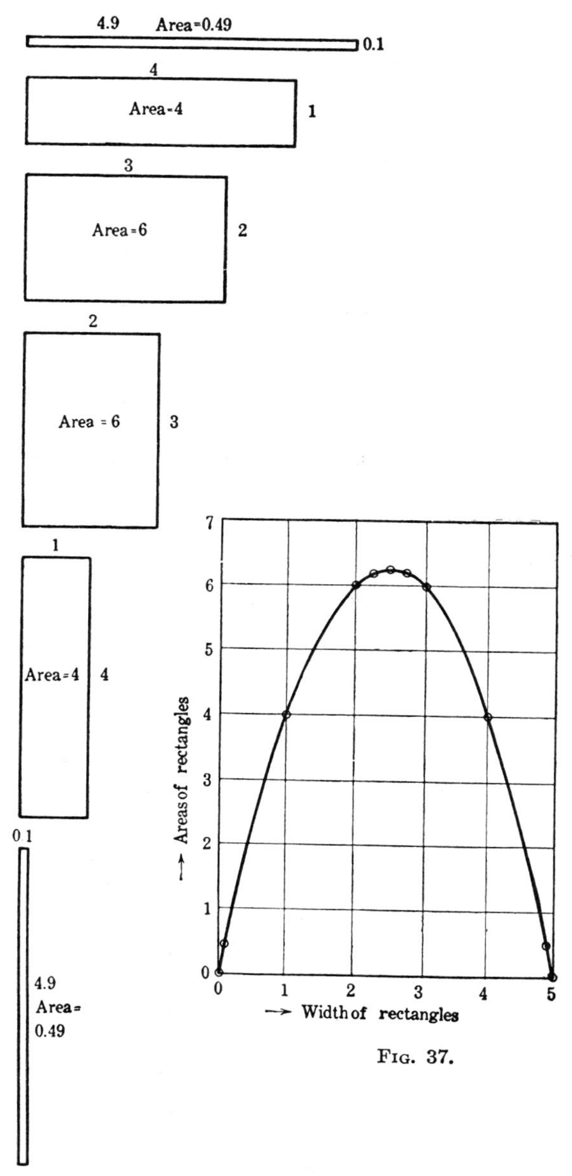

FIG. 37.

FIG. 36.

FIG. 36.—Illustration of a maximum.

Areas of rectangles of different widths, each rectangle having a perimeter of 10 cm.

FIG. 37.—A graphical method for determining what value of x will give to y a maximum value. The areas of the rectangles of Fig. 36 are plotted against the widths

that value of x will then give to the whole equation a maximum value (or minimum value).

In the problem of the rectangle under discussion, x is one side and $5 - x$ is the other, and y is the area. Then $y = x(5 - x) = 5x - x^2$. Differentiating this expression to get the tangent of the curve, it is found that $dy/dx = 5 - 2x$. Now to find what value of x will make dy/dx become zero and give a horizontal tangent, the value of the derivative is set equal to zero and solved for x. Thus if $5 - 2x = 0$, $x = 2.5$. When x has the value 2.5, dy/dx becomes zero and the tangent at this point is horizontal, indicating that the curve is then at the maximum. The maximum value of the function is then obtained by substituting this value of x into the original equation. $y = 5x - x^2 = 5(2.5) - (2.5)^2 = 6.25$.

Taking another example, it is desired to know in the equation $y = 1 + 12x - 2x^2$ what value of x will make y a maximum. $dy/dx = 12 - 4x$. Setting this derivative equal to zero and solving for x,

$$12 - 4x = 0 \text{ and } x = 3.$$

Substituting this value of x into the original equation, $y = 1 + 12x - 2x^2 = 1 + 12 \times 3 - 2 \times 3^2 = 19$. When x is 3, y has its maximum value of 19 and if x is less than 3 or greater than 3, y will be less than 19.

The two functions studied so far exhibit maxima, but it is just as common to have functions which pass through a minimum value. In such a case the curve descends, rapidly at first and then more slowly, passes through a minimum and then ascends. As x increases, the slope is first negative, then zero and then positive. At the minimum, the tangent is horizontal and $dy/dx = 0$, just as in the case of a maximum. It is obvious then, that the procedure given above, in which the derivative is set equal to zero and solved for x, gives either a maximum or minimum, but that it is powerless to distinguish between the two.

In order to decide whether the point in question is a maximum or a minimum, the equation is differentiated a second time. If d^2y/dx^2 is positive, the point is a minimum, and if it is negative, the point is a maximum. It is clear that after passing through a maximum (convex to the X axis) as in Fig. 36, the slope is not only negative, but it becomes more negative as x increases—in other words, the rate of change of the slope, d^2y/dx^2, is negative. In the same way, after passing through a minimum (concave to the X axis) the curve rises with increasing steepness and d^2y/dx^2 is positive.

In the preceding example, $y = 1 + 12x - 2x^2$, $dy/dx = 12 - 4x$ and $d^2y/dx^2 = -4$. The negative value of the second derivative shows that the point $x = 3$, $y = 19$, obtained above, was a maximum.

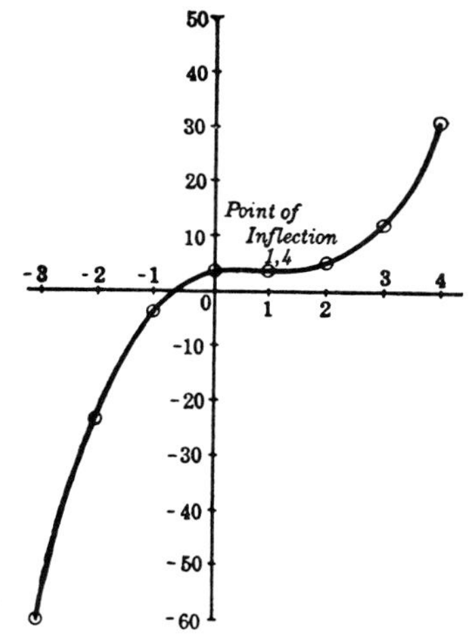

FIG. 38.—An illustration of a point of horizontal inflection. $y = x^3 - 3x^2 + 3x + 3$.

x	-3	-2	-1	0	1	2	3	4
y	-60	-23	-4	3	4	5	12	31

There is still a third case in which dy/dx is zero. When a curve changes from convex to concave, there may be a point on the curve at which the tangent becomes horizontal. This is known as a point of horizontal inflection. The slope becomes smaller and smaller, and passes through zero at the point of inflection and then increases. At a point of inflection of this kind the second derivative as well as the first derivative becomes zero. It must be realized that in changing from one curvature to another the tangent does not pass through a horizontal position in every curve. The equation $y = x^3 - 3x^2 + 3x + 3$, contains a point of horizontal inflection, as shown in Fig. 38.

$$\frac{dy}{dx} = 3x^2 - 6x + 3 = 3(x - 2x + 1) = 3(x - 1)^2$$

When $x = 1$, $(x - 1)$ becomes 0. Therefore dy/dx becomes 0. When $x = 1$, $y = x^3 - 3x^2 + 3x + 3 = 1 - 3 + 3 + 3 = 4$. At the point on the curve where $x = 1$ and $y = 4$, the tangent must be horizontal. Futhermore, $d^2y/dx^2 = 6x - 6$ and substituting for x the value, 1, obtained above, $d^2y/dx^2 = 0$. Since the second derivative is zero at this point, the function does not have a maximum or a minimum value, but it has a point of inflection where the curve changes from convex (to the X axis) to concave; *i.e.*, from a decreasing slope to an increasing slope.

A complete rule then is as follows:

When the derivative of a function is set equal to zero and solved for x, this value of x corresponds to a maximum or a minimum or a point of inflection. If the second derivative is positive for this value of x, there is a minimum; if it is negative, there is a maximum; and if it is neither positive nor negative, but zero, then there may be a point of inflection. (Rule 13)

Another test to distinguish between a maximum and minimum consists in substituting into the original equation values of x a little less and a little greater than the value which makes dy/dx equal to zero. If these values of x give greater values to y, there is a minimum; if they give lesser values to y, there is a maximum; and if one is greater and one is less, there is a point of inflection.

A function may contain both a maximum and a minimum as shown in Fig. 39.

$$y = x^3 - 3x$$

$$\frac{dy}{dx} = 3x^2 - 3$$

When $3x^2 - 3 = 0$,

$$x^2 = 1 \text{ and}$$

$$x = +1 \text{ or } -1.$$

$$\frac{d^2y}{dx^2} = 6x$$

When x is $+1$, d^2y/dx^2 is positive, showing a minimum.

When x is -1, d^2y/dx^2 is negative, showing a maximum.

The minimum value of the function is -2, obtained by substituting $+1$ into the original equation, $y = x^3 - 3x$. This statement does not mean that -2 is the lowest value which the function can have, for when x is -4, y is -52; and indefinitely large values of y can be obtained by increasing x. It means that when x is $+1$, y has the lowest value (-2) in that neighborhood and that larger values of y are obtained by making x larger or smaller than $+1$.

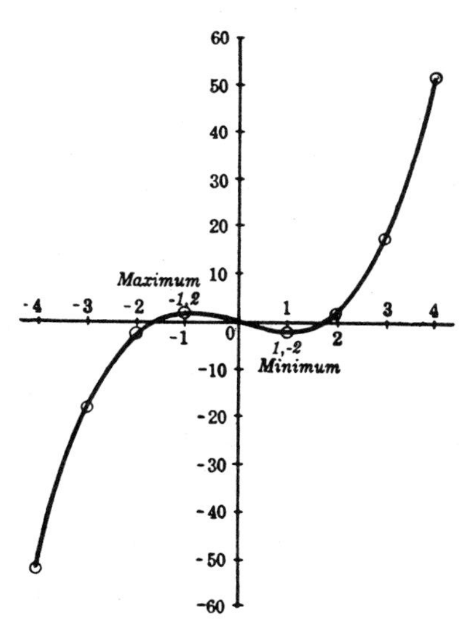

FIG. 39.—A curve exhibiting both maximum and minimum. $y = x^3 - 3x$.

x	-4	-3	-2	-1	0	1	2	3	4
y	-52	-18	-2	$+2$	0	-2	$+2$	18	52

The maximum value of the function, $+2$, is given to the equation when x is -1. Again, $+2$ is not the greatest value which the equation can have, but the greatest in the region where x is -1.

It is evident from the graph that each value of y between $+2$ and -2 may be given by three entirely different values of x, a result which is to be expected from the fact that the original cubic equation has three roots.

It is to be noted that the second derivative of a function of the *second* degree is a constant and can be only positive or only negative or only zero, corresponding to *one* maximum or minimum. In functions with exponents higher than 2, the second derivative may have both positive and negative values, depending on the value of x, and more than one maximum or minimum is possible.

Applications of Maxima and Minima

There are many interesting problems based on maximum and minimum functions. In solving such problems it is necessary first to express the facts with a mathematical formula. The art of expressing physical problems in mathematical language is extremely useful, but, like every art, it comes only with practice.

The drawing of a diagram to illustrate the problem is usually of great help. After the equation for the problem has been found the maximum or minimum values can be readily determined by the methods just described. A few examples are offered to illustrate the procedure.

Into what two parts can the number 4 be divided, so that the sum of the cube of one part and three times the square of the other part shall have a minimum value?

Let $x =$ the first part and $4 - x =$ the second part.
Then

$$y = x^3 + 3(4 - x)^2 = x^3 + 3x^2 - 24x + 48.$$

The facts are now stated in mathematical language and it is desired next to find what value of x will give to y a minimum value.

$$\frac{dy}{dx} = 3x^2 + 6x - 24 = 3(x^2 + 2x - 8) = 3(x + 4)(x - 2)$$

Putting dy/dx equal to zero and solving for x,

$$3(x + 4)(x - 2) = 0, \text{ and } x = 2 \text{ or } -4.$$

When x is 2 or -4, y has a maximum or a minimum value or a point of inflection, and the second derivative is determined in order to decide between the three.

$$\frac{d^2y}{dx^2} = 6x + 6$$

When $x = +2$, d^2y/dx^2 is positive and therefore y is at a minimum
When $x = -4$, d^2y/dx^2 is negative and y is at a maximum.

When 4 is divided into 2 and 2 the conditions of the problem are satisfied. The results should be checked by sketching the curve corresponding to the equation.

In a second example, A is 10 miles east of B. A walks north at 4 miles per hour and B walks east at 2 miles per hour. When are they closest and how far apart are they then?

The problem becomes clearer after drawing a diagram as in Fig. 40. The distances traveled by the two men are different but they are both functions of one variable—time. It is best then to make time the independent variable, and the distance between the men the dependent variable.

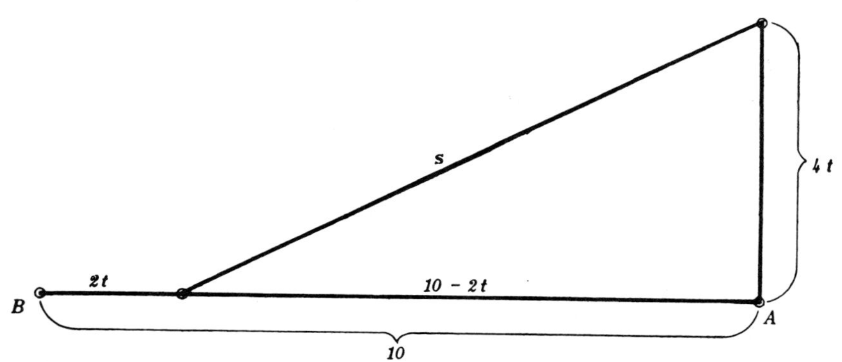

FIG. 40.—A diagram to aid in visualizing a problem involving the calculation of a minimum.

At any time, t, A's position is $4t$ miles north and B's position is $2t$ miles east, since the distance covered is equal to the time multiplied by the velocity. The distance, s, is the hypotenuse of a right-angled triangle and its square is equal to the sum of the squares of the two sides.

$$s^2 = (4t)^2 + (10 - 2t)^2$$

$$s = \sqrt{16t^2 + 100 - 40t + 4t^2} = \sqrt{20t^2 - 40t + 100}$$

$$\frac{ds}{dt} = \frac{40t - 40}{2\sqrt{20t^2 - 40t + 100}}$$

When $t = 1$, $40t - 40 = 0$, and $\frac{ds}{dt}$ becomes zero, giving the condition for a maximum or a minimum. The fact that

$\frac{d^2s}{dt^2}$ is positive shows that y is a *minimum* when $t = 1$. At the end of one hour, therefore, A and B are closest together and after one hour they get farther and farther apart.

When

$$t = 1,\ s = \sqrt{20t^2 - 40t + 100} = \sqrt{20 - 40 + 100} = \sqrt{80} = 8.95 \text{ miles.}$$

In a third example, a projectile is shot up vertically with a velocity of 100 m. per second. What is the greatest height which it will attain, neglecting air resistance?

If it were not for gravity the distance, s, traveled upward in time, t, would be $100t$. If gravity alone were acting the distance through which the projectile fell downward would be $\frac{1}{2}gt^2$ or $4.9t^2$. The actual distance s, above the ground at the time, t, is the resultant of the two distances and $s = 100t - 4.9t^2$. The problem, then, is to find what value of t will give to s a maximum value.

$$\frac{ds}{dt} = 100 - 9.8t$$

When

$$100 - 9.8t = 0, \quad t = \frac{100}{9.8} = 10.2 \quad \left(\text{setting } \frac{ds}{dt} = 0\right)$$

When

$$t = 10.2,\ s = 100 \times 10.2 - 4.9(10.2)^2 = 508.4$$

Therefore the *highest* point that the projectile will attain is 508.4 m. The fact that d^2s/dt^2 is negative proves that the value is a maximum.

This problem is a particularly good one for visualizing the rule used in determining maxima. ds/dt is velocity and the highest point is the point where the velocity is zero. The projectile slows up in its upward flight as gravity has more time to act, becomes stationary, and then falls down with increasing velocity. The downward flight is defined as negative.

In a fourth example the product of the concentration of hydrogen ions, C_{H^+} and hydroxyl ions, C_{OH^-} in any aqueous solution is a constant, 10^{-14}. Mathematically expressed, $C_{H^+} \times C_{OH^-} = 10^{-14}$. At what concentration of hydrogen ions will the sum of the hydrogen ions and hydroxyl ions, y, be a minimum?

The concentration of hydroxyl ions may be expressed in terms of hydrogen ions, thus eliminating one variable.

$$C_{OH^-} = \frac{10^{-14}}{C_{H^+}}$$

$$y = C_{H^+} + C_{OH^-} = C_{H^+} + \frac{10^{-14}}{C_{H^+}}$$

$$\frac{dy}{dC_{H^+}} = 1 - \frac{10^{-14}}{C^2_{H^+}}$$

Setting $1 - \frac{10^{-14}}{C^2_{H^+}}$ equal to zero and solving for C_{H^+},

$$C^2_{H^+} = 10^{-14} \quad \text{or} \quad C_{H^+} = \sqrt{10^{-14}} = 10^{-7}$$

$\frac{d^2y}{dC^2_{H^+}} = +\frac{2 \times 10^{-14}}{C^3_{H^+}}$ and y is a minimum, because $\frac{d^2y}{dC^2_{H^+}}$ is positive.

When the concentration of hydrogen ions is 10^{-7} the number of ions is a minimum. It is also true that at this minimum concentration the number of hydroxyl ions is likewise 10^{-7} since $C_{OH^-} = 10^{-14}/C_{H^+} = 10^{-14}/10^{-7} = 10^{-7}$. When the concentration of hydrogen and hydroxyl ions is just equal the solution is exactly neutral, and the total number of ions is 2×10^{-7}.

Exercises

The following equations are to be plotted and the tangents determined graphically at various points. The values obtained in this way are then compared with the values of dy/dx found by applying the rules for differentiation.

1. $y = 2x + 3$.
2. $y = 2x^2 + 2$.
3. $y = \frac{1}{x^2} + 5$.
4. $y = e^x$.
5. $y = x^2 + 3x + 2$.
6. What is the third derivative of $y = 3x^4 + 2x$? *Ans.:* $\frac{d^3y}{dx^3} = 72x$.
7. What is the fourth derivative of $y = 2x^3 + 3x^2 - 12x + 5$?
8. $y = 2 \log_{10} 2x$. $\frac{d^4y}{dx^4} = ?$ *Ans.:* $\frac{-5.212}{x^4}$.
9. What angle does the tangent to the curve $y = (2x + 3)^2$ make with the X axis at the points $x = 0$, $x = 1$, and $x = 2$?
10. What is the slope of the curve $y = 2x - x^2$ at the points $x = 0$, $x = 1$, and $x = 2$? What is the rate of change of the slope at these points?

Ans.: When $x = 0$, $\frac{dy}{dx} = 2$. $\frac{d^2y}{dx^2} = -2$.

When $x = 1$, $\frac{dy}{dx} = 0$. $\frac{d^2y}{dx^2} = -2$.

What are the maximum or minimum values of the following equations?

11. $y = -x^2 + 2x + 1$. *Ans.:* Maximum of 2 when $x = 1$.
12. $y = x^2 - 3x + 2$.
13. $y = 2x^3 - 3x^2 - 12x + 5$. *Ans.:* Minimum of -15 when $x = 2$, and Maximum of 12 when $x = -1$.
14. $y = x^3 + x^2 - 2x + 3$.
15. $y = 3x^4 - 8x^3 - 6x^2 + 24x - 11$.

Ans.: Maximum of 2 when $x = 1$.
Minimum of -30 when $x = -1$.
Minimum of -3 when $x = 2$.

16. How can the number 20 be divided into two parts such that the sum of four times the reciprocal of one and nine times the reciprocal of the other may be a minimum? *Ans.:* 8 and 12.

17. How can the number 12 be divided into two parts such that the product of one part by the square of the other will have a maximum value? What is this maximum value?

18. A man is in a row boat 3 miles from the nearest point, A, on a straight beach. He wishes to reach a point, B, 5 miles from A, in the shortest possible time. If he can row 3 miles an hour and walk 4 miles an hour, at what point on the beach should he land? How long will it take him to reach B by this route (hint—time = distance/velocity)?

Ans.: He should land at a point 3.4 miles beyond A. He can then make the trip in 1.91 hours. If he had landed at A, it would have taken him 2.25 hours. If he had landed at B, it would have taken him 1.94 hours.

19. What is the relation between the height and radius of a closed cylindrical pint can when the can has the largest volume for a given area (*i.e.*, for a given cost of sheet metal)?

20. The cost, y, of running a certain launch in still water is proportional to the cube of the velocity, v. What is the most economical velocity at which to run the launch up stream against a four mile current? (The cost depends also on the time required to make the trip.)

$$y = kv^3 \text{ and } y = k'\frac{1}{v - 4}$$ *Ans.:* 6 miles per hour.

21. The density, s, of water is given by the equation, $s = s_0(1 + at + bt^2 + ct^3)$ where s_0 is the density at 0° C. and t is the temperature in centigrade degrees.

$a = 5.3 \times 10^{-5}$. $b = -6.53 \times 10^{-6}$.
$c = 1.4 \times 10^{-8}$.

At what temperature does water have a maximum density?

22. The velocity of a certain auto-catalytic reaction is given by the expression $v = kx(a - x)$. The velocity is greatest when $x = \frac{1}{2}a$. x is the amount of material decomposed at time, t, and, a is the amount of material originally present. How can this fact be demonstrated mathematically?

23. A long strip of copper is to have its edges bent up to make a rectangular trough. The strip is 50 cms. wide. How wide should the trough be, and how deep in order that it can carry the greatest amount of water (*i.e.*, the largest cross-section)?

24. The current, I, from a battery is given by $I = \frac{E}{r + R}$, where E is the electromotive force, r the internal resistance, and R the external resistance. The energy utilized in the external circuit is I^2R. If it is desired to utilize as much energy as possible R must be equal to r. How can this fact be demonstrated mathematically?

CHAPTER X

THE DIFFERENTIAL

Theory

The preceding discussions have centered around the derivative, dy/dx. It will be remembered that the derivative is the limit to which the ratio $\Delta y/\Delta x$ approaches as the increase in x (*i.e.*, Δx) becomes smaller and smaller and approaches zero. So far, dy and dx have had no significance when standing by themselves, and they have always been used together in the ratio dy/dx.

It is convenient, however, in many cases to separate the two terms and to use dy and dx as individual quantities. When so used they are called differentials.

The different notations used in calculus must be kept clearly in mind. They may be summarized in the three following definitions, where y is some function of x.

Increment = Δx or Δy = finite increase in x or y.

Derivative = dy/dx = limiting value of $\Delta y/\Delta x$ as Δx becomes smaller and smaller and approaches zero.

Differential = dx(or dy) = one part of the derivative, dy/dx, taken separately. The relation between the differential and the derivative as given by the following equation is important

$$\mathbf{dy = \frac{dy}{dx}dx.} \qquad \text{(Rule 14)}$$

According to this equation *the differential of y is equal to the derivative of y with respect to x, times the differential of x.*

The relation may be shown by first considering increments, which are finite numbers, and to which the ordinary laws of algebra can be applied without reservation.

The fraction $\Delta y/\Delta x$ has the significance of any fraction such as $\frac{1}{2}$ or $0.0001/0.0002$. The following identical equation may be written

$$\frac{\Delta y}{\Delta x} = \frac{\Delta y}{\Delta x}$$

and it can be transposed by the ordinary rules of algebra to read

$$\Delta y = \frac{\Delta y}{\Delta x} \times \Delta x.$$

A very important deduction from this equation is that dy and dx and dy/dx may be treated in calculus as if they were algebraic quantities. It is extremely fortunate for the practical use of calculus that the differentials and the derivatives can be subjected to ordinary mathematical operations.

These operations include, among others, the separation on different sides of an equation for purposes of integration (p. 115); separation for partial differentiation (p. 175); cancelling for functions of functions as in the operation $\frac{dy}{ds} \times \frac{ds}{dx} = \frac{dy}{dx}$ (p. 83); and division by a third variable as in rate problems as for example where $\frac{dv}{dr} = \frac{dv}{dt} \div \frac{dr}{dt}$.

Any of the rules of differentiation given previously for the derivative may now be transposed for use with differentials, as shown in the few following examples:

If $y = x$, $\frac{dy}{dx} = 1$; or $dy = dx$

If $y = u + v$, $\frac{dy}{dx} = \frac{du}{dx} + \frac{dv}{dx}$; or $dy = \frac{du}{dx}dx + \frac{dv}{dx}dx = du + dv$

If $y = x^n$, $\frac{dy}{dx} = nx^{n-1}$; or $dy = nx^{n-1}dx$

If $y = uv$, $\frac{dy}{dx} = u\frac{dv}{dx} + v\frac{du}{dx}$, or $dy = u\frac{dv}{dx}dx + v\frac{du}{dx}dx = udv + vdu$

If $y = \frac{u}{v}$, $\frac{dy}{dx} = \frac{v\frac{du}{dx} - u\frac{dv}{dx}}{v^2}$, or $dy = \frac{vdu - udv}{v^2}$

If $y = e^x$, $\frac{dy}{dx} = e^x$, or $dy = e^x dx$

If $y = \log_e x$, $\frac{dy}{dx} = \frac{1}{x}$, or $dy = \frac{dx}{x}$. Also $d \log_e x = \frac{dx}{x}$.

Practical Approximation

Another application of the equation $dy = \frac{dy}{dx}dx$ lies in the easy solution of practical problems where an approximation is permissible. It is clear that dy is not equal to Δy but it is also clear that the difference between dy and Δy may be made negligible by making Δy sufficiently small. The following problem illustrates the application of this approximation.

A square, 2 cm. on a side, has an area of 4 sq. cm. What is the increase in the area if the sides are increased by 0.01cm. (*i.e.* by ½ of 1 per cent)?

$$y = x^2. \quad \text{(Where } y = \text{area and } x = \text{side).}$$

$$\frac{dy}{dx} = 2x = 4. \quad \text{(Substituting } x = 2. \text{ by statement of problem.)}$$

$$dx = 0.01. \quad \text{(By statement of problem.)}$$

$$dy = \frac{dy}{dx}dx = 4(0.01) = 0.0400. \quad \text{(By rule for differentials.)}$$

By differential calculation, then, the area has increased by 0.040 sq. cm. By direct calculation the area has increased by $(2.01)^2 - 2^2 = 4.0401 - 4.000 = 0.0401$ sq. cm. The error introduced by the approximation is 1 part in 400 or 0.25 per cent.

If the sides had increased from 2 to 2.1 (an increase of 5 per cent) the area would have increased by $(2.1^2 - 2^2)$ or 0.41 sq. cm. The error introduced by using the differential would have been one part in forty or 2.5 per cent.

If the sides had increased by 0.001 (0.05 per cent) the error introduced by using the differential would have been only 0.025 per cent.

It is clear that the calculation with differentials is more accurate when the changes are small. These results may be deduced directly, also, from the table on page 71 where the ratio $\Delta y/\Delta x$ is compared with dy/dx. In the simple problem given above, the direct calculation of the increase is fully as easy if not easier than the calculation by differentials. In many problems however the use of differentials effects a considerable saving of time.

Exercises

1. $y = 3x^2 - \frac{2}{x} - 5. \quad dy = ?$
2. $y = 2x^3 + \frac{4}{x^2}. \quad dy = ?$ *Ans.:* $dy = \left(6x^2 - \frac{8}{x^3}\right)dx.$
3. $y = ae^x x^2. \quad dy = ?$
4. $y = 5 \log_e x. \quad dy = ?$ *Ans.:* $dy = \frac{5}{x}dx.$
5. $y = 2xae^x + ax^2e^x. \quad dy = ?$

6. A soap bubble is expanding in volume. When the radius is 5 cm., approximately how much does the volume increase, when the radius increases by 0.1 cm.? The volume, v, of a sphere is $\frac{4}{3}\pi r^3$.

$$\frac{dv}{dr} = 4\pi r^2$$

$$dv = \frac{dv}{dr}\, dr = 4\pi r^2 dr = 4\pi 5^2 \times 0.1 = 31.4 \text{ cc.}$$

7. When a soap bubble, 50 cc. in volume, has 1 cc. more of air blown into it, how much does the radius increase (approximately)?

$$v = \tfrac{4}{3}\pi r^3 \text{ and } r = \sqrt[3]{\frac{3v}{4\pi}}. \quad dr = \frac{dr}{dv}\, dv.$$

8. A stone is falling at a rate such that the distance s is related to the time, t, by the equation $s = \frac{1}{2}gt^2$, where $g = 9.8$ m. per second. Approximately, how far will it fall between the eleventh and the twelfth seconds?

$$\frac{ds}{dt} = gt \text{ and } ds = gt\, dt = g \times 11.5 \times 1 = 11.5g = 112.7$$

How far will it fall in a second, after it has been falling for a minute? How long will it take in falling from 100 m. to 102 m. below the starting point?

9. The heat, Q, required to raise the temperature, T, of a certain liquid 1° is given by the equation

$$Q = a + bT + cT^2.$$

How much heat is required to heat the material from 300.0 to 300.1° K?

CHAPTER XI

INTEGRAL CALCULUS

Theory

If a number, y, is divided into a large number of small parts, each part being Δy, then the sum of the small parts will give back the original number y. Mathematically expressed, $\Sigma \Delta y = y$ where the symbol Σ stands for the process of summation or addition. If the small parts are made exceedingly small, so small in fact, that making them smaller does not make any difference in a calculation, they are called dy's and again their sum will give back the original y. In this case a very large number of parts must be added together and the process is called integration instead of summation. The symbol for this integration process is the integral sign, $\int$, which is nothing but a distorted S, standing for summation. Then,

$$\int dy = y.$$

If y is a variable which depends for its value on the value of the independent variable, x, then $y = f(x)$, and dy depends on dx. The object of differential calculus is to find the relation between dy and dx when the relation between y and x is known. The object of integral calculus, on the other hand, is to find the relation between y and x when the relation between dy and dx is known. When dy/dx is known, the equation $y = f(x)$ can be solved by calculus, and it can be ascertained what function of x, y is. Integration makes it possible then to determine the general relation between two variables from a knowledge of the way in which they are varying over very small ranges. For example, if the velocity at two different times is known, it is possible to find the general expression which gives the distance of a falling body as a function of time. Also, the vapor pressure of a liquid may be calculated at any temperature if it is known at two temperatures.

Integration is the reverse of differentiation, just as extracting the square root is the reverse of squaring. When $x^4/4$ is differentiated x^3 results, and so it follows that integration of x^3dx gives $x^4/4$. Only those expressions can be integrated which have been found previously by differentiation. For example, no one has yet found a function which on differentiation will give 2^{-x^3} and so the integration of $2^{-x^3}dx$ cannot be done directly. In differential calculus dy and dx are usually kept together as a ratio, but in integral calculus dy is placed by itself so that on summing up all the dy's the value of y can be obtained at once.

If $y = x^4/4$, then $dy/dx = 4x^3/4 = x^3$, and in the notation of differentials (Rule 14) $dy = x^3dx$. Reversing the process, if $dy = x^3dx$, then $\int dy = \int x^3dx$, since the sums of the small parts are equal if the small parts are always equal. It follows next that $y = x^4/4$, since $\int dy = y$, and since $\int x^3dx = x^4/4$ because the differentiation of $x^4/4$ gave x^3. This process can be stated in various ways as follows:

1. If $dy = x^3dx, \quad y = \dfrac{x^4}{4}$
2. If $\dfrac{dy}{dx} = x^3, \quad y = \dfrac{x^4}{4}$
3. The integral of x^3dx is $\dfrac{x^4}{4}$
4. The integration of x^3dx gives $\dfrac{x^4}{4}$
5. $\int x^3dx = \dfrac{x^4}{4}$

The expression in x to be integrated must always contain the term dx. In other words, the integral sign must be followed by a differential.

Now the statement that the integration of x^3 gives $x^4/4$, is not strictly correct; for $y = \dfrac{x^4}{4} - a$, or $y = \dfrac{x^4}{4} - 2$, or $y = \dfrac{x^4}{4} +$ any constant, all give the same result on differentiation, namely $dy/dx = x^3$ (Rule 3). Since all the added constants dropped out on differentiation, a constant term must be put into the integrated expression to allow for them, when the process is reversed. Thus if $dy = x^3dx$, $y = \frac{x^4}{4} + C$ rather than $y = \frac{x^4}{4}$, where C is a constant called the integration constant. The elimination of the integral sign is accompanied by the addition of the integration constant.

The fact that the integration is the reverse of differentiation may be shown clearly by reference to the curve $y = \frac{x^2}{5} + 2$, shown previously in Fig. 31, and again in Fig. 41. The full line, A, is the graph of the equation, and the values of $\Delta y/\Delta x$, represented by triangles with dotted lines were determined at various values of x on page 93. It was the object of differential calculus to find these dotted triangles $\Delta y/\Delta x$, when the triangles were made very small—in other words, it was desired to find the slope of the true tangents. Now in integral calculus the dotted triangles are given and it is required to build them up to give the full line.

The line shown at B is obtained by building up large triangles; and the line shown at C is obtained with the smaller triangles which were shown in Fig. 31. When the Δx's are made exceedingly small, they can be built up into a perfectly smooth curve as shown at A. The curve of A, then, is a limit to which the $\Delta y/\Delta x$'s approach as the Δx's become smaller and smaller.

The value of y for any value of x can be determined by adding up all the Δy's (when the starting point is known). It is given exactly by $\int dy$. Now by trigonometry, dy is equal to dx multiplied by the tangent of the angle, or $dy = \tan \alpha dx = \left(\frac{dy}{dx}\right)dx$. But dy/dx for this particular set of triangles was always $0.4x$.

$$dy = \frac{dy}{dx}dx = 0.4xdx \quad \text{(substituting)}$$

$$\int dy = \int 0.4xdx \quad \text{(taking integrals)}$$

$$y = \frac{x^2}{5} + C. \quad \left(\text{Because } \frac{x^2}{5} + C \text{ gives } 0.4xdx \text{ on differentiation}\right)$$

The curve can be built up from its little triangles and the equation, $y = \frac{x^2}{5} + C$, can be constructed from the known values of dy/dx.

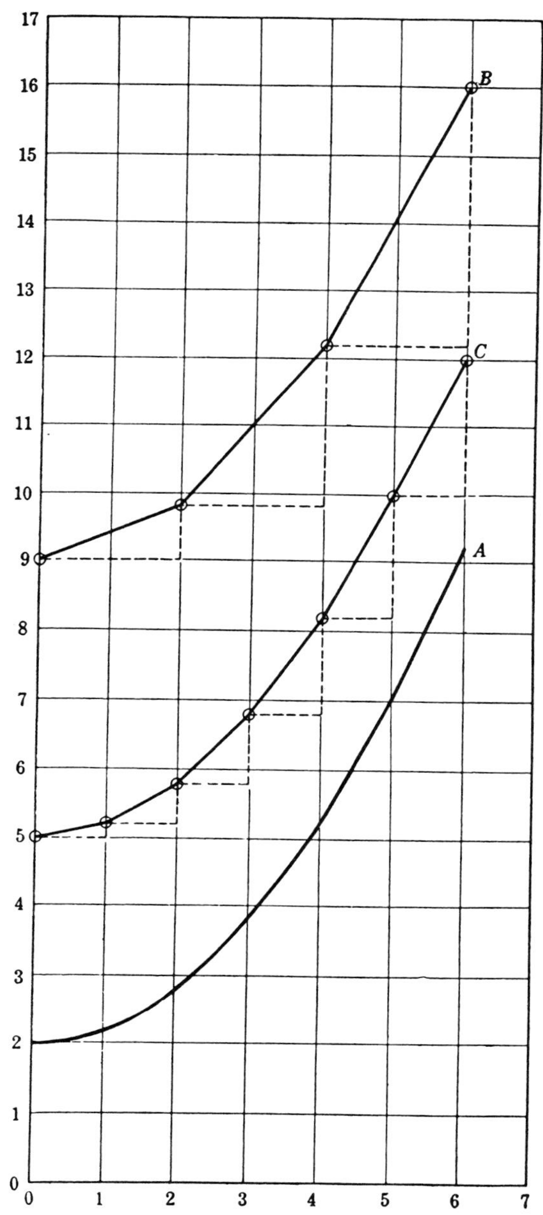

FIG. 41.—The construction of a curve by building up triangles ($\Delta y/\Delta x$). The smaller the values of Δx the smoother is the curve. The triangles of curve C are shown also in Fig. 31.

but the value of C is unknown and so the curve cannot be located definitely and the various ordinates cannot be determined.

The fact that the integrated equation contains an unknown constant, and the fact that there is nothing to tell where to start the curve are both expressions of the same lack of definiteness. When C is known, the curve can be started, for the integration constant C is the intercept of the curve on the Y axis. It is evident that C is the intercept on the Y axis because when $x = 0$, $y = C$; and when $x = 0$ on a graph, the function lies on the Y axis.

The intercept on the Y axis has nothing to do with the slope of the curve but it has a great deal to do with the numerical value of y. In Fig. 41, the series of triangles, B and C, were started at random from various points on the Y axis, but the curve A, was plotted directly from the equation $y = \frac{x^2}{5} + 2$. The integration constant is 2, and the intercept on the Y axis is 2. It is necessary to know the integration constant before the numerical value of an integration can be obtained, and methods for evaluating it, or eliminating it will be described in a later section. In an ordinary integration process the constant, C, should be added, *always*, as a reminder that more information is necessary before a numerical answer can be obtained.

RULES

The rules for integration follow directly from the rules for differentiation, and no special proofs are necessary. The integration of an expression consists in finding a second expression which on differentiation will give the first. As a rule, integration is a more difficult operation than differentiation.

The first rule of integration has already been given. It follows directly from the definitions of the integral and the differential that,

$$\int \mathbf{dx} = \mathbf{x + C}. \qquad \text{(Rule 15)}$$

According to Rule 1, if $y = x^{(n+1)}$, $dy/dx = (n + 1)x^n$ and by Rule 14, $dy = (n + 1)x^n dx$. Since the differential of $x^{(n+1)}$ is $(n + 1)x^n dx$ it follows that the integral of $(n + 1)x^n dx$ is x^{n+1}.

The coefficient $(n + 1)$ may be eliminated by introducing a multiplied constant which will remain unchanged after differentiation and cancel out the undesired coefficient. Thus, if

$$y = \frac{1}{n+1} x^{n+1}, \frac{dy}{dx} = \frac{n+1}{n+1} x^n, \text{ and } dy = x^n dx.$$

This differentiation shows that the integral of $x^n dx$ is

$$\frac{x^{n+1}}{n+1}.$$

This rule may be stated in words as follows: A variable raised to a constant power may be integrated by raising the exponent by 1 and dividing by the increased exponent.

$$\int \mathbf{x^n dx} = \frac{\mathbf{x^{n+1}}}{\mathbf{n+1}} + \mathbf{C}. \qquad \text{(Rule 16)}$$

For example,

$$\int x^3 dx = \frac{x^4}{4} + C, \text{ and}$$

$$\int x^{-2} dx = \frac{x^{-1}}{-1} + C = \frac{-1}{x} + C, \text{ and}$$

$$\int x^{1/2}\, dx = \frac{x^{3/2}}{3/2} = \frac{2}{3}\sqrt{x^3} + C$$

Reversing Rule 4, it follows that if a multiplied constant, k, remains unchanged on differentiation, it will remain unchanged also on integration. Since the multiplied (or divided) constants are not affected by the process of integration it is convenient to transfer them to a position in front of the integral sign and ignore them until the integration is finished.

$$\int \mathbf{kf(x)dx} = \mathbf{k}\int \mathbf{f(x)dx}. \qquad \text{(Rule 17)}$$

For example,

$$\int 2x^4 dx = 2\int x^4 dx = \frac{2x^5}{5} + C$$

$$\int \frac{x^2}{a} dx = \frac{1}{a}\int x^2 dx = \frac{x^3}{3a} + C$$

Reversing Rule 2, it follows that the integral of a sum (or difference) is equal to the sum of the integrals.

$$\int (\mathbf{f}_1(\mathbf{x}) + \mathbf{f}_2(\mathbf{x}) + \mathbf{f}_3(\mathbf{x}))\mathbf{dx} = \int \mathbf{f}_1(\mathbf{x})\mathbf{dx} + \int \mathbf{f}_2(\mathbf{x})\mathbf{dx} + \int \mathbf{f}_3(\mathbf{x})\mathbf{dx}. \quad \text{(Rule 18)}$$

For example,

$$\int \left(3x^2 + \frac{a}{x^3} - 1\right)dx = \int 3x^2dx + \int \frac{a}{x^3}dx - \int dx = \frac{3x^3}{3} + \frac{ax^{-2}}{-2} - x = x^3 - \frac{a}{2x^2} - x + C.$$

When several integration processes are grouped into one, the various integration constants are all combined in the one C.

An added (or subtracted) constant remains after integration, but it is multiplied by the independent variable. This fact follows directly from the two preceding rules. The integral sign is always accompanied by the differential which multiplies the whole expression, including any constant term.

$$\int (\mathbf{a} + \mathbf{f}(\mathbf{x}))\mathbf{dx} = \int \mathbf{adx} + \int \mathbf{f}(\mathbf{x})\mathbf{dx} = \mathbf{ax} + \int \mathbf{f}(\mathbf{x})\mathbf{dx}. \quad \text{(Rule 19)}$$

For example,

$$\int (2 + x^6)dx = \int 2dx + \int x^6dx = 2\int dx + \int x^6dx = 2x + \frac{x^7}{7} + C$$

Since e^x remains unchanged on differentiation, it must remain unchanged also on integration. Reversing Rule 8,

$$\int \mathbf{e}^x\mathbf{dx} = \mathbf{e}^x + \mathbf{C}. \quad \text{(Rule 20)}$$

The application of Rule 16 to the special case of $x^{-1}dx$, leads to difficulty, for

$$\int x^{-1}dx = \frac{x^{-1+1}}{-1+1} = \frac{x^0}{0} = \frac{1}{0}.$$

Obviously, the rule will not work for this particular expression, since the differentiation of $1/0$ does not give $1/x$. The problem resolves itself into the question "What, on differentiation, will give $1/x$?" This is the ultimate method in all integration.

On looking over the rules for differentiation, it is found that one rule gave $1/x$ as a result. According to Rule 9 the derivative of $\log_e x$ is $1/x$ and so it follows that the integral of $\frac{1}{x}\,dx$ is $\log_e x$.

$$\int \frac{dx}{x} = \log_e x + C. \qquad \text{(Rule 21)}$$

This integration is of the utmost importance in physical chemistry and in all science.

a^x on differentiation gives $a^x \log_e a$ by Rule 12. Reversal of this process shows then that the integral of $a^x \log_e a$ is a^x. But the integral of a^x, rather than $a^x \log_e a$ is desired. The extra constant, $\log_e a$, may be cancelled by putting into the expression $1/\log_e a$. Then,

$$\frac{d(a^x/\log_e a)}{dx} = \frac{a^x \log_e a}{\log_e a} = a^x$$

Now an expression has been found which on differentiation will give a^x, and so

The integral of a^x is a^x divided by $\log_e a$.

$$\int a^x dx = \frac{a^x}{\log_e a} + C. \qquad \text{(Rule 22)}$$

The integral of $\log_e x$ is not obvious at once, for it is necessary to do a little rearranging to find an expression which will give $\log_e x$ on differentiation. It will be remembered that the differentiation of a product results in the sum of two terms which contained the original functions as multiplied constants. It should be possible then, by a judicious choice of terms, to make the undesired factors drop out. Trying $x \log_e x$, the differential is found to be $x\frac{(1)}{x} + \log_e x(1)$, or $1 + \log_e x$. This result is close to the desired result but it is necessary to get rid of the extra 1. If an extra term is put into the original expression such that it will yield -1 on differentiation, it will cancel out the undesired 1. The differentiation of $-x$ will give the required -1.

If $y = x \log_e x - x, \frac{dy}{dx} = 1 + (\log_e x) - 1 = \log_e x$; and

the integral of $\log_e x dx$ is therefore $x \log_e x - x$.

$$\int \log_e x dx = x(\log_e x - 1) + C. \qquad \text{(Rule 23)}$$

The rules for the logarithms to other bases follow directly from conversion into $\log_e$. Since $\log_{10} x = \log_{10} e \log_e x = 0.4343 \log_e x$, and since 0.4343 is a multiplied constant which remains unchanged it follows that

$$\int \log x dx = \int \log_{10} x dx = 0.4343x(\log_e x - 1) + C. \quad \text{(Rule 24)}$$

Likewise for the more general case,

$$\int \log_a x dx = (\log_a e)(x)(\log_e x - 1) + C. \quad \text{(Rule 25)}$$

It may have been noticed that the rules for products and quotients were skipped. The rules for differentiation of these functions cannot be reversed, and in fact there are no general rules for integrating products or quotients. Certain methods for handling them under special conditions will be given in a later chapter.

The rules for integration are much less complete than those for differentiation and a considerable amount of mathematical ingenuity is necessary at times. The situation is not unlike that found in qualitative chemical analysis, where a student can follow directions for a separation of the metals; but when he comes to the separation of acids he finds that he is thrown more on his own resources. Integration is an art and proficiency in it requires a great deal of practice.

Integration between Limits

In every integration considered so far, the unknown integration constant, C, has been included as part of the answer. It has been shown that C is equal to the value of the integral obtained by setting x equal to zero and sometimes C can be evaluated by finding, experimentally, the value of y when $x = 0$. Often, however, it is difficult to evaluate the integration constant and still more difficult to attach the proper physical significance to it. Sometimes considerable progress has been made as a result of finding the correct physical interpretation of an integration constant.

In many practical problems one is interested only in the values of the expression between two different values of x. Under these conditions the integration constant conveniently cancels out. The two values between which the integration is to be carried out are known as the limits of the integration. In case the limits are specified the integral is known as a definite integral, and when they are not specified the integral is known as an indefinite integral. All the integrals given up to this point have been indefinite integrals. It is to be noted that the integration constant is not evaluated in a definite integral—it is merely eliminated by cancellation.

$\int \frac{x}{2}\,dx = \frac{x^2}{4} + C$ is an indefinite integral, but if one is concerned only with the value of this expression between the values $x =$ 1 and $x = 4$, it is written:

$$\int_{x=1}^{x=4} \frac{x}{2}dx \quad \text{or} \quad \int_1^4 \frac{x}{2}\,dx.$$

This definite integral is evaluated as follows:

$$\int^4 \frac{x}{2}\,dx = \left[\frac{x^2}{4}\right]_1^4 = \frac{16}{4} - \frac{1}{4} = 3\frac{3}{4}.$$

The small number at the top of the integral sign is called the upper limit and the one at the bottom is the lower limit. These numbers signify that they are to be substituted into the integrated expression, and that the value obtained by substitution of the lower limit is to be subtracted from the value obtained by substitution of the upper limit. (Rule 26)

It is customary to enclose the integrated expression in square brackets with the upper and lower limits designated. The two values are written down and the subtraction is carried out to give the final numerical answer.

A few examples will serve to show how this simple process of integrating between limits is carried out:

1. $\int_0^{10} x dx = \left[\frac{x^2}{2}\right]_0^{10} = \frac{100}{2} - \frac{0}{2} = 50.$
2. $\int_2^6 x dx = \left[\frac{x^2}{2}\right]_2^6 = \frac{36}{2} - \frac{4}{2} = \frac{32}{2} = 16.$
3. $\int_6^2 x dx = \left[\frac{x^2}{2}\right]_6^2 = \frac{4}{2} - \frac{36}{2} = -\frac{32}{2} = -16.$
4. $\int_{-1}^6 x dx = \left[\frac{x^2}{2}\right]_{-1}^6 = \frac{36}{2} - \frac{(+1)}{2} = \frac{35}{2} = 17\frac{1}{2}.$

5. $\int_0^2 e^x dx = \left[e^x\right]_0^2 = e^2 - e^0 = 2.71828^2 - 1 = 7.3891 - 1 = 6.3891.$

6. $\int_{v=1}^{v=100} RT\frac{dv}{v} = RT\int_{v=1}^{v=100}\frac{dv}{v} =$

$$RT\left[\log_e v\right]_1^{100} = RT\,[\log_e 100 - \log_e 1] =$$

$$RT2.303[2 - 0] = 4.606RT$$

This is the expression for the work done by a perfect gas in expanding isothermally from 1 liter to 100 liters. R and T are constants.

7. $\int_{100}^{1} RT\frac{dv}{v} = RT\left[\log_e v\right]_{100}^{1} = RT2.303[\log_{10} 1 - \log_{10} 100] =$
$-4.606RT.$

The result in example (7) has the same numerical value as in (6) but the sign is reversed. The physical significance of the negative sign is that work is done on the gas by the surroundings. Positive work is arbitrarily defined as work done *by* the gas (*i.e.*, expansion), so negative work is work done *on* the gas (compression).

It is evident that reversing the limits reverses the sign of the resulting answer but does not change its numerical value.

Exercises

1. $\frac{dy}{dx} = x^5 + 5.$ $\quad y = ?$

   Ans.: $y = \frac{x^6}{6} + 5x + C.$

2. $dy = 9x^3dx.$ $\quad y = ?$

3. Integrate $ax^9dx.$ *Ans.:* $\frac{ax^{10}}{10} + C.$

4. What is the integral of $10x^{-2}dx$?

5. The integral of $\sqrt[3]{x^4}dx$ is what? *Ans.:* $\frac{3}{7}\sqrt[3]{x^7} + C.$

6. $\int(5x^3 + \sqrt{x} - 2x)dx = ?$

7. $\int(4x^3 - 5x^2)dx = ?$ *Ans.:* $x^4 - \frac{5x^3}{3} + C.$

8. Integrate $x\left(6x^2 + \frac{7}{x^4} + 4\right)dx.$ (Simplified by multiplication)

9. $\int(3x^3 + 5x^2)dx = ?$ *Ans.:* $\frac{3x^4}{4} + \frac{5x^3}{3} + C.$

10. $\int\frac{(e^x - \sqrt[4]{x})}{2}\,dx = ?$

11. $\int \left(\frac{2}{x} - 5\sqrt[4]{x^7}\right) dx = ?$ *Ans.:* $2 \log_e x - \frac{20\sqrt[4]{x^{11}}}{11} + C.$

12. $\int 2a \log_e x dx = ?$

13. $\int 5^x dx = ?$ *Ans.:* $\frac{5^x}{\log_e 5} + C.$

14. $\int \frac{2^z}{3} dz = ?$

15. $\int \frac{\log y}{10} dy = ?$ *Ans.:* $0.04343y(\log_e y - 1) + C.$

16. $\int (\log_5 x + 1) dx = ?$

17. $\int \frac{(x^3 - x^2 - 2x - 1)}{x^2} dx = ?$ (simplified by division)

Ans.: $\frac{x^2}{2} - x - 2 \log_e x + \frac{1}{x} + C.$

18. $\int \left(\frac{1}{x} + \log x - 2x\right) dx = ?$

19. $\int (ax^{-23} - \log_e x) dx = ?$ *Ans.:* $\frac{-a}{22x^{22}} - x(\log_e x - 1) + C.$

20. $\int (x^2 + 2)^3 dx = ?$

21. $\int_0^{10} \frac{x^2}{3} dx.$ *Ans.:* 111.1.

22. $\int_{10}^{0} \frac{x^2}{3} dx.$ 23. $\int_{-10}^{10} \frac{x^2}{3} dx.$ *Ans.:* 222.2.

24. $\int_0^5 (6 + 8x - 3x^2) dx.$

25. $\int_b^a (x^2 + 3) dx.$ *Ans.:* $\frac{a^3 - b^3}{3} + 3(a - b).$

26. $\int_b^{2a} \left(x^3 - \frac{1}{\sqrt[3]{x}}\right) dx.$ 27. $\int_0^2 e^x dx.$ *Ans.:* 6.39.

28. $\int_{3.3}^{1.2} 2e^x dx.$ 29. $\int_1^2 5^y dy.$ *Ans.:* 12.4.

30. $\int_1^{1,000} \log x dx.$ 31. $\int_b^1 2 \log_e z dz.$ *Ans.:* $-2 - 2b(\log_e b - 1).$

32. $\int_1^{10} RT \frac{dv}{v}$ (R and T are constants)

33. $\int_{100}^{5} RT d \log_e v.$ *Ans.:* $-3RT.$

CHAPTER XII

THE SIGNIFICANCE OF "e"

The rules for integration and differentiation already studied are sufficient to meet most of the needs of a first course in physical chemistry. It is a striking fact that although physical chemistry uses calculus constantly, its requirements are usually met with the very simplest and easiest parts of calculus. The problems of this chapter, which have been selected from various branches of physical chemistry bear out the truth of this statement.

Definition of "*e*." The Compound Interest Law

Natural logarithms (logarithms to the base e) are found throughout physical chemistry, and a considerable proportion of the calculus used is concerned with the quantity e. Whenever a quantity is changing at a rate proportional to the magnitude of the quantity, one is justified in looking at once for a mathematical relation based on e. The reason for this relation lies, of course, in the remarkable property of e^x, that it remains unchanged on differentiation ($de^x/dx = e^x$, p. 85). The most common expression of this relation in chemistry is the so-called "mass law," according to which the amount of chemical reaction depends on the effective concentration of the reacting materials. This law finds expression in reaction rates and chemical equilibria, including the phenomena of solubility, vapor pressure, electromotive force, distribution, and others.

The quantity e is so important that its full meaning is studied here a little more in detail, although it has been used already in various mathematical operations;

e is defined as the limit of the series $\left(1 + \frac{1}{n}\right)^n$ where n is made larger and larger, and its numerical value is 2.71828 . . .

Table I shows how this value is calculated:

TABLE I.—CALCULATIONS OF $\left(1+\frac{1}{n}\right)^n$ FOR INCREASING VALUES OF n

n	$\left(1+\frac{1}{n}\right)^n$	Numerical value
1	$(1 + 1/1)^{1}$	2.000
2	$(1 + 1/2)^{2}$	2.250
3	$(1 + 1/3)^{3}$	2.369
5	$(1 + 1/5)^{5}$	2.489
10	$(1 + 1/10)^{10}$	2.594
20	$(1 + 1/20)^{20}$	2.653
40	$(1 + 1/40)^{40}$	2.684
50	$(1 + 1/50)^{50}$	2.691
100	$(1 + 1/100)^{100}$	2.705
1,000	$(1 + 0.001)^{1,000}$	2.717
10,000	$(1 + 0.0001)^{10,000}$	2.718
∞	....	2.71828...

It is evident that the value of the series becomes more nearly 2.71828 . . . as the number of terms is increased. This fact is brought out clearly by reference to Fig. 42. The extrapolation illustrates clearly the meaning of the term "approaching as a limit," for 2.71828 . . . is the limit to which this series approaches as n approaches infinity. No value of n can give to the expression a larger value than 2.71828 . . .

The law which governs the increase of a property in this series is sometimes called the "compound interest law" or the "law of organic growth." Both names emphasize the fact that the increase is proportional to the size of the thing which is increasing,—the interest depends on the amount of money in the bank and the increase in weight of a plant over a period of time depends on the size of the plant. It is more properly called the "law of exponential growth."

The workings of the law can be illustrated concretely by considering a dollar which has been invested at 10 per cent interest (per year). If it is invested at simple interest, one-tenth of the original capital will be added each year and at the end of ten years there will be $1 in interest in addition to the original capital of $1. The $1 will have grown to $2.

If it is invested at compound interest, the interest at the end of the first year will be one-tenth of \$1 or 10 cents. The interest at the end of the second year will be one-tenth of \$1.10 or 11 cents; and at the end of the third year it will be one-tenth of the original dollar, plus the 10 cents of the first year, plus the

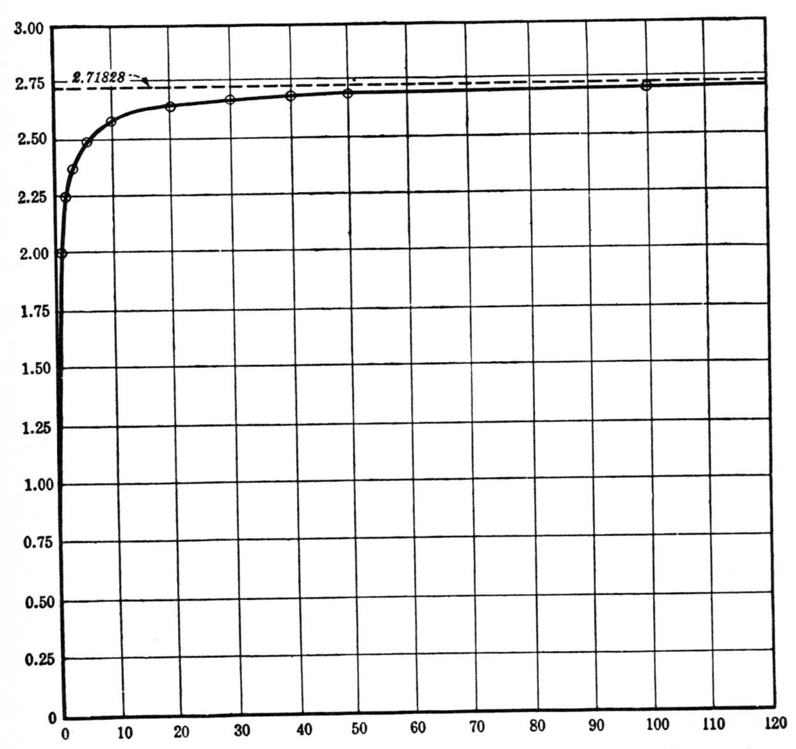

Fig. 42.—The values of e obtained by extrapolating $(1 + 1/n)^n$ to large values of n. The curve approaches the value 2.71828 . . . as a limit.

11 cents of the second year, *i.e.* one-tenth of \$1.21 which is 12 cents. Each year the capital will increase by one-tenth; that is, the capital, C, will become equal each year to $(1 + 1/10)$ times the capital of the preceding year. This fact is represented by the equation

$$C = 1(1 + 1/10)(1 + 1/10)(1 + 1/10) \cdots$$

If this process is carried out for 10 years the capital, C, $=1(1 + 1/10)^{10}$ and reference to Table I shows that this equation has the value of 2.594. The dollar will grow to $2.59 in 10 years at compound interest.

When the interest is compounded annually as in the preceding example, part of the interest is not drawing interest. If the interest is compounded semi-annually, the capital will grow more rapidly because the interest of the first half-year will then be drawing interest during the second half. Instead of increasing by 10 per cent each year the capital will increase by 5 per cent each half-year, a fact which is represented by the equation

$$C = 1(1 + 1/20)(1 + 1/20)(1 + 1/20) \cdots$$

If this is carried out for 10 years or 20 payments,

$$C = 1(1 + 1/20)^{20} = 2.653$$

The dollar will grow to $2.65 in 10 years, when the interest is compounded semi-annually.

If the interest is compounded quarterly, with forty payments in 10 years, the dollar will grow to $2.68. Finally, if the interest is compounded every instant with a very large number of payments, the dollar will grow to $2.718.

This last method of compounding interest is impractical, of course, but it is typical of many processes found in the natural sciences. As a matter of fact the processes do involve changes by jumps, corresponding to appreciable periods of time for compounding interest, but the jumps are so small as to give the appearance of continuity. A glance at the table will show that if there are only 1,000 stages in a process e will have the value of 2.717 as compared with the value of 2.71828 corresponding to an infinite number of stages (perfect continuity). If a chemical substance is decomposing, it is decomposing one molecule at a time; and if an electrical condenser is discharging, it is discharging one electron at a time; while even a hot body which was formerly supposed to be cooling at a perfectly uniform rate, is without doubt throwing out its radiant energy one quantum at a time. But one molecule or one electron or one quantum is so very small in comparison with any change which can be measured by physical or chemical means that the *true* compound interest value,

2.71828 . . . , can be applied with accuracy. It is to be noted that there are *approximately* 10^{23} molecules connected with a gram molecule of material and that with only 10^4 molecules the deviation from the true value of e is less than $\frac{1}{100}$ of 1 per cent.

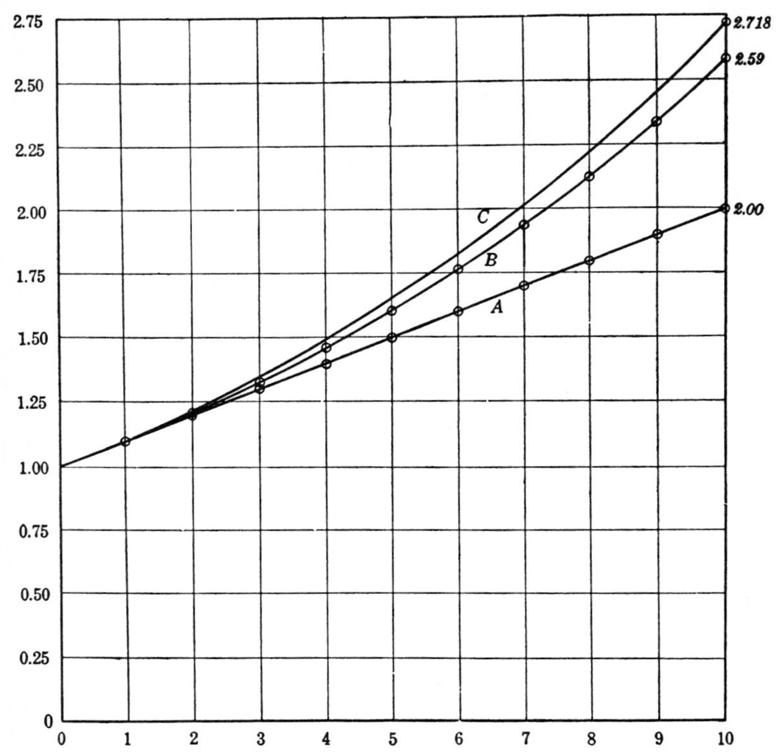

FIG. 43.—A graphical representation of the law of exponential growth. The growth of $1 at interest for 10 years at 10 per cent per year.

Curve A: Simple interest—giving $2.
Curve B: Compound interest, compounded every year—giving $2.59.
Curve C: Compound interest, compounded every instant—giving $2.718.

A graphical presentation of the ideas under consideration is helpful. Figure 43A gives the increase of capital with time according to simple interest. Each year the capital is increased by one-tenth of the original capital. Each ordinate increases by 10 per cent. The tenth ordinate has a value of 2. The middle curve of Fig. 43B shows the increase according to compound interest, where each ordinate increases by 10 per cent of the immediately preceding ordinate, rather than 10 per cent of the original ordinate. The tenth ordinate has a value of 2.594. If the number of ordinates had been twenty instead of ten and each one had been increased by one-twentieth of the preceding one, the curve would have been slightly steeper and the last ordinate would have been 2.653 instead of 2.594. When a very large number, n, of ordinates is taken, each one being $1/n$th higher than the preceding one, the upper curve of Fig. 43*C* is obtained. The last ordinate here has a value of 2.71828 . . . , which is the value of e.

It has been shown before (p. 86) that e raised to the x power is $\left(1 + \frac{1}{n}\right)^{nx}$, when n is very large, and that this in turn is equal to the series $1 + x + \frac{x^2}{1 \times 2} + \frac{x^3}{1 \times 2 \times 3} + \frac{x^4}{1 \times 2 \times 3 \times 4} - \cdots$ This series remained unchanged on differentiation and so it was proved that the derivative of e^x is e^x (Rule 8). This fact was the basis of the differentiation and integration of logarithms and exponents.

These considerations led to the rule that $\frac{d \log_e x}{dx} = \frac{1}{x}$ or that $d \log_e x = dx/x$. This relation is so important that it should have further amplification. The fact of the relationship can be shown by specific examples without any reference to calculus, as follows:

Let x = any number, 6.000 for example.

Let $x + \Delta x$ = a slightly larger number, 6.03 for example.

$$\Delta x = 6.03 - 6.00 = 0.03 \qquad \text{Log}_e\ 6.03 = 1.79670$$

$$\frac{\Delta x}{x} = \frac{0.03}{6.00} = 0.005 \qquad \text{Log}_e\ 6.00 = 1.79170$$

$$\Delta\ \text{Log}_e x = 0.005$$

It is evident that $\Delta x/x = \Delta \log_e x$ in this case where the calculation is carried out only to 3 places. The same relation would have been found no matter what number had been taken. If Δx is taken larger the relation will not hold so well. On the other hand the smaller Δx is taken the more nearly true does the relation become; and it is exactly true that $dx/x = d \log_e x$, for

$\frac{d \log_e x}{dx}$ is the limit to which $\frac{\Delta \log_e x}{\Delta x}$ approaches as Δx approaches zero. Integration of this equation gives $\int \frac{dx}{x} = \log_e x + C$. (p. 121)

Applications of the Compound Interest Law to Physical Chemistry

A unimolecular reaction constitutes one of the best examples of the compound interest law in chemistry. It is an example of the "mass law." Experiments show that the amount of material (hydrogen peroxide, for example) decomposing per unit of time is proportional to the amount present at that time. If c is the concentration at time, t, $-dc/dt$ is the rate of decrease in concentration. The minus sign indicates that the concentration (of hydrogen peroxide) is *decreasing* with an increase in time. Now the fact that this rate of decrease is proportional to the amount of material present is expressed mathematically by the equation

$$-\frac{dc}{dt} \propto c, \quad \text{or} \quad -\frac{dc}{dt} = kc^*$$

It will be remembered that when one quantity is proportional to another, it can be set equal to it if a multiplied constant is introduced. This equation can be transposed, giving $-dc/c = kdt$, and integrating between the limits c_2 at the time t_2 and c_1 at t_1, where t_2 is greater than t_1,

$$\int_{c_1}^{c_2} -\frac{dc}{c} = \int_{t_1}^{t_2} kdt$$

$$-[\log_e c]_{c_1}^{c_2} = [kt]_{t_1}^{t_2}$$

$$(-\log_e c_2) - (-\log_e c_1) = k(t_2 - t_1)$$

$$k = \frac{1}{t_2 - t_1} \log_e \frac{c_1}{c_2}$$

$$k = \frac{2.303}{t_2 - t_1} \log \frac{c_1}{c_2}.$$

This equation can be somewhat simplified by taking, t, the earlier time as zero time. Then c_0 is the concentration at zero time and c is the concentration at any time, t. It is understood

* With physical problems where conventional terms have been already established, it is not always possible to confine the symbols for variables to the last letters of the alphabet as suggested on page 71.

that experimental measurements may be started at any time and that "zero time" is purely arbitrary. The equation then becomes,

$$k = \frac{2.303}{t} \log \frac{c_0}{c}.$$

In some books on physical chemistry the concentration at zero time is denoted by a, and the concentration at time, t, is then $a - x$, where x is the amount decomposed at time, t. The equation is then written,

$$k = \frac{2.303}{t} \log \frac{a}{a - x}.$$

Applying the equation to a specific case, it is supposed that there are 1,000 molecules at the beginning of the measurement and that after 1 hour 500 have decomposed, leaving 500. In this case,

$$k = \frac{2.303}{1} \log \frac{1,000}{500} = 2.303 \times 0.3010 = 0.6932.$$

Having evaluated k, it is now possible to calculate the concentration, c, at any time, t. For example, if it is desired to know how many molecules will remain undecomposed at the end of 4 hours, $t = 4$; and

$$0.6932 = \frac{2.303}{4} \log \frac{1,000}{c}$$

$$\frac{4 \times 0.6932}{2.303} = \log 1000 - \log c$$

$$\log c = 3 - 1.204 = 1.796$$

$$c = \text{antilog } 1.796 = 62.5$$

Now if the concentration at the start had been ten times greater, ten times as many molecules would have decomposed and the ratio would have been unchanged. It is a necessary deduction from this formula for a unimolecular reaction that the *fraction* of the molecules decomposing in a given period of time is independent of the number present. k is known as the reaction velocity constant—or better, as the specific reaction rate. It is defined as the fraction of the molecules present which decompose in a unit of time, provided that k is so small that the total number of molecules is essentially unchanged during the time interval.

The statement that the number of molecules decomposing in a unit of time is proportional to the amount present may be visualized with the help of a homely illustration. A corn popper is divided up into equal areas. If the temperature is uniform and if the corn has been thoroughly mixed the number of corn kernels which pop per minute is the same in each particular area. If two areas are taken the number popping per minute is twice as large and if n areas are taken the number is n times as large, etc. If, for example, one-third of the number pop in one area in a minute and two areas are taken, twice as many will pop but there are twice as many kernels present, and the fraction popping remains unchanged. It is evident that the fraction decomposing is independent of the number taken.

In the particular example given above the numbers were chosen so that the answer can be checked by mental arithmetic. In 1 hour 1,000 molecules changed to 500, or half of them decomposed. This establishes the rate of decomposition and since the rate is independent of the number of molecules, one-half of the remaining decompose also in the second hour, leaving one-half of one-half of 1,000 or 250. At the end of the third hour there are left 125 and at the end of the fourth there are one-half of 125 or 62.5.[1]

The constant, k, may be expressed in any unit of time— eciprocal minutes, hours, or seconds, for example. In theoretical work the second is the best unit of time. The numerical value of k will, of course, be sixty times as great when expressed in minutes as when expressed in seconds.

The example just given is so simple that it may lead to a misconception. How many molecules remain after one-half hour? The student is warned against jumping to the conclusion that if one-half of the number decompose in one hour, one-quarter will decompose in half an hour, leaving 750 out of the original 1,000. More than one-quarter will decompose in half an hour, because the number decomposing is proportional to the number present and there are more molecules present in the first half-hour

[1] Obviously it is impossible to have half a molecule in this sense, but of course any measurable quantity of material contains billions and billions of molecules instead of the one thousand of the hypothetical problem. Even if there were only 62.5, it would be realized that all the physical-chemical laws are statistical averages and that the number 62.5 stands for a series of results, sometimes 62 and sometimes 63.

than in the second. The rate of decomposition is changing continuously, but integration over a period of an hour gives a total decomposition of one-half of the number present at the beginning of the hour. To get the total number decomposing during half an hour the expression must be integrated over the period of one-half hour, giving

$$\int_{1,000}^{c} \frac{-dc}{c} = \int_{0}^{0.5} k dt.$$

$$0.6932 = \frac{2.303}{0.5} \log \frac{1,000}{c} \quad \text{(integrating and substituting } k = 0.6932 \text{ as determined on p. 133)}$$

$$\frac{0.6932}{4.606} = 3 - \log c \text{ (simplifying and setting } \log 1,000 = 3)$$

$$\log c = 3 - 0.1550 = 2.8450$$

$$c = 700$$

After half an hour, 700 molecules out of 1,000 are left. This is appreciably different from the number 750 obtained on the false assumption that the rate of decomposition is uniform over the full hour and that the number decomposed in half an hour will be half of the number decomposed in an hour.

Passing next to an example which is not so obvious as the first one—the number of atoms of radium emanation (radon) decreases to one-fifth of the number in 12,800 min. What is the value of k, in reciprocal minutes?

$$k = \frac{2.303}{1.28 \times 10^4} \log \frac{1}{0.2} = \frac{2.303}{1.28 \times 10^4} \log 5 = \frac{2.303 \times 0.699}{1.28 \times 10^4} = 1.26 \times 10^{-4}.$$

This equation means that a little over $\frac{1}{100}$ of 1 per cent of the total number of atoms, present in any sample of radon, decompose in a minute. What fraction will be left after 1 hour?

$$1.26 \times 10^{-4} = \frac{2.303}{60} \log \frac{1}{x}$$

$$-\log x = \frac{1.26 \times 10^{-4} \times 60}{2.303} = 0.00328$$

x = antilog (-0.00328) = antilog $\bar{1}.99672 = 0.9925$ (p. 11)
After 1 hour 99.25 per cent of the atoms are still unchanged.

How long will it take for half of the atoms to disintegrate, *i.e.*, after what time will the number of atoms be 0.5?

$$1.26 \times 10^{-4} = \frac{2.303}{t} \log \frac{1}{0.5} = \frac{2.303}{t} \log 2$$

$$t = 5{,}500 \text{ (minutes)}$$

A similar formula applies also in optical problems. The absorption of light (the decrease in intensity per unit of thickness) is proportional to the intensity of the light. Expressed mathematically, $-dI/dl = kI$. In this equation I is the intensity of light after passing through a layer of absorbing material l units thick.

$$-\frac{dI}{I} = kdl$$

Integrating between the limits I and I_0, and l and 0, the following equation is obtained. I is the intensity of light after passing through l cm. (or any other unit) of absorbing material, and I_0 is the intensity of light when $l = 0$.

$$\int_{I_0}^{I} -\frac{dI}{I} = \int_0^l kdl$$

$$-\Big[\log_e I\Big]_{I_0}^{I} = k\Big[l\Big]_0^l$$

$$k = \frac{2.303}{l}\Big[\log I_0 - \log I\Big] = \frac{2.303}{l} \log \frac{I_0}{I}$$

Applying this formula again to a specific problem: A beam of light is reduced by 10 per cent in passing through 1 cm. of a certain solution, giving an intensity of 1.00 − 0.1 = 0.9. What will be its intensity after passing through 10 cm.?

$$k = \frac{2.303}{1} \log \frac{1.00}{0.9} = 2.303 \log 1.1111 =$$

$$2.303 \times 0.0457 = 0.10525$$

Having established the value of k from two experimental observations (intensities at two different depths,—zero and l) the intensity, I, at any value of l may be calculated.

When $l = 10$, for example,

$$0.10525 = \frac{2.303}{10} \log \frac{1.00}{I} = -0.2303 \log I$$

$$\log I = \frac{-0.10525}{0.2303} = -0.457$$

$$\text{Intensity} = I = \text{antilog } \bar{1}.543 = 0.349$$

The intensity is reduced to 34.9 per cent of its original value after passing through 10 cm. of the material.

Through how much material would it have to pass before its intensity is reduced by 90 per cent; that is, reduced to an intensity of 10 per cent?

$$0.10525 = \frac{2.303}{l} \log \frac{1.00}{0.1}$$

$$l = \frac{2.303}{0.10525} \times 1 = 21.9 \text{ cm.}$$

Exponential Equations

Now there is another way of working out problems of this kind based on exponents instead of logarithms. The logarithmic formulas seem to be easier for chemists, and in fact there is no surer way to make the average chemist skip a page of a scientific article than to place on it a couple of exponential equations. Physicists, on the other hand, usually prefer the exponential equations. The two are equivalent, of course, and both give identical results. As a matter of fact log tables are usually more accessible and more accurate than tables of exponents. It is absolutely essential, however, to acquire familiarity with both methods.

It has been shown before that $d(e^x)/dx = e^x$, and if $y = e^x$, then $dy/dx = y$. The statement that $dy/dx = y$, means that the rate of change of the function is numerically equal to the function itself. But the situation in which the rate of change of the function is *proportional* to the function itself is much more common. Now this more general condition of proportionality is satisfied if $y = ae^{kx}$ where a and k are constants. In spite of these extra constants it is still true that $\frac{dy}{dx} \propto y$, or that $\frac{dy}{dx} =$ (constant) (y); for if $y = ae^{kx}$, $\frac{dy}{dx} = k(ae^{kx}) = ky$.

Whenever it is found that the value of a property, y, is increasing with some variable at a rate which is proportional to the magnitude of the property, the fact can be stated mathematically by the equation $y = ae^{kx}$, for the derivative of ae^{kx} is proportional to ae^{kx}. If the quantity is *decreasing* at a rate which is proportional to the magnitude of the property, then $-dy/dx$

is proportional to y. In this case $y = ae^{-kx}$, for then $dy/dx = -k(ae^{-kx})$ and dy/dx is proportional to $-y$.

This a has a special significance. It is equal to the function, y, when $x = 0$, for then e^{kx} or e^{-kx} is equal to 1. It will be remembered that any number raised to the zero power is 1. The formula, therefore, is usually written $y = y_0e^{-kx}$, where y_0 is a constant, which is the value of y when $x = 0$. y and x are any variables.

The fact that the decrease in concentration of reacting material, c, per unit of time, t, is proportional to the concentration, is expressed by the equation,

$$c = c_0e^{-kt}.$$

c_0 is the concentration where $t = 0$; that is, at the beginning of the experiment.

The fact that the decrease in intensity, I, of light, is proportional to the depth, l, of absorbing material is given by the equation,

$$I = I_0e^{-kl}.$$

I_0 is the intensity of light when $l = 0$, that is, before passing through any of the absorbing material.

The fact that the pressure of the atmosphere, p, decreases at a rate proportional to the altitude, h, is given by the equation,

$$p = p_0e^{-kh},$$

where p_0 is the pressure when the height, h, is zero; *i.e.*, at sea level.

It is obvious that the exponential equation and the logarithmic equation derived before, are equivalent. Both were derived from the expression that dy/dx is proportional to y,—the exponential form by finding the most general expression which on differentiation would be proportional to itself; and the logarithmic form by straight integration. The first is changed into the second by taking $\log_e$ of both sides, thus

$$\begin{aligned} y &= y_0e^{-kx}. \quad \text{(Exponential equation)} \\ \log_e y &= \log_e y_0 + \log_e e^{-kx} = \log_e y_0 - kx \\ kx &= \log_e y_0 - \log_e y \\ k &= \frac{2.303}{x} \log \frac{y_0}{y}. \quad \text{(Logarithmic equation)} \end{aligned}$$

This expression is identical with the one derived on page 133.

The exponential equations can be solved by reference to tables of e^x and e^{-x} given on page 286, or they may be first converted to the logarithmic formula and solved in the manner just given.

The problems of this chapter previously solved by logarithms should now be solved by reference to the tables of e^x and e^{-x}.

One thousand atoms of a radio active element decompose at such a rate that they are half gone in an hour. How many are remaining after 4 hours?

$$c = c_0 e^{-kt};\ 500 = 1{,}000 e^{-k(1)};\ e^{-k} = 0.5$$

Reference to the table on page 287 shows that when $e^{-k} = 0.5$, $k = 0.694$. Having evaluated k, the next problem is to find c when t is 4.

$$c = 1{,}000 e^{-0.694 \times 4}$$

$$\frac{c}{1{,}000} = e^{-2.776}$$

The table shows that when the exponent of e is -2.776, the value of the function is 0.0625, and $c = 1{,}000 \times 0.0625 = 62.5$.

In the other problem solved by integration on page 135, one-fifth of the radon atoms are left after 12,800 min.

$$I = I_0 e^{-kt}$$

$$e^{-k \times 12{,}800} = 0.2$$

In the tables it is found that when $e^{-x} = 0.2$, x has the value of 1.609. Then $12{,}800k = 1.609$, or $k = 1.26 \times 10^{-4}$.

What fraction will be left after 60 min.?

$$I = I_0 e^{-kt}$$

$$\frac{I}{I_0} = e^{-1.26 \times 10^{-4} \times 60} = e^{-0.00756}$$

In the tables, it is found that when $x = 0.00756$, $e^{-x} = 0.9927$. Then $I/I_0 = 0.9927$, and 99.27 per cent of the atoms are left after 60 min.

How long will it take for half of the atoms to decompose?

$$\frac{I}{I_0} = e^{-kt} \quad 0.5 = e^{-1.26 \times 10^{-4} t}$$

The tables show that when $e^{-x} = 0.5$, $x = 0.694$. The two exponents must be equal since the numbers are equal and $1.26 \times 10^{-4}t = 0.694$. Then $t = 5{,}508$.

Exercises

1. A beam of light is reduced 1 per cent in intensity, I, by passing through a certain pane of glass. How much light is absorbed by passing through 100 panes?

Ans.: The intensity of the incident light is 100 and that of the transmitted light is 36.61 and the absorption is therefore 63.39 per cent.

How many panes are necessary to reduce the light to one-half its intensity?

Ans.: 68.9 panes.

How much light is absorbed in passing through 10 panes?

How many panes are necessary to reduce the light to 1 per cent of its intensity?

2. The time taken for half of a given quantity of material to disintegrate is known as the period of "half-life." It is related to the specific reaction rate as follows:

$$\text{Period of half-life} = \frac{2.303}{k} \log \frac{1}{1 - \frac{1}{2}} = \frac{2.303}{k} \log 2 = \frac{0.6933}{k}.$$

If a unimolecular reaction is one-quarter completed in 100 min., what is the specific reaction rate, k, and the period of half-life?

Ans.: $k = 0.002{,}88$; half-life = 241 minutes.

How long will it be until one-tenth of the material is left?

3. What is the value of k in a unimolecular reaction if the reaction is one-quarter completed in 60 minutes?

How long will it be until 5 per cent is remaining?

4. A certain radio-active material disintegrates at such a rate that 90 per cent remains after a period of one minute. The radio active transformations follow the unimolecular equation.

a. What is the value of the radio-activity constant?

Ans.: $k = 0.1053$ min.$^{-1}$.

b. What per cent will be left at the end of a second?

Ans.: 99.82 per cent.

c. What per cent at the end of an hour?

d. What per cent at the end of a day?

5. The period of half-life $(0.6932/k)$ of Uranium is 4.67×10^9 years. How many atoms of uranium out of a gram atom of uranium disintegrate in a year?

One gram atom contains 6.06×10^{23} atoms.

Ans.: $k \times 6.06 \times 10^{23} = 9 \times 10^{13}$ atoms.

6. Radon (radium emanation) disintegrates at such a rate that after one hour 99.25 per cent is left; after 3 days 18 hours and 45 minutes 50 per cent is left, and after 25 days and 12 hours 1 per cent is left. These relations are to be fitted with a mathematical equation expressing per cent decomposed as a function of time.

7. A quartz plate 1 cm. in thickness absorbs 16.4 per cent of the ultraviolet light passing through, at a wave length of 203 $m\mu$. How much is absorbed by a plate 2 cm. in thickness? 10 cm. in thickness? 1 mm. in thickness?

Ans: 30.11 per cent by 2 cm.

8. A certain red solution 15 cm. in thickness absorbs 20 per cent of the blue light passing through. What per cent will be absorbed by a layer 3.2 cm. in thickness? By a layer 25 cm. in thickness?

9. A beaker of water is now 60° above room temperature and 20 min. ago it was 70° above room temperature. Assuming that the rate of cooling is proportional to the difference in temperature between the beaker and room temperature, what will the temperature be after 15 min.? After 2 hours? How long will it take for the water to cool to a temperature 10° above room temperature? *Ans.:* 53.71° after 15 min.

10. What will be the temperature of a beaker of water in an hour's time, if it was boiling (100°) 15 min. ago and is now 80° C.? A graph is to be plotted showing the temperature every 10 min.

11. Some of the preceding problems are to be worked with exponential equations as well as with logarithmic equations.

CHAPTER XIII

DIFFERENTIATION AND INTEGRATION OF TRIGONOMETRICAL FUNCTIONS[1]

In calculus, angles are always measured in radians rather than in degrees. The radian measure makes it possible to reduce the expression for the differential of the sine of an angle to a very simple quantity, and the differentiation and integration of all the other trigonometrical functions are then obtained from the differential of the sine.

It will be remembered that the ratio arc/radius gives the circular or radian measure of an angle (p. 269). When the arc is equal to the radius, the ratio becomes 1 and the angle intercepting such an arc is an angle of one radian. The total circle of 360° is 2π radians, and one radian $= 360/2\pi = 57.295°$. Also 1 degree $= 2\pi/360 = 0.0175$ radians.

The definitions of the sine, cosine, and tangent of an angle must be kept clearly in mind throughout this chapter (p. 269). Some of the more important theorems of trigonometry, also, are reviewed in the appendix (p. 270).

Differentiation of Trigonometrical Functions

The derivative of the sine of an angle is the cosine of the angle.

$$\text{if } y = \sin x, \quad \frac{dy}{dx} = \cos x. \qquad \text{(Rule 27)}$$

As usual in proving a rule for the differentiation of a quantity, x is increased by a small amount, Δx, whereupon y increases automatically by the small amount Δy. Then Δy may be found by subtracting the original equation as follows:

$$\begin{aligned} y + \Delta y &= \sin (x + \Delta x) \\ -(y &= \sin x) \\ \hline \Delta y &= \sin (x + \Delta x) - \sin x \end{aligned}$$

[1] This chapter is not used so much in chemistry and it may be omitted better than the chapters which follow.

This expression may be simplified considerably by applying some of the standard rules of trigonometry. There is a rule for subtracting one sine from another (p. 270) which leads to the following expression:

$$\Delta y = \sin (x + \Delta x) - \sin x = 2 \cos \frac{(x + \Delta x + x)}{2} \sin \frac{(x + \Delta x - x)}{2}$$

$$\Delta y = 2 \cos \left(\frac{2x}{2} + \frac{\Delta x}{2}\right) \sin \frac{\Delta x}{2} = \frac{\sin \ \Delta x/2}{\frac{1}{2}} \cos \left(x + \frac{\Delta x}{2}\right)$$

$$\frac{\Delta y}{\Delta x} = \frac{\sin \Delta x/2}{\Delta x/2} \cos \left(x + \frac{\Delta x}{2}\right) \text{ (dividing by } \Delta x)$$

Now as Δx approaches zero, $\Delta y/\Delta x$ approaches dy/dx, as a limit,

$$\frac{\sin \Delta x/2}{\Delta x/2} \text{ approaches 1, as a limit,}$$

$$\cos \left(x + \frac{\Delta x}{2}\right) \text{ approaches } \cos x \text{ as a limit,}$$

and therefore $dy/dx = \cos x$.

The reason that $\frac{\sin \Delta x/2}{\Delta x/2}$ approaches 1 as a limit when Δx approaches zero is evident from Fig. 44.

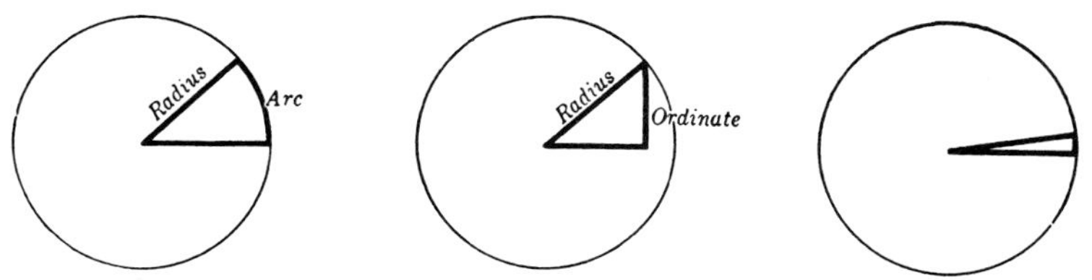

FIG. 44.—Angle (radians) = arc/radius. Sine of angle = ordinate/radius. For very small angles, the arc and the ordinate nearly coincide and the ratio, (sine angle) / angle, approaches 1 as a limit.

As the angle approaches zero the arc and the ordinate tend to become identical and the ratio ordinate/arc approaches the limit 1. Dividing both numerator and denominator of this fraction by the radius does not change the value of the ratio and $\frac{\text{ordinate}}{\text{radius}} \div \frac{\text{arc}}{\text{radius}}$ also approaches the limit 1. But ordinate/radius is the sine of the angle and arc/radius is the angle, α, and $(\sin \alpha)/\alpha$ approaches 1 as α approaches zero. In this case the angle α is taken as $\Delta x/2$.

It is evident also that $\cos \left(x + \frac{\Delta x}{2}\right)$ approaches $\cos x$ as a limit when Δx approaches zero, for Δx or $\Delta x/2$ drops out and $x + \frac{\Delta x}{2}$ becomes x.

The derivative of the cosine of an angle is the negative sine of the angle,

$$\textbf{if } \mathbf{y = \cos x, \quad \frac{dy}{dx} = -\sin x.} \qquad \text{(Rule 28)}$$

This rule for differentiation depends on a rule of trigonometry which states that the cosine of an angle is equal to the sine of a right angle $\left(\frac{\pi}{2}\text{ radians}\right)$ minus the angle. It is shown on page 270 that

$$\cos x = \sin\left(\frac{\pi}{2} - x\right) \text{ and that}$$

$$\sin x = \cos\left(\frac{\pi}{2} - x\right)$$

Substituting for the cosine of an angle, its value in terms of the sine

$$\text{if } y = \cos x, \frac{dy}{dx} = \frac{d(\cos x)}{dx} = \frac{d \sin\left(\frac{\pi}{2} - x\right)}{dx}$$

Differentiating $\sin\left(\frac{\pi}{2} - x\right)$ and remembering that it is a function of a function,

$$\frac{d \sin\left(\frac{\pi}{2} - x\right)}{dx} = \cos\left(\frac{\pi}{2} - x\right)\frac{d\left(\frac{\pi}{2} - x\right)}{dx} = \cos\left(\frac{\pi}{2} - x\right)(-1)$$

After substituting for $\cos\left(\frac{\pi}{2} - x\right)$ its equivalent $\sin x$, the expression becomes,

$$\frac{dy}{dx} = \cos\left(\frac{\pi}{2} - x\right)(-1) = -\sin x.$$

The derivative of the tangent of an angle is the square of the secant of the angle.

$$\textbf{If } \mathbf{y = \tan x,} \quad \frac{\mathbf{dy}}{\mathbf{dx}} = \mathbf{\sec^2 x} \qquad \text{(Rule 29)}$$

In deriving this rule, use is made of the relation $\tan x = \frac{\sin x}{\cos x}$, given on page 270.

Differentiating $y = \frac{\sin x}{\cos x}$ as a quotient and applying the rules just given for differentiating sines and cosines,

$$\frac{dy}{dx} = \frac{\cos x \frac{d \sin x}{dx} - \sin x \frac{d \cos x}{dx}}{\cos^2 x} = \frac{\cos x \cos x - \sin x\,(-\sin x)}{\cos^2 x} =$$

$$\frac{\cos^2 x + \sin^2 x}{\cos^2 x} = \frac{1}{\cos^2 x}.$$

But the reciprocal of the cosine is the secant so if $y = \tan x$, $dy/dx = \sec^2 x$.

The cotangent, the secant, and the cosecant are not discussed. These functions are seldom necessary in solving chemical problems. It is an excellent exercise for the student, however, to prove that

$$\frac{d(\cot x)}{dx} = -\operatorname{cosec}^2 x$$

$$\frac{d(\sec x)}{dx} = \frac{\sin x}{\cos^2 x}$$

$$\frac{d(\operatorname{cosec} x)}{dx} = \frac{-\cos x}{\sin^2 x}$$

Inverse Trigonometrical Functions

The inverse trigonometrical functions at first sight would appear to be unimportant. They are unusual functions which do not occur often in chemistry. It happens, however, that their differentiations give rather simple algebraic expressions. At once, then, they become valuable for integration formulas. If some one in the past had not discovered that the derivative of the arc sin of x is $1/\sqrt{1 - x^2}$ it would not now be possible to integrate $dx/\sqrt{1 - x^2}$. Various other unusual functions which appear complex and unimportant are also valuable for purposes of integration. The hyperbolic sine, sinh, is a case in point (p. 261).

The inverse trigonometrical functions are called arc sine, arc cosine, etc, and sometimes written $\sin^{-1}$ or $\cos^{-1}$.

The derivative of the arc sine of x is the reciprocal of the square root of $(1 - x^2)$.

$$\textbf{If } \mathbf{y = \text{arc sin } x,\ \frac{dy}{dx} = \frac{1}{\sqrt{1 - x^2}}} \qquad \text{(Rule 30)}$$

If y is the arc sine of x, then y is the angle whose sine is x. For example, the sine of 45° (0.7875 radians) is 0.7071, and the arc sine of 0.7071 is 45° or 0.7875 radians. In general, if $y = \text{arc sin } x$,

$$x = \sin y, \text{ and } \frac{dx}{dy} = \cos y.$$

Inverting,

$$\frac{dy}{dx} = \frac{1}{dx/dy} = \frac{1}{\cos y}.$$

But

$$\cos y = \sqrt{1 - \sin^2 y} \quad (\text{since } \cos^2 y + \sin^2 y = 1)$$

$$\frac{dy}{dx} = \frac{1}{\sqrt{1 - \sin^2 y}} = \frac{1}{\sqrt{1 - x^2}}. \quad (\text{Substituting } x \text{ for } \sin y)$$

The derivative of the arc tangent of x is the reciprocal of the quantity $(1 + x^2)$.

$$\text{If } y = \text{arc tan } x, \frac{dy}{dx} = \frac{1}{1 + x^2} \qquad \text{(Rule 31)}$$

This rule is derived as follows:

$$x = \tan y \text{ and } \frac{dx}{dy} = \sec^2 y, \text{ or } \frac{dy}{dx} = \frac{1}{\sec^2 y}$$

Since $\sec^2 y = 1 + \tan^2 y$ by trigonometry, it is evident that

$$\frac{dy}{dx} = \frac{1}{1 + \tan^2 y} = \frac{1}{1 + x^2}$$

The other trigonometric functions may be differentiated for practice. The rules are as follows:

If $y = \text{arc cosine } x$,

$$\frac{dy}{dx} = -\frac{1}{\sqrt{1 - x^2}}$$

If $y = \text{arc cot } x$,

$$\frac{dy}{dx} = -\frac{1}{1 + x^2}$$

If $y = \text{arc sec } x$,

$$\frac{dy}{dx} = \frac{1}{x\sqrt{x^2 - 1}}$$

If $y = \text{arc cosec } x$,

$$\frac{dy}{dx} = -\frac{1}{x\sqrt{x^2 - 1}}$$

Integration of Trigonometrical Functions

The rules just derived may now be reversed to give the integrals of the trigonometric functions.

Since the derivative of the cosine is the negative sine (Rule 28) it follows that the integral of the negative sine is the cosine. Therefore, on reversing signs, the integral of the sine is the negative cosine.

$$\int \sin x dx = -\cos x + C. \qquad \text{(Rule 32)}$$

Since differentiation of $\sin x$ gave $\cos x$, it follows that integration of $\cos x$ gives $\sin x$.

$$\int \cos x dx = \sin x + C. \qquad \text{(Rule 33)}$$

Comparing Rules 27 and 28 with Rules 32 and 33 it is evident that the signs of the integrals of the sine and cosine are just opposite to those of the derivatives.

In finding a function which will give the tangent on differentiation, it will be remembered that the tangent is equal to sine/cosine.

It is possible to put the sine into the numerator by differentiating the cosine; and the cosine may be put into the denominator by introducing $\log_e$ cosine.

$$\text{If } y = -\log_e (\cos x), \text{ then } \frac{dy}{dx} = \frac{\sin x}{\cos x} = \tan x.$$

Reversing the process,

$$\int \tan x dx = -\log_e \cos x + C. \qquad \text{(Rule 34)}$$

The integral of the tangent is the negative logarithm of the cosine.

Since the inverse trigonometrical functions gave simple algebraic functions (Rules 30 and 31), the integration of these algebraic functions must give the inverse trigonometrical functions. In reversing these rules for purposes of integration they may be made more general by substituting a^2 for 1 where a is any constant. The rules for these integrations may be found on page 293.

Exercises

In the following problems the derivative dy/dx is to be found in each case:

1. $y = \sin^2 x = (\sin x)^2$.

Let $s = \sin x$ and $y = s^2$.

$$\frac{ds}{dx} = \cos x \cdot \frac{dy}{ds} = 2s.$$

$$\frac{dy}{ds} \times \frac{ds}{dx} = 2s \cos x = 2 \sin x \cos x.$$

2. $y = \sin x \cos x$.

$$\frac{dy}{dx} = \sin x(-\sin x) + \cos x \cos x =$$

$\cos^2 x - \sin^2 x$.

3. $y = \sqrt[3]{a + 2\tan^2 x}$.

Let $s = 2\tan^2 x$, then $y = (a + s)^{1/3}$.

$$\frac{ds}{dx} = 2(2\tan x)\frac{d\tan x}{dx} = 4\tan x \sec^2 x.$$

$$\frac{dy}{ds} = \frac{1}{3}(a + s)^{-2/3}.$$

$$\frac{dy}{dx} = \frac{dy}{ds} \times \frac{ds}{dx} = \frac{4\tan x \sec^2 x}{3\sqrt[3]{(a + 2\tan^2 x)^2}}.$$

4. $y = b\cos x$.

5. $y = \frac{2}{a}\sin x$. *Ans.:* $\frac{2}{a}\cos x$.

6. $y = 5\tan^2 x$.

7. $y = 2x(\cos x)$. *Ans.:* $-2x\sin x + 2\cos x$.

8. $y = \sin(2x^3 + 5)$.

9. $y = 5\tan 3x$. *Ans.:* $15\sec^2 3x$.

10. $y = \log_e \sin x$.

11. $y = \sin x^a$. *Ans.:* $ax^{a-1}\cos x^a$.

12. $y = \cos^3 \sqrt[3]{x}$.

13. $y = \sin^{-1}\frac{2x - 5}{3}$. *Ans.:* $\frac{1}{\sqrt{-x^2 + 5x - 4}}$.

14. $y = \sin x$.

What are the first, second, third, and fourth derivatives?

15. A particle vibrates according to the equation $y = k\sin(at - c)$ where a, k and c are constants.

What is the velocity dy/dt at any instant? *Ans.:* $ak\cos(at - c)$.

16. What value of x will make $\sin x \cos x$ a maximum (in radians and in degrees)?

What are the integrals of the following:

17. $\int \sin 2x\,dx = ?$ *Ans.:* $-\frac{1}{2}\cos 2x + C$.

It might be expected that the application of Rule 32 would give $-\cos 2x$ but the expression is a function of a function and the derivative of $-\cos 2x$ is $2\sin 2x$. It is necessary to introduce something to cancel this undesired factor 2 which multiplies $\sin 2x$. Multiplication by the term $\frac{1}{2}$ will effect this cancellation. Then since $\frac{d(-\frac{1}{2}\cos 2x)}{dx} = \left(-\frac{1}{2}\right)(-2\sin 2x) = \sin 2x$, it follows that $\int \sin 2x\,dx = -\frac{1}{2}\cos 2x + C$.

18. $\int \cos\frac{x}{3}dx = ?$

19. $\int_0^\pi \sin x\,dx = ?$

Ans.: $\Big[-\cos x\Big]_0^\pi = (-\cos\pi) - (-\cos 0) = 1 + 1 = 2.$

20. $\int(\frac{1}{2}\cos x + 3\sin x + 1)dx = ?$

CHAPTER XIV

INTEGRATION

It was pointed out before that there are no general rules for integrating the more complicated expressions such as products and quotients and functions of functions. Each case must be handled as a separate problem. In this chapter, various aids to integration are described. Through these devices it is possible, frequently, to express the problem in a form that is recognized as having been obtained by a previous differentiation.

Algebraic Simplification

It is advantageous to change a product into a sum whenever possible. For example,

$$\int (a + bx)x^2dx = \int (ax^2 + bx^3)dx = \int ax^2dx + \int bx^3dx = \frac{ax^3}{3} + \frac{bx^4}{4} + C.$$

In another example,

$$\int (x^2 + a)^2dx = \int (x^4 + 2ax^2 + a^2)dx = \int x^4dx + \int 2ax^2 + \int a^2dx = \frac{x^5}{5} + \frac{2ax^3}{3} + a^2x + C.$$

Sometimes it is possible to simplify a quotient by carrying out the indicated division. If the highest exponent in the denominator of a fraction is lower than, or equal to the highest exponent in the numerator the quotient may be changed into a sum. For example,

$$\int \left(\frac{x^3 + 3x^2 + 2}{x}\right)dx = \int \left(x^2 + 3x + \frac{2}{x}\right)dx = \frac{x^3}{3} + \frac{3x^2}{2} + 2 \log_e x + C.$$

In another example,

$$\int \frac{x + 1}{x - 1}dx = \int \left(1 + \frac{2}{x - 1}\right)dx = x + 2 \log_e (x - 1) + C.$$

Integration by Substitution

Substitution is one of the most valuable helps, just as it is in differentiation. The method is somewhat similar to that used in the differentiation of a function of a function, although less general. When the expression contains a function of a function, one of the functions is set equal to s and differentiated with respect to x. dx is then obtained from the derivative ds/dx and substituted into the integrand. The integrand is the expression after the integral sign, which is to be integrated.

For example,

$$\int \sin 3x dx = ?$$

$$s = 3x, \text{ and then } ds/dx = 3, \text{ and } dx = ds/3$$

The original expression may now be changed by substitution into an expression containing only a single function.

$$\int (\sin 3x)(dx) = \int (\sin s)\frac{(ds)}{3} = \frac{1}{3}\int \sin s ds = -\frac{1}{3}\cos s = -\frac{1}{3}\cos 3x + C.$$

The integration of $\sin x \cos x dx$, may be accomplished in a similar manner by letting $\sin x = s$.

$$\text{If } s = \sin x, \frac{ds}{dx} = \cos x, \text{ or } dx = \frac{ds}{\cos x}.$$

Substituting

$$\int (\sin x)(\cos x)(dx) = \int (s)(\cos x)\frac{(ds)}{\cos x} =$$

$$\int s ds = \frac{s^2}{2} = \frac{\sin^2 x}{2} + C.$$

This method is applicable if the derivative of one function is a constant as in the first case; or if the derivative of one function cancels with the other part as in the second case. If such a simplification cannot be effected by the substitution the method is useless. For example, if it is desired to integrate $\sin x \log x dx$ and the same procedure is followed,

$$s = \sin x \text{ and } \frac{ds}{dx} = \cos x, \text{ and } dx = \frac{ds}{\cos x}.$$

$$\int \sin x \log_e x dx = \int s \frac{\log_e x}{\cos x} ds.$$

The method of substitution cannot be used in this case for it leads to an integral of mixed variables s and x, which is more complicated than the original. In this case some other method must be found if the expression is to be integrated (integration by parts for example).

Still other examples are given below to show the wide application of this method.

$$\int x^3\sqrt{2x^4 + 5}\,dx.$$

Let

$$s = 2x^4 + 5. \quad \text{Then } \frac{ds}{dx} = 8x^3 \text{ and } dx = \frac{ds}{8x^3}$$

Substituting,

$$\int x^3\sqrt{2x^4 + 5}\,dx = \int \frac{x^3 s^{1/2} ds}{8x^3} = \frac{1}{8}\int s^{1/2} ds.$$

Integrating,

$$\frac{1}{8}\int s^{1/2} ds = \frac{1}{8}\frac{s^{3/2}}{(3/2)} = \frac{2}{24}\sqrt{s^3} = \frac{1}{12}\sqrt{(2x^4 + 5)^3} + C$$

This problem could have been solved also by letting $s = \sqrt{(2x^4 + 5)}$.

Then

$$\frac{ds}{dx} = \frac{1}{2}(2x^4 + 5)^{-1/2}(8x^3),$$

and

$$dx = \frac{2\sqrt{2x^4 + 5}}{8x^3}ds = \frac{sds}{4x^3}$$

Substituting,

$$\int x^3\sqrt{2x^4 + 5}\,dx = \int (x^3)(s)\frac{sds}{4x^3} = \frac{1}{4}\int s^2 ds$$

Integrating,

$$\frac{1}{4}\int s^2 ds = \frac{1}{4}\frac{s^3}{(3)} = \frac{1}{12}\sqrt{(2x^4 + 5)^3} + C.$$

The two results are identical but the first substitution is the simpler. Care should be exercised in the choice of substitutes, to render the mathematical operations as simple as possible.

$$\int e^{ax} dx$$

Let $\quad s = ax. \quad \text{Then } \dfrac{ds}{dx} = a, \text{ and } dx = \dfrac{ds}{a}$

$$\int e^{ax} dx = \int e^s \frac{ds}{a} = \frac{1}{a}\int e^s ds = \frac{1}{a}(e^s) = \frac{e^{ax}}{a} + C$$

This answer could have been obtained by inspection for the derivative of $e^{ax} + C$ is ae^{ax} and to cancel this undesired a in front of the e, it is necessary to include the term $1/a$. The differentiation of $\frac{e^{ax}}{a} + C$ gives $\frac{a}{a}e^{ax} = e^{ax}$, and therefore the integral of $e^{ax}dx$ is $\frac{e^{ax}}{a} + C$.

Integration by Parts

Integration by parts is based on the rule for differentiating a product, according to which

$$d(uv) = udv + vdu$$

Integrating this equation,

$$\int d(uv) = \int udv + \int vdu$$

or

$$uv = \int udv + \int vdu$$

and

$$\int udv = uv - \int vdu$$

If it is desired to integrate a product, one part of the product may be set equal to u and the other part equal to dv. Integration of the term which is equivalent to dv gives v and differentiation of u gives du. The last term, $\int vdu$, must then be integrated. If the integral of vdu cannot be obtained readily, the method is not applicable. In general, integration by parts is used when the product to be integrated consists of one part which can be easily differentiated and one part which can be readily integrated. Judgment is necessary in putting equal to dv the factor which is most readily integrated, and substituting for u the factor which is easily differentiated. The following examples illustrate the method:

$$\int x \sin x dx = ?$$

Let $x = u$, and then $du = dx$

Let $\sin x dx = dv$, and then $v = \int dv = \int \sin x dx = -\cos x$.

$$\int udv = uv - \int vdu$$

$$\int x(\sin x dx) = x(-\cos x) - \int (-\cos x)dx \text{ (substituting)}$$

$$= x(-\cos x) - (-\sin x)$$

$$= \sin x - x \cos x + C$$

The fact that this integration is correct may be proved (as in every case) by differentiating the result.

$$\frac{d(\sin x - x\cos x) + C}{dx} = \cos x - (-x\sin x + \cos x) = x\sin x$$

In another example,

$$\int x\log_e x dx = ?$$

$$\text{Let } xdx = dv;\text{ and then } v = \int dv = \int xdx = \frac{x^2}{2}$$

$$\text{Let } \log_e x = u;\text{ and then } \frac{du}{dx} = \frac{1}{x} \text{ or } du = \frac{1}{x}dx$$

$$\int udv = uv - \int vdu$$

$$\int x(\log_e x)(dx) = (\log_e x)\left(\frac{x^2}{2}\right) - \int\left(\frac{x^2}{2}\right)\left(\frac{1}{x}dx\right)$$

$$= \frac{x^2}{2}\log_e x - \frac{x^2}{4} + C$$

This result can be checked by differentiation. It should be noted that if ($\log_e xdx$) had been put equal to dv the integration would have been considerably more difficult for $\log_e xdx$ gives a simpler expression on differentiation than on integration. The result would have been correct, however. Different ways of integrating a given expression may give results which look quite different, but they must give the same answer on differentiation. Usually they are equivalent except for an added constant.

Sometimes it is necessary to carry out the integration by parts twice or more before the final result is obtained.

$$\int x^2\cos x\, dx = ?$$

$$\text{Let } x^2 = u;\text{ and then } du = 2x\, dx$$

$$\text{Let } \cos xdx = dv;\text{ and then } v = \int\cos xdx = \sin x$$

$$\int x^2\cos xdx = x^2\sin x - \int(\sin x)(2xdx)$$

$$= x^2\sin x - 2\int x\sin xdx$$

The last term must be handled as a separate problem in integration bv parts, and it has been shown already that it is equal to $\sin x - x\cos x$.

Then

$$\int x^2\cos x\, dx = x^2\sin x - 2(\sin x - x\cos x)$$

$$= (x^2 - 2)\sin x + 2x\cos x + C$$

It is by no means possible to use integration by parts successfully for every product as, for example, in the following case,

$$\int xe^{x^2}dx = e^{x^2}\frac{x^2}{2} - \int \frac{x^2}{2}2xe^{x^2}dx = \frac{e^{x^2}x^2}{2} - \int x^3e^{x^2}dx$$

The expression is no nearer solution than the original. Although integration by parts fails in this case, integration by substitution can be applied.

Integration by Partial Fractions

In some cases an integrand with a denominator may be split up advantageously into a sum of simpler expressions by the method of partial fractions.

The splitting up of an expression into partial fractions is the reverse of adding fractions to give a common denominator. Fractions are added as in the following example:

$$\frac{1}{x+2} + \frac{2}{x-2} = \frac{1(x-2) + 2(x+2)}{(x+2)(x-2)} = \frac{x - 2 + 2x + 4}{x^2 - 4} = \frac{3x+2}{x^2-4}$$

and a fraction may be split up into partial fractions with numerators A and B which are evaluated in the manner of the following example:

$$\frac{3x+2}{x^2-4} = \frac{A}{x+2} + \frac{B}{x-2} = \frac{A(x-2) + B(x+2)}{(x^2-4)}$$

$A(x - 2) + B(x + 2) = 3x + 2$ (The numerators are equal in an identical equation.)

To evaluate A, the B term is made to drop out by setting $x = -2$ for then the coefficient of B becomes zero.

$$\begin{aligned} A(-2-2) + B(-2+2) &= 3(-2) + 2 \\ -4A + 0 &= -6 + 2 \\ A &= 1 \end{aligned}$$

In the same way by putting $x = +2$, A is made to drop out.

$$\begin{aligned} A(+2-2) + B(2+2) &= 3(2) + 2 \\ 0 + 4B &= 6 + 2 \\ B &= 2 \end{aligned}$$

Having evaluated A and B the original fraction is split up into its partial fractions,

$$\frac{3x+2}{x^2-4} = \frac{A}{x+2} + \frac{B}{x-2} = \frac{1}{x+2} + \frac{2}{x-2}$$

The expression $\frac{3x + 2}{x^2 - 4}$ can be more easily integrated after splitting it up into its partial fractions as follows:

$$\int \frac{3x + 2}{x^2 - 4}dx = \int \frac{A}{x + 2}dx + \int \frac{B}{x - 2}dx = \int \frac{1}{x + 2}dx + \int \frac{2}{x - 2}dx =$$
$$\log_e (x + 2) + 2 \log_e (x - 2) = \log_e (x + 2)(x - 2)^2 + C$$

The method may be used when there are three or more factors as in the following example. More complicated expressions are beyond the scope of this book but they depend only on the rules for partial fractions.

$$\int \frac{2x + 1}{x^3 + x^2 - 2x}dx = ?$$

$$\frac{2x + 1}{x^3 + x^2 - 2x} = \frac{A}{x} + \frac{B}{x - 1} + \frac{C}{x + 2} =$$
$$\frac{A(x - 1)(x + 2) + B(x)(x + 2) + C(x)(x - 1)}{x(x - 1)(x + 2)}$$
$$A(x + 2)(x - 1) + B(x)(x + 2) + C(x)(x - 1) = 2x + 1$$

When $x = 0$, B and C drop out and $-2A = 1$, when $x = 1$, $(x - 1)$ becomes zero and A and C drop out leaving B equal to 1. When $x = -2$, $(x + 2)$ becomes zero and A and B drop out leaving $6C$ equal -3. Then $A = -\frac{1}{2}$, $B = 1$ and $C = -\frac{1}{2}$.

$$\int \frac{2x + 1}{x^3 + x^2 - 2x}dx = -\frac{1}{2}\int \frac{dx}{x} + 1\int \frac{dx}{x - 1} - \frac{1}{2}\int \frac{dx}{x + 2} = -\frac{1}{2} \log_e x +$$
$$\log_e (x - 1) - \frac{1}{2} \log_e (x + 2) = \log_e \frac{x - 1}{x^{1/2}(x + 2)^{1/2}} = \log_e \frac{x - 1}{\sqrt{x^2 + 2x}} + C$$

Exercises

1. $\int \frac{(x^2 + 3x)^2}{5}dx = ?$ *Ans.:* $\frac{x^5}{25} + \frac{3x^4}{10} + \frac{3x^3}{5} + C.$

2. $\int x^2(3x^5 + 5)(3x^5 - 5)dx = ?$

3. $\int (2x^2 + x)^3dx = ?$ *Ans.:* $\frac{8x^7}{7} + 2x^6 + \frac{6x^5}{5} + \frac{x^4}{4} + C.$

4. $\int_0^2 (x + 1)(x^2 - 3)dx = ?$

5. $\int \frac{x^2 + 2x + 1}{x + 1}dx = ?$ *Ans.:* $\frac{x^2}{2} + x + C.$

6. $\int \frac{x^5 + 2x^4 - x^2 + 3}{x^2}dx = ?$

7. $\int \frac{(2 + x - 3x^2)}{\sqrt{x}}dx.$ *Ans.:* $4\sqrt{x} + \frac{2}{3}\sqrt{x^3} - \frac{6}{5}\sqrt{x^5} + C.$

8. $\int \frac{x^2}{\sqrt{x}}(1 + 2x - 3x^2)dx = ?$

9. $\int x^2\sqrt{x^3 - a}\,dx = ?$ *Ans.:* $\frac{2}{9}\sqrt{(x^3 - a)^3} + C.$

10. $\int \frac{dx}{3x - 2} = ?$

11. $\int \cos ax dx = ?$ *Ans.:* $\frac{\sin ax}{a} + C.$

12. $\int x\sqrt{5 - x^2}\,dx = ?$

13. $\int \frac{dx}{\sqrt{2x + 5}} = ?\quad s = 2x + 5.$ *Ans.:* $\sqrt{2x + 5} + C.$

14. $\int x\sqrt{a + x}\,dx = ?\quad s = a + x$ and $dx = ds$. Also $x = s - a$.

Therefore

$$\int x\sqrt{a + x}\,dx = \int (s - a)s^{1/2}ds$$

15. $\int e^{-x/5}dx = ?$ *Ans.:* $-5e^{-x/5} + C.$

16. $\int \frac{x + 3}{x^2 + 6x + a}dx = ?$

17. $\int xe^x dx = ?$ *Ans.:* $e^x(x - 1) + C.$

18. $\int x \cos x dx = ?$

19. $\int x^2e^{-x}dx = ?$ *Ans.:* $-e^{-x}(x^2 + 2x + 2) + C.$

20. $\int e^x \cos e^x dx = ?$

21. $\int 2x^2 \log_e x dx = ?$ *Ans.:* $\frac{2x^3}{3}\left(\log_e x - \frac{1}{3}\right) + C.$

22. $\int (x + 1) \log_e (x + 2)dx = ?$

23. $\int \frac{dx}{x^2 - 1} = ?$ *Ans.:* $\log_e \sqrt{\frac{x - 1}{x + 1}} + C.$

24. $\int \frac{dx}{x^2 - 4} = ?$

25. $\int \frac{dx}{(x - a)(x - b)} = ?$ *Ans.:* $\frac{1}{a - b} \log_e \frac{x - a}{x - b} + C.$

26. $\int \frac{5x - 3}{x^2 - 9}dx = ?$

27. $\int \frac{dx}{x^2 - 3} = ?$ *Ans.:* $\frac{1}{2\sqrt{3}} \log \frac{x - \sqrt{3}}{x + \sqrt{3}} + C.$

28. $\int \frac{3}{x^2 - 5}dx = ?$

29. $\int \frac{dx}{x^2(x - 1)} = ?$

30. $\int \frac{dx}{(a - x)(b - x)(c - x)} = ?$

CHAPTER XV

THE USE OF INTEGRATION TABLES

The integrations which have been studied in the previous pages are comparatively simple ones, yet they show that integration is an art, requiring practice. It seems to be an art, too, which is quickly lost when left unused. It is beyond the scope of this introductory book to go farther into the art of integration. Fortunately, the various formulas for integration have been generalized and collected into standard tables of integration. On page 289 is given the table of Peirce and Carver[1] comprising 154 different formulas.

The expression to be integrated is first classified as follows:

Rational algebraic integrals, *i.e.*, containing no fractional exponents	Formulas 1–31
Integrals involving $\sqrt{ax + b}$	Formulas 32–43
Integrals involving $\sqrt{x^2 + a^2}$ and $\sqrt{a^2 - x^2}$	Formulas 44–67
Integrals involving $\sqrt{ax^2 + bx + c}$	Formulas 68–87
Miscellaneous integrals with fractional exponents	Formulas 88–90
Logarithmic integrals	Formulas 91–94
Exponential integrals	Formulas 95–98
Trigonometric integrals	Formulas 99–154

In all these formulas x and $\underline{X}$ represent variables and all other letters represent constants.

$\underline{X}$ stands for $ax^2 + bx + c$, and $q = b^2 - 4ac$. a is the coefficient of x^2, b is the coefficient of x and c is a constant. These expressions occur so frequently that a considerable saving is effected by the use of the symbols $\underline{X}$ and q.

[1] PEIRCE and CARVER: "Handbook of Formulas and Tables for Engineers," McGraw-Hill Book Company, Inc., New York.

A few problems serve to illustrate the use of the tables.

$$\int \frac{dx}{9x^2 + 4} = ? \tag{1}$$

This integral belongs in the group of rational algebraic integrals and it resembles Formula 16 which is

$$\int \frac{dx}{x^2 + a^2} = \frac{1}{a} \tan^{-1} \frac{x}{a}$$

The problem may be changed into this form by taking the constant $\frac{1}{9}$ outside the integral giving $\frac{1}{9}\int \frac{dx}{x^2 + 4/9}$. $\frac{4}{9}$ corresponds to a^2. When the integration is complete, the answer must be multiplied by the constant $1/9$.

$$\frac{1}{9}\int \frac{dx}{x^2 + 4/9} = \frac{1}{9}\left(\frac{1}{2/3} \tan^{-1} \frac{x}{2/3}\right) = \frac{1}{6} \tan^{-1} \frac{3x}{2} + C.$$

$$\int x\sqrt{3x + 2}\, dx = ? \tag{2}$$

This expression involves $\sqrt{ax + b}$ and comes under Formula 32.

$$a = 3 \text{ and } b = 2.$$

$$\int x\sqrt{ax + b}\, dx = \frac{2(3ax - 2b)\sqrt{(ax + b)^3}}{15a^2} \qquad \text{(Formula 32)}$$

$$\int x\sqrt{3x + 2}\, dx = \frac{2[(3 \times 3x) - (2 \times 2)]\sqrt{(3x + 2)^3}}{15 \times 3^2} = \frac{(18x - 8)\sqrt{(3x + 2)^3}}{135} + C.$$

$$\int \frac{dx}{\sqrt{(5 - 3x + 2x^2)^3}} = \int \frac{dx}{\sqrt{(2x^2 - 3x + 5)^3}} = ? \tag{3}$$

This expression contains the term $\sqrt{ax^2 + bx + c}$. Formula 86 is applicable.

$$\underline{a} = 2 \qquad b = -3 \qquad c = 5$$

$$\underline{X} = ax^2 + bx + c \text{ and } q = b^2 - 4ac$$

$$\int \frac{dx}{\underline{X}\sqrt{\underline{X}}} = \frac{-2(2ax + b)}{q\sqrt{\underline{X}}} = \frac{-2(2ax + b)}{(b^2 - 4ac)\sqrt{\underline{X}}} \qquad \text{(Formula 86)}$$

$$\int \frac{dx}{(2x^2 - 3x + 5)\sqrt{2x^2 - 3x + 5}} = \frac{-2(2 \times 2x - 3)}{((-3)^2 - 4 \times 2 \times 5)\sqrt{2x^2 - 3x + 5}} = \frac{-(8x - 6)}{(9 - 40)\sqrt{2x^2 - 3x + 5}} = \frac{8x - 6}{31\sqrt{2x^2 - 3x + 5}} + C$$

The following example is taken from the trigonometric functions:

$$\int \sin^4 x dx = ? \tag{4}$$

Formula 102 applies in this case, followed by Formula 100.

$$\int \sin^n x dx = -\frac{\sin^{n-1} x \cos x}{n} + \frac{n-1}{n}\int \sin^{n-2} x dx \qquad \text{(Formula 102)}$$

$$\int \sin^4 x dx = -\frac{\sin^{4-1} x \cos x}{4} + \frac{4-1}{4}\int \sin^2 x dx$$

$$= -\frac{\sin^{4-1} x \cos x}{4} + \frac{4-1}{4}\frac{(x - \sin x \cos x)}{2} \qquad \text{(Formula 100)}$$

$$= -\frac{1}{4}\sin^3 x \cos x + 3/8(x - \sin x \cos x) + C$$

Frequently an algebraic rearrangement is necessary to put the expression into a form which corresponds to a rule given in the table. For example,

$$\int \sqrt{1 + \frac{b}{x}}\, dx = ?$$

There is no direct rule for integrating this expression but multiplication of both numerator and denominator by x gives

$$\sqrt{\frac{(x+b)}{x}\frac{x}{x}}\, dx, \text{ or } \frac{\sqrt{bx + x^2}}{\sqrt{x^2}}\, dx, \text{ or } \frac{\sqrt{bx + x^2}}{x}\, dx.$$

The last expression comes under Formula 83.

As another example, the integration of

$$\sqrt{1 + 4x^2}\, dx$$

may be effected as follows:

$$\sqrt{\frac{(1 + 4x^2)4}{4}} dx = \sqrt{\left(\frac{1}{4} + x^2\right)}\sqrt{4} = 2\sqrt{\left(\frac{1}{2}\right)^2 + x^2}$$

The integration of $\sqrt{(1/2)^2 + x^2}\, dx$ is accomplished easily with the help of Formula 44.

The integrations already performed in the earlier parts of the book should now be carried out with the help of the tables. The tables will also help in the integration of the material of the next chapter and later problems.

Exercises

1. $\int \frac{dx}{5x + 8x^3} = ?$ *Ans.:* $\frac{1}{10} \log_e \frac{x^2}{5 + 8x^2} + C.$

2. $\int \frac{dx}{9 + x^2} = ?$

3. $\int_{-1}^{1} \frac{dx}{x^2(2 + 3x)} = ?$ *Ans.:* $\left[\frac{-1}{2x} + \frac{3}{4} \log_e \frac{2 + 3x}{x}\right]_{-1}^{1} = 0.20708.$

4. $\int \frac{xdx}{(a - 2x)^2} = ?$

5. $\int \frac{dx}{\sqrt{1 + 3x + x^2}} = ?$ *Ans.:* $\log_e (\sqrt{1 + 3x + x^2} + x + 3/2) + C.$

6. $\int \sin^5 xdx = ?$

7. $\int \frac{2xdx}{x + 2x^2 + 3x^3} = ?$ *Ans.:* $\frac{2}{\sqrt{2}} \tan^{-1} \frac{(3x + 1)}{\sqrt{2}} + C.$

8. $\int_1^2 x^3 \log_{10} xdx = ?$

9. $\int_0^2 \frac{x^2dx}{1 + 2x} = ?$ *Ans.:* 0.70118.

10. $\int \sqrt{(1 - 2x - 3x^2)^3}\, dx = ?$

CHAPTER XVI

GEOMETRICAL APPLICATION OF INTEGRAL CALCULUS

Areas

In Fig. 45 is shown a rectangle 50 units long and 40 units wide. Obviously its area is 2,000 square units. A triangle is formed by drawing the diagonal. The area of the triangle is known by geometry to be 1,000 square units but it is instructive to calculate the area by the methods of calculus.

The area may be estimated roughly by inscribing a number of small rectangles in the triangle and adding up the area of each rectangle. In the second diagram there are four such rectangles and the area is only 800, whereas it should be 1,000. The reason for this discrepancy is obvious, for the little triangles above the rectangles have not been included in the calculation. In the third diagram the area of these little neglected triangles is smaller and the area of all the rectangles is 900, a value nearer to the true area. In the fourth diagram the large triangle is divided up into 24 small rectangles and the total area is found to be 962. This area is more nearly correct than the others because the error introduced by neglecting the area of the little triangles above the rectangles is less. If the number of rectangles is increased still further, the area will be still more nearly correct. They can be increased, except for mechanical difficulties of graphing, to give any desired degree of accuracy. In other words, they may be increased to such a large number that a further increase makes no appreciable difference in the total area, and then the total area is the true area of the triangle. This fact is shown graphically in Fig. 46 where extrapolation to zero width gives an area of 1,000. The true area is the limit to which the sum of the areas of the rectangles approaches, as Δx approaches zero.

As the number of rectangles is increased the width of each decreases, and when the rectangles are so narrow that making them narrower does not appreciably change the total area the area must have its true value. The height, y, of each rectangle is given by its intersection with the hypothenuse of the triangle.

The area of the sum of the rectangles is given mathematically by the expression, $\sum_{x=0}^{x=50} y\Delta x$. As Δx becomes smaller the area of the rectangles becomes more nearly equal to the area of

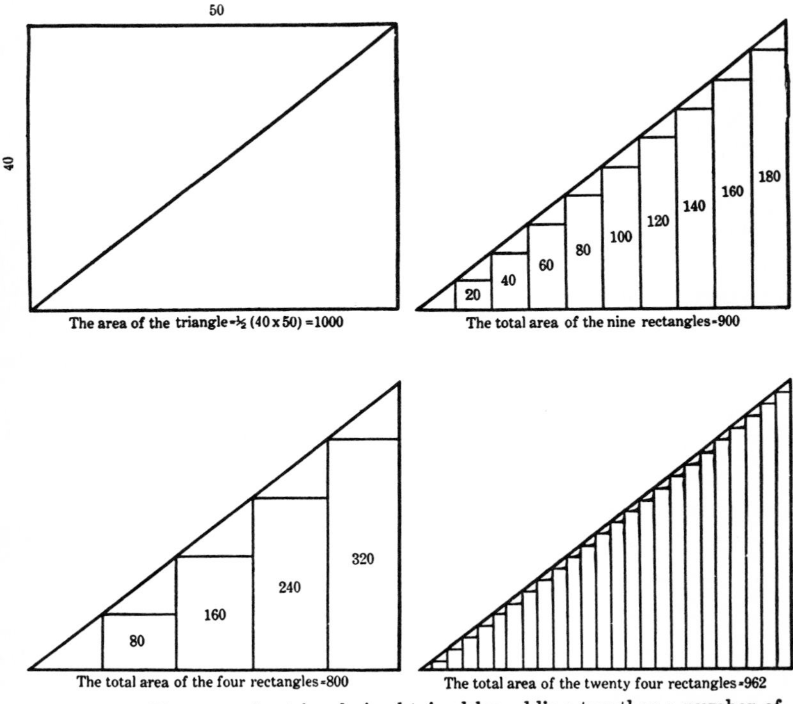

FIG. 45.—The area of a triangle is obtained by adding together a number of rectangles. The value is only approximate, because the areas of the small triangles are not included. The error becomes less as the number of rectangles is increased.

the large triangle. The limit to which $\sum_0^{50} y\Delta x$ approaches as Δx approaches zero is $\int_0^{50} ydx$, and this limit gives the true area of the triangle.

The integration of ydx gives then the correct area but the integration cannot be carried out unless more information is available. It is necessary to express y as a function of x before the expression can be integrated. In the case of the triangle under discussion, the equation of the line is known, for the slope is $4/5$ and the line passes through the origin. The value of

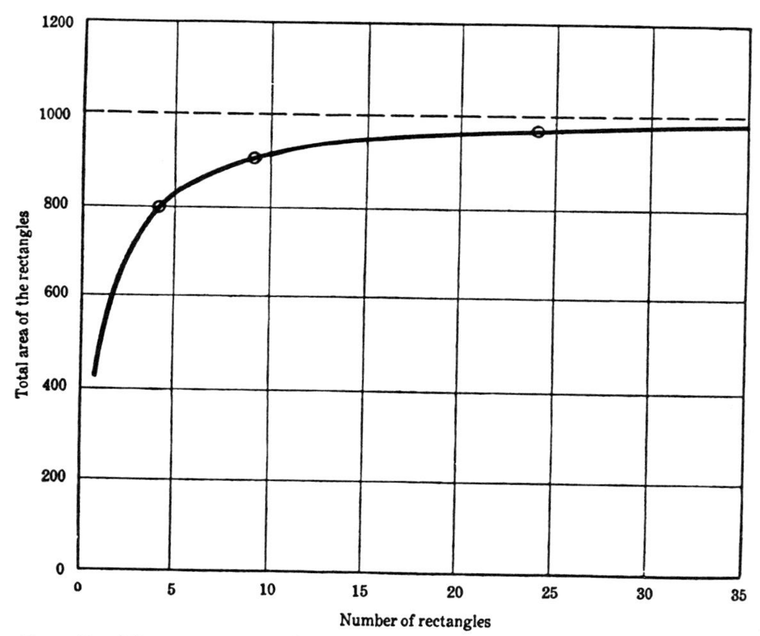

FIG. 46.—The true area of the triangle of Fig. 45 may be found by plotting the sum of the areas of all the rectangles against the number of rectangles and extrapolating to a very large number of rectangles. In this case, the limiting value of the area is 1,000. The limiting value may be found much more easily with integral calculus.

every point on the line then is such that $y = 0.8x$. Substituting into the original equation,

$$\text{area} = \int_0^{50} y\,dx = \int_0^{50} 0.8x\,dx = \left[\frac{0.8x^2}{2}\right]_0^{50} = (0.4 \times 50^2) - (0) = 1{,}000.$$

This value checks, of course, with the area calculated by geometry.

The area of a pressure-volume diagram is of great importance in physical chemistry. When a piston is moved against a force by an expanding gas the work done is equal to the pressure multiplied by the increase in volume.

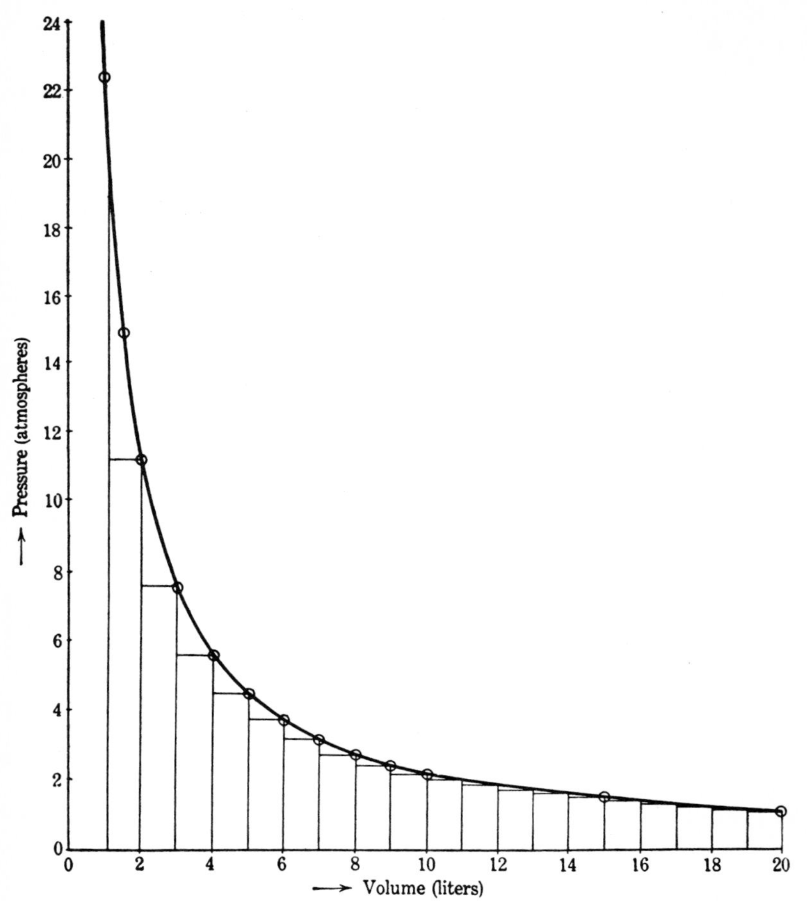

FIG. 47.—The area of this figure between the volumes 1 and 10 is 51.6 liter-atmospheres. It represents the maximum work done by a mole of gas expanding from 1 to 10 liters. The area may be determined approximately by adding up the areas of the small rectangles. It may be calculated exactly with integral calculus.

In Fig. 47 is given the volume of a gram molecule of a perfect gas at different pressures, when the temperature is kept constant at 0° C. The points are given in Table I.

TABLE I.—VOLUME OF A GRAM MOLECULE OF A GAS AT DIFFERENT PRESSURES

Volume, liters	Pressure, atmospheres
0.01	2,240.
0.10	224.0
0.50	44.80
1.00	22.40
1.50	14.90
2.00	11.20
3.00	7.47
4.00	5.60
5.00	4.48
6.00	3.73
7.00	3.20
8.00	2.80
9.00	2.49
10.00	2.24
15.00	1.49
20.00	1.12
50.00	0.448
100.0	0.224
1,000.	0.022

If it is desired to find the area under this curve between $v = 10$ and $v = 1$, the area may be divided up into rectangles as before, each one being Δv wide and p high. Each has a different height, p, depending on the value of v. The approximate area $= \sum_{v=1}^{v=10} p\Delta v$, and the exact area $= \int_1^{10} pdv$. This area has the physical significance that it is the work in liter atmospheres done by a gram molecule of gas in expanding from 1 liter to 10 liters.

Now this integration cannot be carried out unless the mathematical relation between p and v is known. Fortunately, laboratory experiments have established the fact that pressure varies inversely as the volume, *i.e.*, $p = k\frac{1}{v}$. When pressure is expressed in atmospheres and the volume in liters, $p = 22.4/v$. Substituting $22.4/v$ for p in the integration,

$$\text{Area} = \int_1^{10} \frac{22.4}{v}\,dv = 22.4\int_1^{10} \frac{dv}{v} = 22.4 \times 2.303\Big[\log_{10} v\Big]_1^{10} =$$
$$51.6(1 - 0) = 51.6 \text{ liter-atmospheres.}$$

In another example it is desired to find the area between the curve $y = 6 + 8x - 3x^2$ and the X axis, from $x = 0$ to $x = 5$. This area may be found, as before, by integrating; *i.e.*, by finding the area of the very large number of very narrow rectangles which make up the area, each rectangle being y high and dx wide.

$$\text{Area} = \int_0^5 (6 + 8x - 3x^2)dx = 6\int_0^5 dx + 8\int_0^5 xdx - 3\int_0^5 x^2dx =$$

$$\left[6x + \frac{8x^2}{2} - \frac{3x^3}{3}\right]_0^5 = \left[6x + 4x^2 - x^3\right]_0^5 = (30 + 100 - 125) - 0 = 5.$$

The area is five square units; for example, 5 sq. cm. if y and x are in centimeters or 5 sq. in. if y and x are in inches.

Volumes

If a horizontal line is rotated 360° around the X axis, a cylinder is generated. If a slanting line is rotated, a cone is formed and if a semi-circle is rotated through 360° a sphere is produced. Now, as these lines are swung around, every point of the line describes a complete circle and the total volume of the solid which is produced in this way may be considered as composed of a pile of a large number of discs, each dx wide and πy^2 in area. The volume of each disc is then $\pi y^2 dx$, and the integral of this expression between the proper limits of x gives the true volume. The integration is the adding together of a very large number of very small discs, and the discs are so small that no error is introduced by neglecting the exceedingly small volume between the edge of the discs and the actual surface of the solid. The integration gives the limit to which the summation process approaches as Δx approaches zero and the number of terms added together approaches infinity.

The expression $\int \pi y^2 dx$ can be solved only when the mathematical relation between y and x is known. If this relation cannot be established from geometrical considerations or from laboratory experiments the problem cannot be solved by integration formulas. In such a case the volume can be obtained approximately with the help of mathematical series (p. 200); or by measuring in the laboratory the volume of liquid displaced when the solid is completely submerged in an inert liquid.

The following examples show how volumes may be calculated exactly by integral calculus.

What is the volume of a right-angled cone which is 25 cm. high and 10 cm. wide at the base?

It is considered that the cone has been generated by revolving a straight line through a complete circle around the X axis, as shown in Fig. 48. One end of the straight line is fixed at the origin, and at a distance of 25 cm. along the X axis, the line describes a circle 5 cm. in radius. This circle of 5 cm. radius is

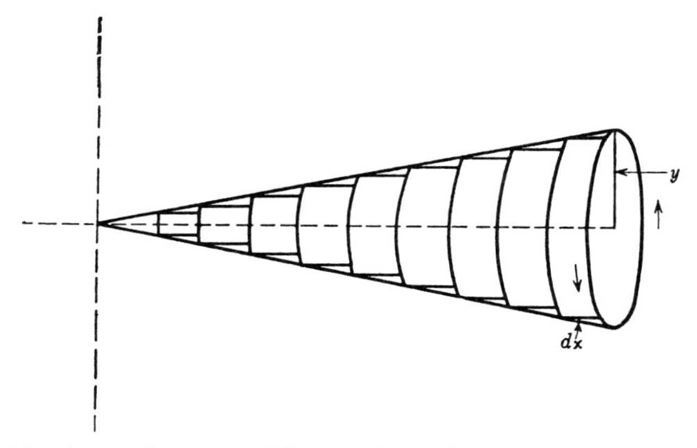

Fig. 48.—A cone is generated by rotating a line around the X axis. Its volume is composed of a large number of discs each πy^2 in area and dx in thickness. The true volume may be obtained by integral calculus when the equation of the line is known.

in fact the base of the cone. All sections of the line will describe circular discs which are less than 5 cm. in radius, and they become smaller the nearer they are to the origin. The sum of the several discs gives the volume of the cone, as shown in Fig. 48, except for the little triangular volumes which surround the discs. By making the discs sufficiently thin, however, the triangular rings can be made negligible in comparison with the total volume, no matter how great an accuracy is demanded. The radius of each disc is the height above the X axis, or in other words the value of y. Each disc then has the volume of $\pi y^2 dx$, and the approximate volume of the cone $= \sum_{x=0}^{x=25} \pi y^2 \Delta x$. The true volume is $\int_0^{25} \pi y^2 dx$. It is the limit to which the expression $\sum_0^{25} \pi y^2 \Delta x$ approaches as Δx approaches zero.

This integral can be solved because the relation between y and x may be obtained from the straight line which generated the cone. This line goes through the origin where $x = 0$ and $y = 0$; and the ordinate, corresponding to an abscissa of 25, is 5. The equation of the line is $y = x/5$. Substituting into the equation given above, the volume of the cone is

$$\int_0^{25} \pi y^2 dx = \int_0^{25} \pi \left(\frac{x}{5}\right)^2 dx = \frac{\pi}{25}\int_0^{25} x^2 dx = \frac{\pi}{25}\left[\frac{x^3}{3}\right]_0^{25} =$$

$$\frac{\pi \times 25^3}{25 \times 3} - 0 = \frac{625\pi}{3} = 654.3 \text{ cu. cm.}$$

It is important to realize that the volume of a solid may be calculated by adding up the volumes of a number of discs stacked along the Y axis, as well as if they are stacked along the X axis as in the example just given. The cone can be set anywhere in space and it is just as satisfactory to place it in a vertical position. In this case, the equation of the line is $y = -5x + 25$ or $x = -\frac{y}{5} + 5$, and the volume of the cone is

$$\int_0^{25} \pi x^2 dy = \int_0^{25} \pi\left(-\frac{y}{5} + 5\right)^2 dy = 654.3 \text{ cu. cm.}$$

The answer is the same but the mathematical operations are more tedious.

The volume of a sphere may be calculated by considering it as a volume which is generated by the revolution of a semicircle around the X axis through 360°. It is then composed of a stack of thin discs each of which is dx wide and of a cross section of πy^2. The volume of each disc is then $\pi y^2 dx$ and the volume of the sphere is the integral of $\pi y^2 dx$. From the equation of the circle, $x^2 + y^2 = r^2$, which was revolved to give the sphere, it follows that $y^2 = (r^2 - x^2)$. The expression is then integrated between the limits $x = r$ and $x = -r$.

Volume of sphere is

$$\int_{-r}^{r} \pi y^2 dx = \int_{-r}^{r} \pi(r^2 - x^2)dx = \pi r^2 \int_{-r}^{r} dx - \pi \int_{-r}^{r} x^2 dx =$$

$$\left[\pi r^2 x - \frac{\pi x^3}{3}\right]_{-r}^{r} = \left(\pi r^3 - \frac{\pi r^3}{3}\right) - \left(-\pi r^3 + \frac{\pi r^3}{3}\right) = \frac{4}{3}\pi r^3$$

LENGTHS OF CURVES

The straight line distance between two points is given by the theorem of the right triangle as shown on page 34. The distance, l, between two points along a curved line can be determined with the help of calculus. If the curve is divided up into short chords Δl, the sum of the chords $\sum \Delta l$ gives approximately the distance

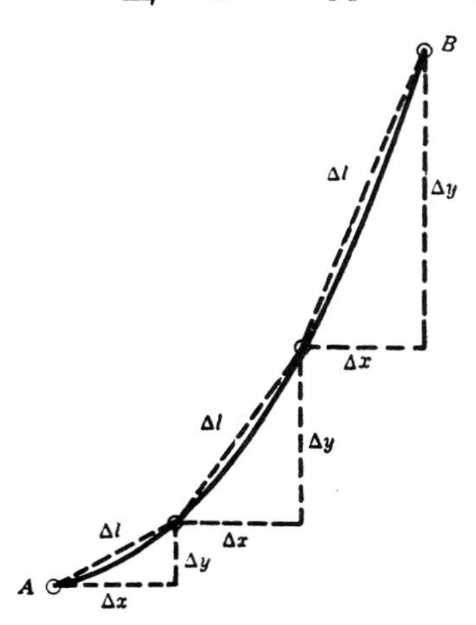

FIG. 49.—The length of a line between two points A and B is determined approximately by adding up all the segments as follows:

$$\sum_A^B \Delta l = \sum_A^B \sqrt{(\Delta x)^2 + (\Delta y)^2}$$

The exact length is given by

$$\int_A^B \sqrt{1 + (dy/dx)^2}\, dx.$$

l, as shown in Fig. 49. The shorter the chords are made the more nearly correct is the answer, and $\int_A^B dl$ gives an absolutely correct value for the distance between a and b along the curve.

But $(\Delta l)^2 = (\Delta x)^2 + (\Delta y)^2$ because Δl is the hypothenuse of a right-angled triangle.

Then dividing by (Δx^2),

$$\frac{(\Delta l)^2}{(\Delta x)^2} = \frac{(\Delta x)^2}{(\Delta x)^2} + \frac{(\Delta y)^2}{(\Delta x)^2} = 1 + \frac{(\Delta y)^2}{(\Delta x)^2}$$

and

$$\frac{\Delta l}{\Delta x} = \sqrt{1 + \left(\frac{\Delta y}{\Delta x}\right)^2} \text{ or } \Delta l = \sqrt{1 + \left(\frac{\Delta y}{\Delta x}\right)^2}\, \Delta x$$

The length of the line between A and B may now be found in terms of dy and dx, by passing to limits

$$\text{Length of line, } l, = \int_A^B dl = \int_{x=A}^{x=B} \sqrt{1 + \left(\frac{dy}{dx}\right)^2}\, dx$$

A similar formula may be derived by dividing $(\Delta l)^2 = (\Delta x)^2 + (\Delta y)^2$ by $(\Delta y)^2$ to give,

$$l = \int_{y=A}^{y=B} \sqrt{1 + \left(\frac{dx}{dy}\right)^2}\, dy$$

To illustrate the application of the formula, it is desired to find the length, l, of the line $y = 3x + 2$ between the points for which $x = 1$ and $x = 4$.

$$l = \int_1^4 \sqrt{1 + \left(\frac{dy}{dx}\right)^2}\, dx = \int_1^4 \sqrt{1 + 9}\, dx = \sqrt{10}\Big[x\Big]_1^4 =$$
$$3.16\,(4 - 1) = 9.48$$

This is a straight line, and the distance may be checked by the formula given on page 34.

The statement of the formulas is very simple but sometimes the solution of the integral is complicated. The integration tables are of great help in such cases.

What is the length, l, of the curve

$y^2 = 12x$ between $x = 0$ and $x = 3$

$$y = \sqrt{12x} \text{ and } \frac{dy}{dx} = \frac{\sqrt{12}}{2\sqrt{x}}. \quad \text{Also } \left(\frac{dy}{dx}\right)^2 = \frac{12}{4x} = \frac{3}{x}$$

$$l = \int_0^3 dl = \int_0^3 \sqrt{1 + \left(\frac{dy}{dx}\right)^2}\, dx = \int_0^3 \sqrt{\left(1 + \frac{3}{x}\right)}\, dx = \int_0^3 \sqrt{\frac{x^2 + 3x}{x^2}}\, dx$$

This integration can be performed with the help of Formulas 83 and 68 of the integration tables.

$$\int_0^3 \frac{\sqrt{x^2 + 3x}}{x}\, dx = \left[\sqrt{x^2 + 3x} + \frac{3}{2}\log_e\left(\sqrt{x^2 + 3x} + \frac{2x + 3}{2}\right)\right]_0^3$$
$$= 4.2426 + 3.2527 - 0.6087 = 6.88$$

SURFACES

The area of the surface generated by the rotation of a line around an axis may be determined by calculus in a manner similar to that used in the calculation of volumes. Each little segment of the line, l, forms a ring which is dl wide and $2\pi y$ in circumference. The sum of all the little rings, then, gives the area of the surface.

$$\text{Surface area} = \int 2\pi y\, dl.$$

But $\sqrt{1 + (dy/dx)^2}dx$ may be substituted for dl as shown in the preceding section.

$$\text{Surface area between } x_2 \text{ and } x_1 = 2\pi \int_{x_1}^{x_2} y\sqrt{1 + \left(\frac{dy}{dx}\right)^2}dx.$$

For example, the lateral area of a cone 25 cm. in height and 10 cm. in diameter at the base may be calculated. Setting it along the X axis with its point on the origin, the cone is generated by sweeping the line $y = x/5$ around the X axis through a complete circle. Each little section of the line sweeps out its ring and the total area is

$$2\pi \int_0^{25} y\sqrt{1 + \left(\frac{dy}{dx}\right)^2} dx = 2\pi \int_0^{25} \frac{x}{5} \sqrt{1 + \left(\frac{1}{5}\right)^2} dx =$$
$$\frac{2\pi\sqrt{1.04}}{5} \int_0^{25} x dx = \left[\frac{2\pi \times 1.02}{5} \frac{x^2}{2}\right]_0^{25} = 400.47.$$

Exercises

1. What is the area of the right triangle which has 8 cm. for a height and 6 cm. for a base? *Ans.:* 24 sq. cm.

2. What is the area of the right triangle which has a base of 10 cm. and a height of 10 cm.?

3. What is the area under the curve $y = x^2 + 4$, from $x = 0$ to $x = 4$? *Ans.:* 37.33.

4. What is the area bounded by the X axis, the line $y = \frac{x}{2} + 3$ and the two vertical lines $x = 1$ and $x = 5$?

5. What is the area under the curve $y = x^2 + 2x + 3$ from $x = 3$ to $x = -2$? *Ans.:* 31.66.

6. What is the area under the curve $y = \sqrt{1 + x}$ between $x = 9$ and $x = 0$?

7. What is the area between $y = 9 - x^2$ and the X axis? The points of intersection must serve as the limits of integration. They are found by setting $y = 0$ and solving for x. *Ans.:* 36.

8. What is the area under the curve $y = x^3 - 2x^2 + \frac{3x}{2} + \frac{1}{2}$ between the points $x = 3$ and $x = 1$?

9. What is the area under the curve $y = 2x^2 - x + 4$ between $x = 6$ and $x = 2$? *Ans.:* 138.7.

10. What is the area of the ellipse $\frac{x^2}{a^2} + \frac{y^2}{b^2} = 1$? It is easiest to calculate the area of one quadrant and then multiply later by 4. The area is integrated between $x = a$ and $x = 0$. $y = \frac{b}{a}\sqrt{a^2 - x^2}$.

11. What is the area of a circle?

The area under the curve in one quadrant may be found with the help of the equation, $x^2 + y^2 = r^2$. In a simpler process the circle may be divided up into rings of increasing size. Each ring is dx wide and $2\pi x$ long.

$$\int_{x=0}^{x=r} 2\pi x dx = \pi r^2.$$

12. What is the area of the parabola, $y^2 = 4x$, out to $x = 4$?

13. What is the closed area bounded by $y^2 = 9x$ and $y = 3x$?

In problems of this kind the areas under each curve are determined between the limits set by the points of intersection. The points of intersection are found by solving the equations simultaneously. Then the smaller area above the X axis is subtracted from the larger area.

Ans.: 0.5.

14. What is the area enclosed by the two lines $x + y = 5$ and $xy = 4$?

15. How much work is done when a gram molecule of a perfect gas expands from 5 liters to 15 liters at 25° C. (298° K.)?

$$\text{Work} = \int_5^{15} \frac{RT}{v} dv = \int_5^{15} \frac{0.082 \times 298}{v} dv. \quad (R = 0.082 \text{ liter-atmospheres})$$

16. How much work must be done when a gram molecule of a perfect gas is compressed from 15 liters to 5 liters at 298° K.?

The physical significance of the negative sign is that work must be done on the gas, whereas in the previous problem the expanding gas did work on the surroundings. Work done by a system on its surroundings is arbitrarily defined as positive work.

It should be noted that reversing the order of the limit always changes the sign of the answer without changing its magnitude.

17. What is the volume of a right-angled cone which is 10 cm. in height and 10 cm. in diameter at the base? *Ans.:* 262 cc.

18. What is the volume between $x = 0$ and $x = 5$ generated by the line $y = 2x$ as it revolves about the X axis?

19. What is the volume between $x = 0$ and $x = 4$, generated by the curve $y = \sqrt{1 + x^2}$ as it revolves about the X axis?

$$\int_0^4 \pi y^2 dx = \int_0^4 \pi(1 + x^2)dx = 25\tfrac{1}{3}\pi$$

20. What is the volume generated by the revolution of the curve $y = x^2 - 4$ around the X axis, between the points of intersection of the curve on the X axis?

21. What is the volume of the solid obtained by revolving the ellipse $\frac{x^2}{a^2} + \frac{y^2}{b^2} = 1$ around the X axis?

$$y^2 = \frac{b^2}{a^2}(a^2 - x^2)$$

$$\frac{\text{Volume}}{2} = \pi \int_0^a \frac{b^2}{a^2} (a^2 - x^2)dx = \frac{2\pi ab^2}{3}$$

It is more convenient to work only with the volume at the right of the Y axis. The answer must be multiplied by 2 to give the volume of the whole.

22. What is the volume generated by revolving the ellipse $\frac{x^2}{4} + \frac{y^2}{16} = 1$ around the Y axis?

23. The curve $p = 22.4/v$ is revolved about the V axis. What is the volume of the figure between $v = 1$ and $v = 10$? *Ans.:* 1418.7.

24. What is the distance along the line $y = 2x + 3$ from the point where $x = 0$ to the point where $x = 10$?

25. What is the length of the curve $y = x^2$ between the two points for which $x = 0$ and $x = 2$? *Ans.:* 4.577.

26. What is the length of the curve $y = a + bx + cx^2$ between the two points for which $x = 0$ and $x = 10$?

27. What is the circumference of a circle where $x^2 + y^2 = r^2$? The length in one quadrant $= \int_0^r \sqrt{1 + \left(\frac{dy}{dx}\right)^2}\, dx = \int_0^r \frac{r\, dx}{\sqrt{r^2 - x^2}}$ *Ans.:* $2\pi r$.

28. What is the length of the hyperbola $p = 22.4/v$ from $v = 1$ to $v = 10$?

29. What is the lateral area of a cone 10 cm. in height and 10 cm. in diameter at the base? *Ans.:* 175.6.

30. What is the lateral area of a cylinder, generated by revolving the line $y = 5$ around the X axis, between $x = 0$ and $x = 20$?

31. What is the area of a sphere, generated by revolving the semicircle $x^2 + y^2 = r^2$ around the X axis? *Ans.:* $4\pi r^2$.

32. What is the area of a parabolic mirror, produced by rotating the parabola $y^2 = 12x$ around the X axis? The mirror does not extend along the X axis beyond $x = 3$.

33. What is the lateral area from $x = 0$ to $x = 10$ produced by revolving the curve $y = a + bx + cx^2$ around the X axis?

CHAPTER XVII

PARTIAL DIFFERENTIATION

Partial differentiation is very important in thermodynamics and in any mathematical treatment of advanced physics or physical chemistry.

In all the discussions so far there has been but one independent variable. y has varied when x changed, or both u and v have varied when x changed, but always the variables depended on one independent variable for their values.

In many physical and chemical phenomena there are two or more *independent* variables which both together determine the value of a dependent variable. For example, an increase of temperature increases the volume of a gas, and an increase of pressure decreases the volume of the same gas. The pressure change and the temperature change do not depend on each other and they can be applied at different times, or simultaneously. According to a general principle of science the total change is the sum of all the independent changes no matter how they are applied. In this chapter some of the rules of partial differentiation are given by which phenomena of this type may be treated mathematically.

A new symbol is necessary to denote partial differentiation. ∂ is the symbol for partial differentiation just as d is the symbol for complete differentiation. Thus if $u = f(x \text{ and } y)$, $(\partial u/\partial x)_y$ means that y is treated as a constant while x and u are varied. Likewise $(\partial u/\partial y)_x$ means that x is treated as a constant while y and u are varied. If three independent variables are involved x, y and z the partial derivatives are written $(\partial u/\partial x)_{y,z}$ and $(\partial u/\partial y)_{x,z}$ and $(\partial u/\partial z)_{x,y}$. If there are more variables than x, y and z, they must be included also as subscripts. In general, however, if no parenthesis and subscripts are given it is understood that all other quantities, except the two involved in the derivative, are kept constant. Thus $\partial u/\partial x$ is equivalent to $(\partial u/\partial x)_{y,z}$ in the example just given.

The Fundamental Theorem

The fundamental theorem of partial differentiation is given by the following equation where u is a function of the two independent variables x and y,

$$du = \left(\frac{\partial u}{\partial x}\right)_y dx + \left(\frac{\partial u}{\partial y}\right)_x dy.$$

This equation states that the total differential is the sum of the partial differentials. The derivation of this expression may be

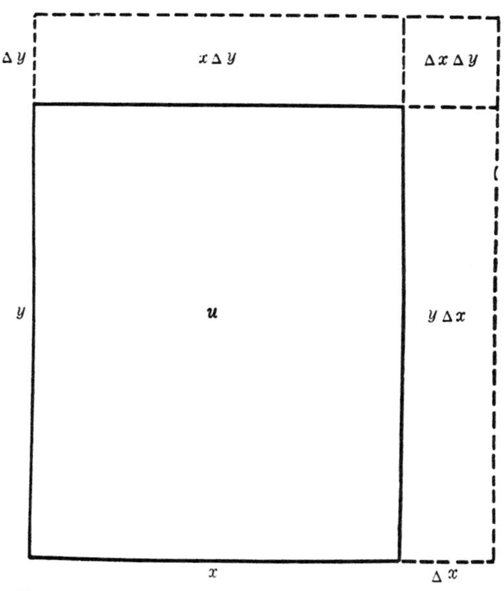

Fig. 50.—A diagram illustrating the derivation of the fundamental equation of partial differentiation.

$$\Delta u = y\Delta x + x\Delta y + \Delta x\Delta y$$

and

$$du = \left(\frac{\partial u}{\partial x}\right)_y dx + \left(\frac{\partial u}{y}\right)_x dy.$$

understood with the help of Fig. 50. Taking a simple case, u is the area of a rectangle which has the sides x and y. Then $u = xy$. Now holding x constant, y is increased by Δy. The corresponding increase in area, Δu is given by $x(\Delta y)_x$. Then $(\Delta u)_x = x\Delta y$ and $(\Delta u/\Delta y)_x = x$.

Holding y constant, x is increased by Δx. The corresponding increase in area is given by $y\Delta x$. Then $(\Delta u)_y = y(\Delta x)_y$ and $(\Delta u/\Delta x)_y = y$.

The total increase in area corresponding to an increase of Δx and Δy is shown by the dotted lines of Fig. 50. It is seen that this increase is made up of three parts, as given by the equation

$$\Delta u = (\Delta u)_y + (\Delta u)_x + \Delta x\Delta y = y\Delta x + x\Delta y + \Delta x\Delta y.$$

Substituting for x and y, their values just obtained,

$$\Delta u = \left(\frac{\Delta u}{\Delta x}\right)_y \Delta x + \left(\frac{\Delta u}{\Delta y}\right)_x \Delta y + \Delta x\Delta y.$$

Now if the increments Δx and Δy are made smaller and smaller, the equation may be written in the notation of differentials according to principles already discussed (p. 110).

$$du = \left(\frac{\partial u}{\partial x}\right)_y dx + \left(\frac{\partial u}{\partial y}\right)_x dy$$

It is an important principle in mathematics that second order differentials drop out in comparison with first order differentials. This fact may be visualized in Fig. 50 by noting that the ratio of the small square $\Delta y\Delta x$ to the total area of the dotted lines, decreases progressively as Δy and Δx are taken smaller and smaller.

If three independent variables x, y, and z are involved, the equation becomes

$$du = \left(\frac{\partial u}{\partial x}\right)_{y,z} dx + \left(\frac{\partial u}{\partial y}\right)_{x,z} dy + \left(\frac{\partial u}{\partial z}\right)_{x,y} dz$$

The use of the formula for partial differentiation is illustrated by the following example: $w = 2x^2 + 3xy + 4y^3$ where x and y are both independent variables and w changes with x and y in a manner shown by the equation. It is desired to find what increase in w is produced by very small increases in x and y when both are acting simultaneously.

$$\text{when } y \text{ is constant, } \left(\frac{\partial w}{\partial x}\right)_y = 4x + 3y$$

$$\text{when } x \text{ is constant, } \left(\frac{\partial w}{\partial y}\right)_x = 3x + 12y^2$$

and when both x and y vary the total change is given by

$$dw = \left(\frac{\partial w}{\partial x}\right)_y dx + \left(\frac{\partial w}{\partial y}\right)_x dy = (4x + 3y)dx + (3x + 12y^2)dy.$$

It is now possible to differentiate a variable raised to a variable power, as, $z = x^y$.

Regarding y as constant,

$$\frac{\partial z}{\partial x} = yx^{y-1}$$

Regarding x as constant,

$$\frac{\partial z}{\partial y} = x^y \log_e x$$

$$dz = \left(\frac{\partial z}{\partial x}\right)_y dx + \left(\frac{\partial z}{\partial y}\right)_x dy = yx^{y-1}dx + x^y \log_e x dy.$$

GEOMETRICAL SIGNIFICANCE OF PARTIAL DIFFERENTIATION

This fundamental equation of partial differentiation is so important in physical chemistry and particularly in thermodynamics that it is now studied in further detail. The volume, V, of a gas depends on the absolute temperature, T, and on the pressure, P. Each variable is independent of the other, and the increase in volume resulting from a simultaneous increase in both temperature and pressure is given by the equation

$$dV = \left(\frac{\partial V}{\partial T}\right)_P dT + \left(\frac{\partial V}{\partial P}\right)_T dP.$$

This equation means in words that for very small changes the total change in volume is equal to,—(*a*) the change in temperature multiplied by the temperature coefficient of volume plus (*b*) the change in pressure multiplied by the pressure coefficient of volume, plus (*c*) a small term, which for small changes in temperature and pressure is relatively insignificant.

In a concrete, numerical example, if 1,000 cc. of gas at 273° K. and 760 mm. pressure, is heated one-tenth of a degree the increase in volume will be $0.1 \times$ (temperature effect on the volume) $= \frac{1}{273} \times 1{,}000 \times 0.1 = 0.366$ cc. If 1,000 cc. of gas at a pressure of 760 mm. mercury is subjected to a pressure increase of 0.1 mm., the volume will be changed by $(-\frac{1}{760}) \times 1{,}000 \times 0.1 = -0.131$ cc. In other words the volume will be decreased 0.131 cc.

If 1,000 cc. of a gas at 273° K. and 760 mm. is heated 0.1° C. and compressed by 0.1 mm. (mercury) simultaneously, the volume will be the same as if the two operations were carried out separately and the total change is the sum of the partial changes, as given by $dV = 0.366$ cc. $-$ 0.131 cc. $=$ 0.235 cc. These changes are taken small and the result is fairly accurate according to the principle of differentials given before (p. 110).

The equation may be represented nicely on a three-dimensional diagram as in Fig. 51 where pressure and temperature are plotted horizontally at right angles to each other and the volume of a gram molecule of gas is plotted vertically. Projections of this space model on the V–T plane are shown in Fig. 51B and projections on the V–P plane are shown in Fig. 51C.

The four points on the surface m, n, o, p, are formed by intersecting planes parallel to the V–T and the V–P planes. The distances between them are fairly large for purposes of illustration but they are to be considered as very close together, corresponding to differentials rather than increments.

The slope of the line mn or op on the V–T diagram is $(\partial V/\partial T)_P$. It is the temperature coefficient of volume, pressure being constant. The change in volume, due to temperature ($V_n - V_m$, or $V_p - V_o$) is equal to the slope multiplied by the temperature increase and is given by $(\partial V/\partial T)_P\, dT$. The slope of the line np or om on the V-P diagram is $\left(\frac{\partial V}{\partial P}\right)_T$. It is the pressure coefficient of volume, temperature being constant. The change[1] in volume due to pressure ($V_p - V_n$ or $V_o - V_m$) is equal to the slope multiplied by the increase in pressure or $\left(\frac{\partial V}{\partial P}\right)_T dP$.

The total change in volume due to both temperature increase and pressure increase is given by $V_p - V_m$, corresponding to the rise along a plane cutting diagonally across the surface $mnpo$ and joining m and p. Instead of cutting directly across the surface the same elevation in V is obtained by passing from m to n and then to p, and $V_p - V_m = (V_n - V_m) + (V_p - V_n)$. Passing around the other way, it is true also that $V_p - V_m = (V_o - V_m) + (V_p - V_o)$.

[1] In the example shown in Fig. 51 the change is negative, because the volume *decreases* when the pressure increases. V_p is smaller than V_n and $(V_p - V_n)$ is negative.

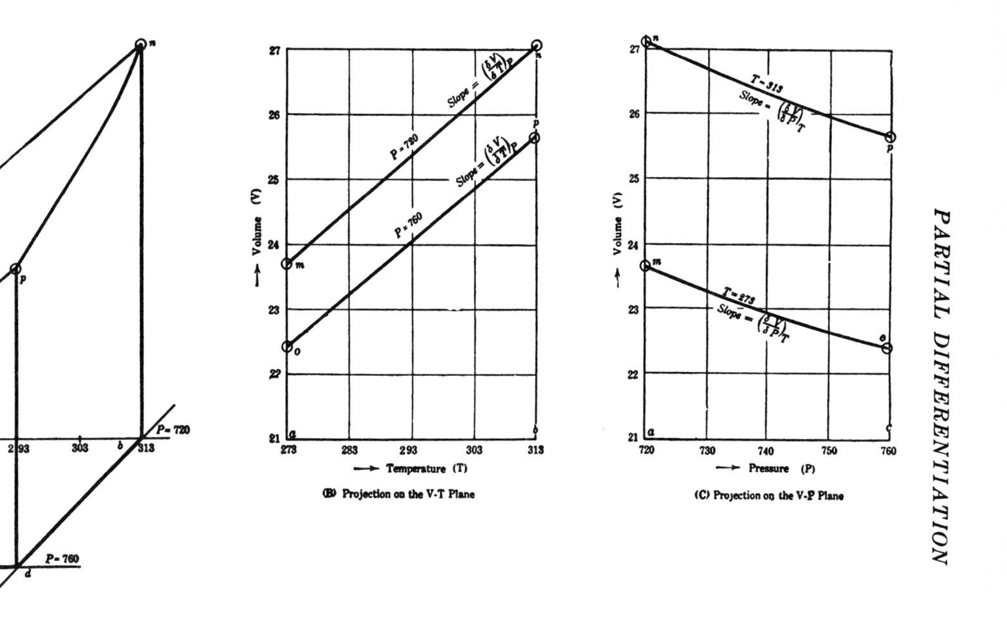

FIG. 51.—A three-dimensional graph in which the volume of a mole of gas, the pressure, and the temperature are plotted at right angles. Two dimensional projections are shown at the right. The figure illustrates the fundamental theorem of partial differentiation

$$dV = (\partial V/\partial T)_P\, dT + (\partial V/\partial P)_T\, dP.$$

The connection with the fundamental equation is now evident for by substitution,

$$V_p - V_m = (V_n - V_m) + (V_p - V_n) \text{ corresponds to}$$
$$dV = \left(\frac{\partial V}{\partial T}\right)_P dT + \left(\frac{\partial V}{\partial P}\right)_T dP.$$

A complete understanding of this three-dimensional diagram is necessary not only in connection with partial differentiation but in connection with certain phase rule diagrams used in physical chemistry. If the student has difficulty with it he is advised to build up the model in three dimensions by tracing over Figs. 51*B* and 51*C* on four different pieces of paper, cutting out with scissors and setting the four pieces together upright in the form of a hollow square. The back of the model is *amnb* and the front is *copd*. The right side of the model is closed by *bnpd* and the left side by *amoc*.

The fundamental equation

$$du = \left(\frac{\partial u}{\partial x}\right)_y dx + \left(\frac{\partial u}{\partial y}\right)_x dy$$

may be changed from the differential notation to the derivative notation by dividing through by another differential, *dt* for example. The equation then becomes

$$\frac{du}{dt} = \left(\frac{\partial u}{\partial x}\right)_y \frac{dx}{dt} + \left(\frac{\partial u}{\partial y}\right)_x \frac{dy}{dt}$$

As a matter of fact the fundamental equation of partial differentiation should have been developed first, through the conception of derivatives and then changed to the differential notation. This procedure would have been more rigorous mathematically but it is less easy to visualize.

Special Cases of the Fundamental Theorem

If one of the quantities is held constant during an operation, certain simplifications are possible. If the volume, V, is always constant, P and T must be adjusted so as to keep V constant. This constancy of volume is represented geometrically by passing a horizontal plane through the surface represented in Fig. 51. On this horizontal plane, every point has the same value of V (since the V axis is the vertical axis) and the intersection of this plane with the surface gives a contour line. The case is exactly analogous to a contour line on a topographical map, which is a line such that all points on the line have the same distance above a given base level.

If V is constant, there can be no change in V and dV becomes equal to zero, and the fundamental equation may be written

$$\left(\frac{\partial V}{\partial T}\right)_P dT + \left(\frac{\partial V}{\partial P}\right)_T dP = dV = 0$$

Rearranging algebraically, the equation becomes

$$\left(\frac{\partial V}{\partial P}\right)_T dP = -\left(\frac{\partial V}{\partial T}\right)_P dT$$

$$\left(\frac{\partial P}{\partial T}\right)_V = -\frac{(\partial V/\partial T)_P}{(\partial V/\partial P)_T} \text{ (dividing by } dT \text{ and by } \left(\frac{\partial V}{\partial P}\right)_T$$

It is necessary to change the left-hand member of the equation, dP/dT, to $(\partial P/\partial T)_V$ because the restriction that V is a constant was imposed on the whole equation as the first step in the development of this special equation.

Expressed in words this equation states that at constant volume, the increase in pressure per degree rise in temperature is equal to the coefficient of thermal expansion divided by the compressibility. The coefficient of thermal expansion is defined as the increase in volume for a small increase in temperature and the compressibility is defined as the decrease in volume corresponding to a small increase in pressure. The *decrease* in volume with pressure is equivalent to $-\partial V/\partial P$ and the negative sign of the equation is thus explained.

The relation between P, V, and T could have been written $dP = \left(\frac{\partial P}{\partial T}\right)_V dT + \left(\frac{\partial P}{\partial V}\right)_T dV$. If P is held constant, by a treatment similar to that just given it is found that

$$\left(\frac{\partial V}{\partial T}\right)_P = -\frac{(\partial P/\partial T)_V}{(\partial P/\partial V)_T}$$

This equation corresponds to a "contour" line produced in Fig. 51 by a vertical plane, parallel to the VT plane, intersecting the surface.

This equation means that the coefficient of thermal expansion at constant pressure is equal to the increase in pressure with temperature at constant volume divided by the decrease in pressure with increase of volume at constant temperature.

It can be shown also that

$$\left(\frac{\partial V}{\partial P}\right)_T = -\frac{(\partial T/\partial P)_V}{(\partial T/\partial V)_P}$$

This equation corresponds to a contour line produced in Fig. 51 by a vertical plane, parallel to the PV plane intersecting the surface.

These various coefficients may be made more definite by taking concrete cases.

The thermal expansion of mercury at constant pressure $(\partial V/\partial T)_P$ is 0.00018 per 1° C. An increase of 1° C. at constant pressure increases the volume of 1 cc. by .00018 cc.

The compressibility of mercury at constant temperature, $(\partial V/\partial P)_T$: is -0.000003 cc. per 1 atmosphere. An increase of one atmosphere at constant temperature causes 1 cc. of mercury to decrease by 0.000003 cc.

The increase in pressure of mercury with an increase of temperature at constant volume, $(\partial P/\partial T)_V$ is 60 per 1° C. It is evident from this statement that a volume of mercury will exert a force of 60 atmospheres on its walls when it is heated 1° C. in a closed vessel (assuming that the vessel does not expand). Substituting these values into the equation

$$\left(\frac{\partial V}{\partial T}\right)_P = -\frac{(\partial P/\partial T)_V}{(\partial P/\partial V)_T};\ 0.00018 = -\frac{60}{-(1/0.000003)} = 0.00018$$

These values are in agreement with the equation.

Another rearrangement of the fundamental equation is useful.

$$dV = \left(\frac{\partial V}{\partial T}\right)_P dT + \left(\frac{\partial V}{\partial P}\right)_T dP$$

$$\frac{dV}{dT} = \left(\frac{\partial V}{\partial T}\right)_P + \left(\frac{\partial V}{\partial P}\right)_T \left(\frac{dP}{dT}\right) \quad \text{(dividing through by } dT\text{)}$$

The difference between the first and second terms of this equation must be kept clearly in mind. The partial derivative, $(\partial V/\partial T)_P$, gives the rate of increase of volume with a small increase in temperature when the pressure is held constant and the temperature is the only independent variable. The total derivative, dV/dT, gives the rate of increase in volume with a very small temperature increase when the pressure and any other factors are allowed to vary also.

This form of the equation is particularly important in thermodynamics where some function of the variables rather than the variables themselves is to be studied. For example, the relation between volume, temperature, and pressure may be desired under conditions such that the energy, E, of the system is always constant. E is a function of T and P. The last equation may then be written

$$\left(\frac{\partial V}{\partial T}\right)_E = \left(\frac{\partial V}{\partial T}\right)_P + \left(\frac{\partial V}{\partial P}\right)_T \left(\frac{\partial P}{\partial T}\right)_E.$$

$(\partial V/\partial T)_E$ represents the slope of a special contour line on the surface shown in Fig. 51. It is a line not of constant volume, nor of constant pressure, nor of constant temperature, but a line of constant energy.

The case is somewhat analogous to the "salt lines" plotted on topographical maps in sanitary surveys near the ocean. The map gives three variables—east-west, north-south, and altitude. Analyses for sodium chloride in the well waters of the district enable one to draw on the map, lines of constant salt percentage in the ground water. The salt concentration is a function of the east-west variable and the north-south variable, and on the surface of the maps then a contour line of constant salt concentration may be drawn, just as a contour line of constant energy may be drawn on the surface shown in Fig. 51.

In a small book entitled "A Condensed Collection of Thermodynamic Formulas," Bridgman[1] has shown that there are ten fundamental quantities in thermodynamics including pressure, absolute temperature, volume, entropy, heat, work, internal energy, heat content, free energy of Helmholtz, and thermodynamic potential of Gibbs or free energy of G. N. Lewis. There are 720 first derivatives involving three variables $(10 \times 9 \times 8)$ for these ten fundamental quantities and eleven billion thermodynamic relations, involving any four derivatives. Since many of these formulas are alike, Bridgman has succeeded in reducing the number of working formulas to 45. With these and some second derivatives all the relations of thermodynamics may be built up in a purely formal, mathematical manner without restrictions of experimental conditions. For ordinary work a much smaller number of equations is sufficient.

[1] P. 264.

Successive Partial Differentiation

The rules of successive differentiation already discussed (p. 97) apply also to partial differentiation, as illustrated by the following example

$$u = x^3 \sin y$$

Holding y constant,

$$\frac{\partial u}{\partial x} = 3x^2 \sin y; \quad \frac{\partial(\partial u/\partial x)}{\partial x} = \frac{\partial^2 u}{\partial x^2} = 6x \sin y;$$

$$\frac{\partial(\partial^2 u/\partial x^2)}{\partial x} = \frac{\partial^3 u}{\partial x^3} = 6 \sin y$$

Holding x constant,

$$\frac{\partial u}{\partial y} = x^3 \cos y; \quad \frac{\partial(\partial u/\partial y)}{\partial y} = \frac{\partial^2 u}{\partial y^2} = -x^3 \sin y;$$

$$\frac{\partial(\partial^2 u/\partial y^2)}{\partial y} = \frac{\partial^3 u}{\partial y^3} = -x^3 \cos y.$$

It is an important property of successive partial differentiation that the order of differentiation is immaterial. $\dfrac{\partial(\partial V/\partial T)_P}{\partial P}$ has the same value as $\dfrac{\partial(\partial V/\partial P)_T}{\partial T}$ for both are equal to $\dfrac{\partial^2 V}{\partial P \partial T}$.

In other words, this equality means that the rate of change with pressure of the coefficient of thermal dilation is equal to the rate of change with temperature of the coefficient of compressibility.

The truth of this proposition can be demonstrated in the specific example given above.

$$u = x^3 \sin y$$

$$\frac{\partial u}{\partial x} = 3x^2 \sin y; \quad \frac{\partial(\partial u/\partial x)}{\partial y} = \frac{\partial^2 u}{\partial x \partial y} = 3x^2 \cos y$$

$$\frac{\partial u}{\partial y} = x^3 \cos y; \quad \frac{\partial(\partial u/\partial y)}{\partial x} = \frac{\partial^2 u}{\partial x \partial y} = 3x^2 \cos y.$$

The result is the same, no matter whether the first derivative of u with respect to x is differentiated with respect to y, or whether the first derivative of u with respect to y is differentiated with respect to x.

Exercises

1. $u = x^2 - 2x^3y + y^3.$ $du = ?$ *Ans.:* $2xdx - 6x^2ydx - 2x^3dy + 3y^2dy.$

2. $u = a + bx + cx^2 + my + ny^2.$ $du = ?$

3. $u = y \log_e x.$ $du = ?$ *Ans.:* $ydx/x + \log_e xdy.$

4. $u = 2x^{ay}.$ $du = ?$

5. $u = x^3e^y \sin z.$ $du = ?$

Ans.: $3x^2e^y \sin zdx + x^3e^y \sin zdy + x^3e^y (\cos z)dz$

6. $u = 2x^3y^2z + \dfrac{x}{y} \log z.$ $du = ?$

7. With the relation $\dfrac{\partial y}{\partial x} = -\dfrac{\partial u/\partial x}{\partial u/\partial y}$, how can it be shown that if

$$u = x^3y^2 + x^2y^3 - 25, \ \frac{dy}{dx} = \frac{-3xy - 2y^2}{2x^2 + 3xy}.$$

8. $u = 2x^2y^3 + \dfrac{e^xy^2}{3}.$ $\dfrac{\partial^2 u}{\partial y \partial x} = ?$

9. $u = \log_e (e^x + e^y + e^z)\cdot$ $\dfrac{\partial^3 u}{\partial x \partial y \partial z} = ?$ *Ans.:* $2e^{x+y+z-3u}.$

10. $y = xz \log \dfrac{x}{z}\cdot$ $\dfrac{\partial^2 y}{\partial x \partial z} = ?$

11. One side of a rectangle is 15 cm. long and decreasing uniformly 1 cm. per minute. The other side is 10 cm. long and increasing 2 cm. per minute. At what rate is the area increasing? *Ans.:* 20 sq. cm. per minute. At what rate is the area increasing after it has been increasing for 5 min.?

12. A kite is moving horizontally along the ground at the rate of 3 ft. a second, and it is rising at the rate of 5 ft. a second. How fast is the string being unwound?

13. A cone having a height, h, and a radius of base, r, has a volume, $V = \frac{1}{3}\pi r^2h$. What is the change in volume corresponding to very small changes in both radius and height when the two are changed simultaneously?

$$\frac{\partial V}{\partial r} = \frac{2\pi}{3}rh$$

$$\frac{\partial V}{\partial h} = \frac{\pi}{3}r^2$$

$$dV = \frac{\partial V}{\partial r}dr + \frac{\partial V}{\partial h}dh = \frac{2\pi}{3}rhdr + \frac{\pi}{3}r^2dh.$$

When the radius is increasing at the rate of 0.2 cm. per minute and the height is increasing at the rate of 0.3 cm. per minute, what is the rate of increase of volume when a cone is 10 cm. in radius and 20 cm. in height?

14. Water is running out of a conical funnel at the rate of 0.02 cc. per minute. The funnel is 10 cm. in height and 6 cm. across the top. How fast is the surface of the water falling when the depth of the water is 5 cm.?

15. According to the simple gas law, $PV = RT$ where P = pressure, V = volume, T = absolute temperature and R is the gas constant. How

can it be shown that for a very small, simultaneous change of temperature and volume, $dP = -\frac{P}{V}dV + \frac{P}{T}dT$?

16. What is the maximum or minimum of the function $u = 2x + 3xy + y$? The necessary (but not sufficient) conditions for a maximum or minimum are that both the partial derivatives shall become zero simultaneously. The treatment in this case is similar to that given on page 102. The two equations after being set equal to zero must be solved simultaneously to find the relation between x and y.

17. How should a number, k, be divided into three parts such that the continued product is as great as possible? This problem is equivalent to finding the values of x and y which make the function, u, a maximum.

$$u = xy(k - (x + y))$$

Ans.: u is a maximum when the three parts are equal.

18. The velocity of a bimolecular reaction is represented by the equation $v = dx/dt = k(a - x)(b - x)$, where x is the concentration of material which has reacted at time, t, and a and b are the original concentrations of the two reacting materials. How can it be proved that the velocity, v, is greatest when $a = b$?

19. If a system changes in volume during heating so that P is constant (but not V) the heat absorbed, Q, is equal to the increase in internal energy dE plus the work done against the atmosphere PdV. $Q = dE + PdV$. The heat capacity, C_P, at constant pressure is equal to the heat absorbed for a very slight increase in temperature

$$C_P = \frac{Q}{dT} = \left(\frac{\partial E}{\partial T}\right)_P + P\left(\frac{\partial V}{\partial T}\right)_P$$

The equation is obtained by dividing by dT and imposing the restriction of constant pressure. This equation is to be expressed in words.

20. Internal energy, E, is a function of two independent variables, volume V, and temperature, T. $dE = \left(\frac{\partial E}{\partial T}\right)_V dT + \left(\frac{\partial E}{\partial V}\right)_T dV$

What steps are necessary to show that

$$\left(\frac{\partial E}{\partial T}\right)_P = \left(\frac{\partial E}{\partial T}\right)_V + \left(\frac{\partial E}{\partial V}\right)_T\left(\frac{\partial V}{\partial T}\right)_P?$$

How may this equation be expressed in words?

CHAPTER XVIII

DIFFERENTIAL EQUATIONS

Any equation which contains a derivative or a differential, is a differential equation. The solution of a differential equation consists in finding by the process of integration an equivalent expression in which the derivatives or differentials have disappeared. For example, $dy/dx = 5$ is a differential equation and its solution is $y = 5x + c$; likewise $y^2 = x^2 + cx$ is the solution of the differential equation $(x^2 + y^2)dx = 2xydy$. Most of the differential equations which are commonly used in physical chemistry are very simple, and in fact many of them have been met and solved on earlier pages of this book, without being referred to as differential equations. The more complex differential equations demand considerable skill for their solution, and they are beyond the scope of the present book. A few examples are given but the student is referred to standard texts in mathematics for details. Chemical applications of differential equations may be found in Mellor's "Higher Mathematics for Students of Physics and Chemistry," and Hitchcock and Robinson's "Differential Equations in Applied Chemistry" (p. 263).

Simple Differential Equations in Physical Chemistry

The very simplest equations are considered in this section, and in all these cases it is possible to place all the y's and dy's on one side of the equation and all the x's and dx's on the other. Only two variables, x and y, are involved, and the equation can be so adjusted that no single term contains both variables.

After placing all the terms in x on one side and all the terms in y on the other, both sides of the equation are integrated, and the differentials disappear. It is necessary to add an integration constant to the equation, unless the integration has been carried out between limits so that the new constant cancels out.

One of the most important differential equations in physical chemistry is $-dy/dx = ky$. This equation expresses the fact that the dependent variable is decreasing at a rate which is proportional at all times to the magnitude of the dependent variable. The solution of this equation and its practical applications which have been discussed at length in Chap. XII, are reviewed briefly:

$$\frac{-dy}{dx} = ky$$
$$\frac{-dy}{y} = kdx$$
$$\int \frac{-dy}{y} = \int kdx$$
$$-\log_e y = kx + c$$

The equation $-\log_e y = kx + c$ is the solution of the differential equation $-dy/dx = ky$. Sometimes it is more useful to integrate between limits; y_1 corresponding to x_1 as determined by experimental measurements, and y_2 corresponding to x_2.

The solution is then as follows

$$\int_{y_1}^{y_2} \frac{-dy}{y} = \int_{x_1}^{x_2} kdx$$
$$\Big[-\log_e y\Big]_{y_1}^{y_2} = \Big[kx\Big]_{x_1}^{x_2}$$
$$-\log_e y_2 + \log_e y_1 = k(x_2 - x_1)$$
$$k = \frac{2.303}{x_2 - x_1} \log_{10} \frac{y_1}{y_2}$$

Expressed as an exponential equation, $y = y_0 e^{-kx}$ (p. 137).

Another example of the simple differential equation often used in physical chemistry is the so-called "reaction isochore" of Van't Hoff which gives the relation between absolute temperature, T, and chemical equilibrium, K, and the heat of reaction, Q. The term chemical equilibrium covers a large number of phenomena, such as vapor pressure, solubility, and ionization as well as the ordinary chemical reactions.

$$\frac{d \log_e K}{dT} = \frac{Q}{RT^2}$$
$$\int d \log_e K = \int \frac{Q}{R} \frac{dT}{T^2}$$

This equation cannot be integrated until the relation between Q and T is established, but usually Q is so nearly independent of temperature that for practical purposes it may be considered a constant. R is a constant, the well-known gas constant, which is expressed in the same units as Q. Assuming Q as well as R to be constant, the equation may be integrated as follows,

$$\int d \log_e K = \frac{Q}{R}\int \frac{dT}{T^2}$$

$$\log_e K = -\frac{Q}{R}\frac{1}{T} + C.$$

Integrating between limits, where K_1 is the equilibrium constant at T_1 and K_2 is the equilibrium constant at the higher temperature T_2, the equation becomes

$$\int_{K_1}^{K_2} d \log_e K = \frac{Q}{R}\int_{T_1}^{T_2} \frac{dT}{T^2}$$

$$\log_e K_2 - \log_e K_1 = \frac{Q}{R}\left[\frac{-1}{T_2} - \frac{-1}{T_1}\right] = \frac{Q}{R}\left[\frac{1}{T_1} - \frac{1}{T_2}\right]$$

$$\log_{10} \frac{K_2}{K_1} = \frac{Q}{2.303R}\,\frac{T_2 - T_1}{T_2T_1}$$

This is the practical working equation which is the solution of the differential equation

$$\frac{d \log_e K}{dT} = \frac{Q}{RT^2}$$

This differential equation is deduced from thermodynamical considerations under conditions where the change in temperature is very small. The integrated expression is true, however, over wide temperature intervals, and it makes possible the use of experimental data at different temperatures. It is limited only by the physical chemical restrictions that Q is constant, and that the gas laws, involving R, are obeyed.

Another important differential equation of physical chemistry is

$$\frac{dx}{dt} = k(a - x)(b - x)$$

This is known in physical chemistry as the equation for a bimolecular reaction or a reaction of the second order. x is the amount of material reacting, a is the initial concentration of one reacting material and b is the initial concentration of a second substance.

Separating the variables,

$$\frac{dx}{(a - x)(b - x)} = kdt$$

Integrating by partial fractions (p. 154) between times, t_2, and t_1 (where t_2 is greater than t_1) and between the corresponding values of x, namely x_2 and x_1

$$\int_{t_1}^{t_2} k\,dt = \int_{x_1}^{x_2} \frac{1/(b-a)}{a-x}dx + \int_{x_1}^{x_2} \frac{1/(a-b)}{b-x}dx$$

$$k(t_2 - t_1) = \left[\frac{1}{a-b}(\log_e(a-x)) + \frac{1}{a-b}(-\log_e(b-x)\right]_{x_1}^{x_2} = \frac{1}{a-b}\left[\log_e \frac{a-x}{b-x}\right]_{x_1}^{x_2}$$

$$k = \frac{2.303}{(a-b)(t_2-t_1)} \log_{10} \frac{(a-x_2)(b-x_1)}{(b-x_2)(a-x_1)}$$

If t_1 is taken as zero time, x_1 is zero also. By making this change and designating t_2 and x_2 as t and x the equation becomes,

$$k = \frac{2.303}{t(a-b)} \log_{10} \frac{b(a-x)}{a(b-x)}.$$

Separable Variables

In the examples just given the y's and x's can be separated easily. The difficulties arise when the variables are mixed up in the equation so that a considerable amount of mathematical rearrangement is necessary before they can be separated for purposes of integration. A few examples will serve to show some of the ways in which differential equations may be prepared for integration.

Example 1.—If $x\dfrac{dy}{dx} = 4y, \quad y = Cx^4.$

$$\frac{dy}{y} = 4\frac{dx}{x}$$

$$\int \frac{dy}{y} = 4\int \frac{dx}{x}$$

$$\log_e y = 4\log_e x + C = \log_e x^4 + \log_e C = \log_e (x^4C)$$

Delogarizing,

$$y = Cx^4$$

C is the constant introduced by the integration. It may be eliminated when the relation between y and x is determined by experiment for any one value of x or y. For example, if it is found that $y = 4$ when $x = 2$, then $4 = C2^4$ and C is equal to $4/16$ or $1/4$. Frequently the value of C may be determined by setting x equal to zero. It is usually customary in differential equations to write $\log_e C$ in the logarithmic equation rather than C, for then the reduction to the exponential form is more easily accomplished.

EXAMPLE 2.—If $2(x + 1)\frac{dy}{dx} - a(y + 2) = 0, \quad (y + 2)^2 = C(x + 1)^a.$

$$\frac{2dy}{y + 2} = \frac{adx}{x + 1}$$

$$2\int\frac{dy}{y + 2} = a\int\frac{dx}{x + 1}$$

$$2 \log_e (y + 2) = a \log_e (x + 1) + \log_e C$$

$$\log_e (y + 2)^2 = \log_e ((x + 1)^a C)$$

$$(y + 2)^2 = C(x + 1)^a$$

EXAMPLE 3.—If $my + n\frac{dy}{dx} = l, \quad y = \frac{l}{m} + Ce^{-\frac{mx}{n}}$

$$\frac{dy}{dx} = \frac{l - my}{n} = \frac{m}{n}\left[\frac{l}{m} - y\right]$$

$$\frac{dy}{y - \frac{l}{m}} = -\frac{m}{n}dx$$

$$\int\frac{dy}{y - \frac{l}{m}} = -\frac{m}{n}\int dx$$

$$\log_e\left(y - \frac{l}{m}\right) = -\frac{m}{n}x + \log_e C$$

$$\log_e\frac{y - \frac{l}{m}}{C} = -\frac{m}{n}x$$

$$\frac{y - \frac{l}{m}}{C} = e^{-\frac{m}{n}x}$$

$$y - \frac{l}{m} = Ce^{-\frac{mx}{n}}$$

$$y = \frac{l}{m} + Ce^{-\frac{mx}{n}}$$

EXAMPLE 4.—If $\quad 2\left(x\frac{dy}{dx} + 3y\right) = xy\frac{dy}{dx}, \quad x^6y^2 = Ce^y$

$$2xdy + 6ydx = xydy$$

$$6ydx = xydy - 2xdy$$

$$6\frac{dx}{x} = dy - \frac{2dy}{y}$$

$$6\int\frac{dx}{x} = \int dy - 2\int\frac{dy}{y}$$

$$6 \log_e x = y - 2 \log_e y + \log_e C$$

$$y = \log_e x^6 + \log_e y^2 - \log_e C$$

$$y = \log_e\frac{x^6y^2}{C}$$

$$\frac{x^6y^2}{C} = e^y$$

$$x^6y^2 = Ce^y$$

Homogeneous Differential Equations

Sometimes it is not possible to effect a separation of x and y, except by the artifice of substituting for y its equivalent xz, where z is a new variable such as to make $y = xz$. In this case

$$dy = xdz + zdx.$$

This substitution of $xdz + zdx$ for dy makes possible the solution of differential equations of the type

$$Mdx + Ndy = 0$$

where M and N are homogeneous functions of x and y having the same degree. An expression is said to be homogeneous if each term has the same dimensions, as in the case of the equation

$$ax^2y^4 + bx^3y^3 + cx^4y^2 = y^6$$

The following differential equation is solved in this manner by substituting xz for y

$$\text{If } y^2 - xy\frac{dy}{dx} + x^2\frac{dy}{dx} = 0, \quad y = Ce^{y/x}$$

$$y^2dx - xydy + x^2dy = 0 \text{ (multiplying by } dx\text{)}$$

$$y^2dx + (x^2 - xy)dy = 0$$

$$x^2z^2dx + (x^2 - x^2z)(zdx + xdz) = 0 \text{ (substituting } xz \text{ for } y\text{)}$$

$$x^2z^2dx + x^2zdx + x^3dz - x^2z^2dx - x^3zdz = 0$$

$$\frac{dx}{x} + \frac{dz}{z} - dz = 0 \text{ (dividing by } x^3z\text{)}$$

$$\int\frac{dx}{x} + \int\frac{dz}{z} = \int dz$$

$$\log_e x + \log_e z = z + \log_e C$$

$$z = \log_e \frac{xz}{C}$$

$$\frac{xz}{C} = e^z$$

$$\frac{xy}{Cx} = e^{\frac{y}{x}} \left(\text{substituting } z = \frac{y}{x}\right)$$

$$y = Ce^{\frac{y}{x}}$$

When an equation is not homogeneous it may often be made homogeneous by the addition of constants. For example x is set equal to $u + h$ and y is set equal to $v + k$. Making these substitutes in a non-homogeneous equation an homogeneous equation is produced which can be solved in the manner outlined above.

Exact Differential Equations

The equation

$$Mdx + Ndy = 0$$

is an exact differential equation if M and N, which are functions of x and y, satisfy the criterion of Euler. According to this criterion $\partial M/\partial y$ must be equal to $\partial N/\partial x$, and equations of this type may be integrated by a direct process. Euler's criterion is developed as follows:

If u is a function of x and y, then

$$du = \frac{\partial u}{\partial x}dx + \frac{\partial u}{\partial y}dy \text{ (partial differentiation)}$$

and

$$du = Mdx + Ndy$$

it follows that

$$M = \frac{\partial u}{\partial x} \text{ and } N = \frac{\partial u}{\partial y}.$$

Differentiating the first with respect to y,

$$\frac{\partial M}{\partial y} = \frac{\partial^2 u}{\partial x \partial y}$$

Differentiating the second with respect to x,

$$\frac{\partial N}{\partial x} = \frac{\partial^2 u}{\partial y \partial x}$$

It follows that

$$\frac{\partial M}{\partial y} = \frac{\partial N}{\partial x}.$$

In other words, when

$$\frac{\partial M}{\partial y} = \frac{\partial N}{\partial x}$$

it is known that the differential equation was obtained by the direct differentiation of some function, u, of x and y, and that the original equation may be regained by direct integration. If some factors were eliminated in the formation of the differential equation, $\partial M/\partial y$ is not equal to $\partial N/\partial x$ and the equation is not exact. In such a case it may sometimes be integrated by putting into the form of an exact differential equation, through multiplication by an integrating factor. Considerable ingenuity is required in finding the proper factor which will convert the equation into an exact differential equation.

The following equation is "exact" (or perfect) as shown by the application of Euler's criterion

$$(3y^2x - x^2)dy + (y^3 - 2xy)dx = 0$$

M is the coefficient of dx and N is the coefficient of dy.

$$\frac{\partial M}{\partial y} = \frac{\partial(y^3 - 2xy)}{\partial y} = 3y^2 - 2x$$
$$\frac{\partial N}{\partial x} = \frac{\partial(3y^2x - x^2)}{\partial x} = 3y^2 - 2x$$

Therefore $\partial M/\partial y = \partial N/\partial x$ and the equation is exact.

The following equation is inexact since $\partial M/\partial y$ is not equal to $\partial N/\partial x$

$$y^2dx + 3xdy = 0$$
$$\frac{\partial M}{\partial y} = \frac{\partial(y^2)}{\partial y} = 2y$$
$$\frac{\partial N}{\partial x} = \frac{\partial(3x)}{\partial x} = 3$$

In equations which satisfy Euler's criterion, M is $\left(\frac{\partial u}{\partial x}\right)_y$ and N is $\left(\frac{\partial u}{\partial y}\right)_x$, u being a function of x and y. Mdx may be integrated assuming y constant and Ndy may be integrated assuming x constant, and the complete solution is obtained by adding to $\int Mdx$ the terms from $\int Ndy$ which are not already in $\int Mdx$ and adding the integration constant. The following formula may be applied

$$\int Mdx + \int\left(N - \frac{\partial(\int Mdx)}{\partial y}\right)dy = C$$

The use of this equation may be illustrated with a specific problem.

$$x(x + 2y)dx + (x^2 - y^2)dy = 0$$
$$Mdx + Ndy = 0$$
$$\frac{\partial M}{\partial y} = 2x \quad \frac{\partial N}{\partial x} = 2x \quad \frac{\partial M}{\partial y} = \frac{\partial N}{\partial x}$$

This equation satisfies Euler's criterion and should respond to direct integration by the formula just given.

$$\int M dx = \int (x^2 + 2xy)dx = \frac{x^3}{3} + \frac{2x^2y}{2} + C \text{ (} y \text{ is constant)}$$

$$\frac{\partial(\int M dx)}{\partial y} = \frac{\partial\left(\frac{x^3}{3} + x^2y + C\right)}{\partial y} = x^2$$

$$N - \frac{\partial(\int M dx)}{\partial y} = (x^2 - y^2) - x^2 = -y^2$$

$$\int M dx + \int \left(N - \frac{\partial(\int M dx)}{\partial y}\right) dy = C$$

$$\frac{x^3}{3} + x^2y + \int -y^2 dy = C$$

$$\frac{x^3}{3} + x^2y - \frac{y^3}{3} = C$$

This is the solution of the exact differential equation

$$x(x + 2y)dx + (x^2 - y^2)dy = 0$$

The solution may be checked by differentiating the solution, calling it u.

$$du = \left(\frac{\partial u}{\partial x}\right)dx + \left(\frac{\partial u}{\partial y}\right)dy$$

$$\frac{\partial u}{\partial x} = x^2 + 2xy \cdot \quad \frac{\partial u}{\partial y} = x^2 - y^2$$

$$du = (x^2 + 2xy)dx + (x^2 - y^2)dy,$$

This is the original differential equation.

Linear Differential Equations

The following differential equation is said to be linear

$$\frac{dy}{dx} + Py = Q$$

where P and Q are functions of x alone, or of constants. For example, $\frac{dy}{dx} + x^2y = 2x$ is a linear equation, but $\frac{dy}{dx} + x^2y = 2xy$ is not linear.

It can be shown that a linear differential equation can be solved by the formula

$$y = e^{-\int P dx}\left(\int Q e^{\int P dx} dx + C\right)$$

Differential Equations of the Second Order

Just as there are second-order derivatives or third-order or nth-order, so there are differential equations of the second order or higher orders. Although such equations are occasionally met with in advanced physics or physical chemistry, they cannot be discussed here. One example of the very simplest type must suffice.

What is the solution of the differential equation $d^3y/dx^3 = x^2$? When the derivative is equal to a function of x or a constant, but not a function of y, the rule is to integrate successively introducing an arbitrary constant for each integration.

$$\frac{d^3y}{dx^3} = \frac{d(d^2y/dx^2)}{dx} = x^2$$

$$\int d\left(\frac{d^2y}{dx^2}\right) = \int x^2dx \quad \text{and} \quad \frac{d^2y}{dx^2} = \frac{x^3}{3} + C_1$$

$$\int d\left(\frac{dy}{dx}\right) = \int \left(\frac{x^3}{3} + C_1\right)dx \quad \text{and} \quad \frac{dy}{dx} = \frac{x^4}{12} + C_1x + C_2$$

$$\int dy = \int \left(\frac{x^4}{12} + C_1x + C_2\right)dx \quad \text{and} \quad y = \frac{x^5}{60} + \frac{1}{2}C_1x^2 + C_2x + C_3$$

as a check, it can be readily shown by differentiation that

$$\text{if } y = \frac{x^5}{60} + C_1x^2 + C_2x + C_3, \ \frac{d^3y}{dx^3} = x^2$$

When there are two or more independent variables the equation is called a partial differential equation to distinguish it from the ordinary differential equation which has but one independent variable, x, and one dependent variable, y. Equations of this type are more difficult to solve. Such equations do occur occasionally in certain studies of reaction rates, but they cannot be treated here.

Exercises

What are the solutions of the following differential equations:

1. $\dfrac{dy}{dx} = 7$? *Ans.:* $y = 7x + C$.
2. $\dfrac{dy}{dx} = 2x$?
3. $dy = x^3dx$? *Ans.:* $y = \frac{1}{4}x^4 + C$.
4. $\cos^2 x = \dfrac{dy}{dx}$?
5. $2\dfrac{dy}{dx} = \dfrac{1}{x}$? *Ans.:* $y = \log_e \sqrt{x} + C$.

6. $udv + vdu = 0$?

7. $4\frac{dy}{dx} + ay = 0$? *Ans.:* $\log_e y = -\frac{a}{4}x + \log_e C$, or $y = Ce^{\frac{-ax}{4}}$.

8. $a^2\frac{dy}{dx} + 2y = 0$?

9. $(x^2y + x)dy + (xy^2 - y)dx = 0$? *Ans.:* $xy + \log \frac{y}{x} = C$.

10. $uvdu - (a + u)(b + v)dv = 0$?

11. $(xy + x^2)\frac{dy}{dx} + y^2 = 0$. *Ans.:* $xy^2 = C(2y + x)$

zx is substituted for y

$$x^2(z + 1)\left(z\frac{dz}{dx} + z\right) + z^2x^2 = 0$$

$$\left(\frac{1}{z} - \frac{1}{2z + 1}\right)dz + \frac{dx}{x} = 0$$

$$\log_e z^2 = \log_e (2z + 1) + \log_e x^2 = \log_e C$$

$$xy^2 = C(2y + x)$$

12. $x^2dy - y^2dx - xydx = 0$?

13. $(p - s)dp + sdp = 0$? *Ans.:* $P = Ce^{-p/s}$.

14. $xy^2dy = (x^3 + y^3)dx$?

Are the following differential equations exact or inexact by Euler's criterion?

15. $(a + 4xy)dx + x\cdot dy = 0$?

$$\frac{\partial M}{\partial y} = 4x\cdot \quad \frac{\partial N}{\partial x} = 2x$$

Therefore the equation is inexact.

16. $(a^2u + b^2)du + (b^3 + a^2u)dv = 0$?

17. $(\sin y + y \cos x)dx + (\sin x + x \cos y)dy = 0$? *Ans.:* Exact.

18. $(2x + 6x^2)dx + 2x^3dy = 0$?

19. $\left(x^2 + \frac{3}{x}\right)dx + (xy + e^x)dy = 0$? *Ans.:* Inexact.

20. The discharge of a condenser follows the law $Q = Q_0e^{-kt}$ where Q_0 is the initial charge and Q is the charge at any time, t, and k is the coefficient of leakage. The charge of a certain condenser is reduced from 0.0020 to 0.0015 mf. in 10 min. How long will it take for the condenser to be half discharged? How much electricity Q will be left in the condenser after 1 hr.? (Other examples of this type may be found on p. 139.)

21. The velocity of a unimolecular reaction is given by the differential equation $dx/dt = k(a - x)$. How can this be integrated to give $k = \frac{2.303}{t} \log_{10} \frac{a}{a - x}$?

22. The velocity of a bimolecular reaction, when the concentrations of the two reacting materials are the same, is given by $dx/dt = k(a - x)^2$. What is the integrated expression?

23. The velocity of a trimolecular reaction is given by $dx/dt = k(a - x)(b - x)(c - x)$. What is the integrated expression?

$$Ans.\ k = \frac{2.303}{t} \frac{\log_{10}\left(\frac{a-x}{a}\right)^{b-c} \log_{10}\left(\frac{b-x}{b}\right)^{c-a} \log_{10}\left(\frac{c-x}{c}\right)^{a-b}}{(a - b)\ (b - c)\ (c - a)}.$$

24. The Clausius-Clapeyron equation $\frac{d \log_e P}{dT} = \frac{Q}{RT^2}$ connects vapor pressure, P, of a liquid with absolute temperature and heat of vaporization per gram molecule. R is the gas constant. Q is assumed to be constant. The equation may be derived with the help of a hypothetical thermodynamic cycle.

How may this differential equation be integrated to give

$$\log_{10} \frac{P_2}{P_1} = \frac{Q}{2.303R} \frac{(T_2 - T_1)}{T_2 T_1}?$$

25. The vapor pressure of water is 4.57 mm. at 0° C. and 23.70 mm. at 25° C. What is the heat of vaporization, as calculated by the Clausius-Clapeyron equation, $\log_{10} P_2 - \log_{10} P_1 = \frac{Q}{2.303R}\left(\frac{T_2 - T_1}{T_2 \times T_1}\right)$. $R = 2$ cal. per degree. When R is expressed in calories Q is obtained in calories. The temperature expressed on the centigrade scale must be converted to absolute temperatures.

$$1.37475 - 0.65992 = \frac{Q \times 25}{2.303 \times 2 \times 298 \times 273}$$

$$Q = 10{,}670 \text{ calories.}$$

From the heat of vaporization calculated above and the vapor pressure at one of the temperatures given above, the vapor pressure at 50° C. is to be calculated.

$$\log_{10} P_2 - \log_{10} 23.7 = \frac{10{,}670}{2.303 \times 2}\left(\frac{323 - 298}{323 \times 298}\right) \qquad \textit{Ans.: } 95.5 \text{ mm.}$$

The experimentally determined vapor pressure at 50° is 94 mm.

26. Benzene boils at 77.9 under a pressure of 700 mm. Its heat of vaporization per gram is 95 cal. What is the boiling point at 760 mm.? (Prob. 25.) Since R refers to a gram molecule, Q must be expressed in calories per gram molecule and the 95 cal. must be multiplied by the molecular weight.

27. How can it be proved fromthe equation $\frac{d \log_e P}{dT} = \frac{Q}{RT^2}$, that a straight line is produced when the logarithm to the base 10 of vapor pressure, P, is plotted against the reciprocal of the absolute temperature, T? Q and R are constants (also p. 63). What is the relation between the slope of this line and Q?

28. Ethyl alcohol boils at 76.1° under a pressure of 700 mm. and at 72.4° C. under a pressure of 600 mm. What is its heat of vaporization? What is its normal boiling point at 760 mm.?

29. The solubility of oxalic acid is 15.9 g. per liter at 30°, and 10.2 at 20° C. What is the heat of solution of a gram molecule of oxalic acid in a saturated solution? What is the solubility at 40°? (Similar to Prob. 25.)

Ans.: 7,873 cal.

30. The equilibrium constant, K, for the dissociation of hydroiodic acid is 0.01984 at 448° C.and 0.01494 at 357°. What is the heat of dissociation, Q, and what is the equilibrium constant at 300°C.? (Similar to Prob. 25.)

31. The temperature coefficient of the rate of a chemical reaction can be expressed by a formula similar to that given for the temperature coefficient of equilibrium. The two are similar, because an equilibrium constant is equal to the ratio of two reaction velocity constants, the forward reaction and the reverse reaction. In the formula $\log \frac{k_2}{k_1} = \frac{E}{2.303R}\left(\frac{T_2 - T_1}{T_2 T_1}\right)$, k_2 is the specific reaction rate at T_2 and k_1 is the specific reaction rate at T_1. E is the energy required to "activate" the molecules. It is called the critical increment.

For the decomposition of nitrogen pentoxide at 0°C., k has been found to be 0.0000472, and at 25° C. it is 0.00203. What is the value of E, and what is the specific reaction rate, k, at 65°? The experimentally determined value at 65° is 0.292.

32. For the rearrangement of ammonium cyanate the specific reaction rate, k, is 0.00141 at 39°C. and 0.0228 at 64.5°. What is the specific reaction rate at 80° C.?

CHAPTER XIX

INFINITE SERIES

A series is a succession of terms which is formed according to some definite law. When the first terms of the series are given it is possible to specify the remaining terms.

For example, the numbers 1, 3, 9, 27 . . . follow the law that each number is obtained by multiplying the preceding number by 3. The next succeeding term, then, is $3 \times 27 = 81$, the next is 243, and the next is 729 and so on. These may be written as $3^0, 3^1, 3^2, 3^3, 3^4, 3^5, 3^6, \ldots 3^{n-1}$, and it is noticed that the exponent of each term is equal to 1 less than the number of the term in the series. In the third term the exponent is $3-1$ or 2, in the fourth term it is $4 - 1$ or 3, and so on; and the exponent of the nth term is $n - 1$.

Series of terms of this kind are of considerable theoretical interest in mathematics, and they are very important in practical computations. Logarithms and trigonometrical functions and other functions are readily computed to as many significant figures as desired, with the help of the proper series. When the laws are unknown or the application of integral calculus is impossible, many of the quantities of physics and chemistry may still be calculated by means of series. Although these calculations are approximations, any desired accuracy may be obtained by taking a sufficient number of terms.

Converging Series

Series may be either convergent or divergent. In a convergent series the sum of the terms approaches closer and closer to a definite numerical value as the number of terms is increased. For example, in the series

$$1 + \frac{1}{2} + \frac{1}{4} + \frac{1}{8} + \cdots \frac{1}{2^{n-1}}$$

the sum of the series, s, depends on the number of terms, n. When $n = 1$, $s = 1$. When $n = 2$, $s = 1\frac{1}{2}$. When $n = 3$, $s = 1\frac{3}{4}$. When $n = 4$, $s = 1\frac{7}{8}$, and so on.

This sum is getting larger continuously as n increases, but the increase in s due to the increase of n becomes continuously smaller. As n approaches infinity, s approaches 2 as a limit (p. 127). It comes nearer and nearer to 2 as the number of terms, n, increases, but it can never be greater than 2. Such a series is a convergent series. If the sum of the first few terms differs only slightly from the sum of all the terms the series is said to be rapidly converging.

In a divergent or non-convergent series no definite limit is approached, as in the series

$$1 + 2 + 3 + 4 + \cdots$$

The sum of the series, s, becomes larger and larger indefinitely as the number of terms, n, is increased.

The convergent series are the only ones which are of value in computations and the more rapidly a series converges the more convenient it is. It is important, therefore, to know whether or not a given series is convergent. Usually it is possible to apply simple tests, such as a comparison with a known series; or Cauchy's ratio test; but sometimes it is a difficult matter to decide.

The following series are given as examples of converging series. Some of them have been discussed on earlier pages.

$$(x + y)^n = x^n + nx^{n-1}y + \frac{n(n-1)}{\underline{|2}} x^{n-2}y^2 + \cdots$$

$$\frac{n(n-1)\ \ldots\ (n-m+1)}{\underline{|m}} x^{n-m}y^m + \cdots \quad (y^2 < x^2)^1$$

$$e = 1 + \frac{1}{1} + \frac{1}{\underline{|2}} + \frac{1}{\underline{|3}} + \frac{1}{\underline{|4}} + \cdots$$

$$e^x = 1 + x + \frac{x^2}{\underline{|2}} + \frac{x^3}{\underline{|3}} + \frac{x^4}{\underline{|4}} + \cdots$$

$$\log_e x = \frac{x-1}{x} + \frac{1}{2}\left(\frac{x-1}{x}\right)^2 + \frac{1}{3}\left(\frac{x-1}{x}\right)^3 + \cdots \quad (x > 1)$$

$$\log_e (1 + x) = x - \frac{1}{2}x^2 + \frac{1}{3}x^3 - \frac{1}{4}x^4 + \cdots$$

$$\log_e x = (x - 1) - \frac{1}{2}(x - 1)^2 + \frac{1}{3}(x - 1)^3 - \cdots \quad (2 > x > 0)$$

$$\cos x = 1 - \frac{x^2}{\underline{|2}} + \frac{x^4}{\underline{|4}} - \frac{x^6}{\underline{|6}} + \cdots$$

$$\tan x = x + \frac{x^3}{3} + \frac{2x^5}{15} + \frac{17x^7}{315} + \frac{62x^9}{2835} + \cdots \quad \left(x^2 < \frac{\pi^2}{4}\right)$$

[1] The symbol $\underline{|m}$ stands for factorial m, that is $1 \times 2 \times 3 \times 4 \times \cdots m$.

Computation by Means of Series

As an example of the use of series in computation, it might be required to find the value of $\frac{1}{1-x}$ when $x = 0.5$. The answer, 2, obtained by simple arithmetic, constitutes a check, but frequently the computation by a series is the only method available. By long division

$$\frac{1}{1-x} = 1 + x + x^2 + x^3 + x^4 \cdots x^{n-1} \frac{(1)}{1-x} x^n$$

and the value of $\frac{1}{1-x}$ may be obtained by substituting the value of x into the series on the right. When $x = 0.5$ and $n = 4$, the first four terms give $\frac{1}{1-x} = 1 + 0.500 + 0.250 + 0.125 \cdots = 1.875$. The computation is in error by 0.125. If, however, the computation is carried out to 8 terms instead of 4 a result of 1.992 is obtained. The computation may be made as accurate as desired by using a sufficiently large number of terms.

Maclaurin's Theorem.—A great many experimental results may be represented accurately by a formula of the form $y = A + Bx + Cx^2 + Dx^3 \ldots$

The expression at the right is a series of increasing powers, and Maclaurin's theorem determines the law for the expansion of a *single* variable in a series of increasing powers of the variable, as follows:

$$\text{If } y = A + Bx + Cx^2 + Dx^3 \tag{1}$$

Successive differentiation gives,

$$\frac{dy}{dx} = B + 2Cx + 3Dx^2 \tag{2}$$

$$\frac{d^2y}{dx^2} = 2C + 2 \times 3Dx \tag{3}$$

$$\frac{d^3y}{dx^3} = 2 \times 3D \tag{4}$$

A, B, C, and D may be evaluated in terms of the derivatives by putting $x = 0$, in each of the preceding equations.

$A = y$ when $x = 0$, or $A = f(0)$. (In (1), Bx, Cx^2 and Dx^3 become zero.)
$B = \frac{dy}{dx}$ when $x = 0$, or $1B = \frac{df(0)}{dx}$. (In (2), $2Cx$ and $3Dx^2$ become zero.)
$1 \times 2C = \frac{d^2y}{dx^2}$ when $x = 0$, or $1 \times 2C = \frac{d^2f(0)}{dx^2}$. (In (3), $2 \times 3Dx$ becomes zero.)
$1 \times 2 \times 3D = \frac{d^3y}{dx^3}$ when $x = 0$, or $1 \times 2 \times 3D = \frac{d^3f(0)}{dx^3}$. (In (4) there is no term in x.)

Then

$$D = \frac{d^3f(0)}{dx^3}\frac{1}{1 \times 2 \times 3} = \frac{d^3f(0)}{dx^3}\frac{1}{\underline{|3}}$$

Substituting the values of A, B, C, and D in Eq. (1) for the particular case when $x = 0$,

$$y = f(0) + \frac{df(0)}{dx}\frac{x}{1} + \frac{d^2f(0)}{dx^2}\frac{x^2}{\underline{|2}} + \frac{d^3f(0)}{dx^3}\frac{x^3}{\underline{|3}}$$

This is known as *Maclaurin's series*, and it is very useful in expanding functions into series, for computation.

For example, $\sin x$ may be evaluated for any numerical value of x after expanding by Maclaurin's theorem.

$$y = \sin x$$

$f(0)$ is the value of y when $x = 0$. In this case $f(0) = 0$ because $\sin 0 = 0$. Then

$$f(0) = 0$$

$\frac{df(0)}{dx}$ is the value of $\frac{dy}{dx}$ when $x = 0$. In this case

$$\frac{dy}{dx} = \cos x, \text{ and } \cos x = 1 \text{ when } x = 0.$$

Then

$$\frac{df(0)}{dx} = 1$$

$\frac{d^2f(0)}{dx^2}$ is the value of $\frac{d^2y}{dx^2}$ when $x = 0$. In this case

$$\frac{d^2y}{dx^2} = \frac{d \cos x}{dx} = -\sin x, \text{ and } -\sin x = 0 \text{ when } x = 0.$$

Then

$$\frac{d^2f(0)}{dx^2} = 0$$

$\frac{d^3f(0)}{dx^3}$ is the value of $\frac{d^3y}{dx^3}$ when $x = 0$. In this case

$$\frac{d^3y}{dx^3} = \frac{d(-\sin x)}{dx} = -\cos x, \text{ and } -\cos x = -1, \text{ when } x = 0$$

Then

$$\frac{d^3f(0)}{dx^3} = -1$$

Putting the values of these terms into MacLaurin's series,

$$y = \sin x = 0 + 1\frac{x}{1} + 0\frac{x^2}{\underline{|2}} - 1\frac{x^3}{\underline{|3}} + 0\frac{x^4}{\underline{|4}} + 1\frac{x^5}{\underline{|5}} =$$

$$x - \frac{x^3}{\underline{|3}} + \frac{x^5}{\underline{|5}} - \frac{x^7}{\underline{|7}} + \frac{x^9}{\underline{|9}} \quad . \; . \; . \; \text{etc.}$$

To illustrate the numerical application of this theorem the sine of an angle of 1 radian is computed.

$$\sin x = \sin 1$$

$$\sin 1 = 1 - \frac{1}{\underline{|3}} + \frac{1}{\underline{|5}} - \frac{1}{\underline{|7}}$$

$$1 = 1.00000 \qquad \frac{1}{\underline{|3}} = 0.16667$$

$$\frac{1}{\underline{|5}} = 0.00833 \qquad \frac{1}{\underline{|7}} = 0.00019$$

$$\overline{1.00833} \qquad \overline{0.16686}$$

$$\sin 1 = 1.00833 - 0.16686 = 0.84147.$$

This answer is correct to four places. It could be made accurate to further decimal places by including more terms of the series. If the next term $1/\underline{|9}$ is included, however, it makes a difference of only 0.000,003.

Other illustrations of the wide application of Maclaurin's theorem may be found among the series listed on page 201.

Taylor's Theorem.—Taylor's theorem is like Maclaurin's theorem, except that it is more general. The derivation by successive differentiation is similar to that of Maclaurin's theorem just given.

Taylor's theorem applies to functions of two or more variables. For example,

$$y = f(x + z) = f(x) + \frac{df(x)}{dx}\frac{z}{1} + \frac{d^2f(x)}{dx^2}\frac{z^2}{\underline{|2}} + \frac{d^3f(x)}{dx^3}\frac{z^3}{\underline{|3}} + \cdots$$

Taylor's theorem is particularly useful in computing logarithms and in computing values of functions containing sums and differences.

Fourier's Series.—Just as certain functions can be expressed in terms of infinite series with increasing powers, certain other functions can be expressed with a series of sines and cosines. In both cases the evaluation is purely empirical and the fact that a set of data can be expressed accurately by a series of powers or

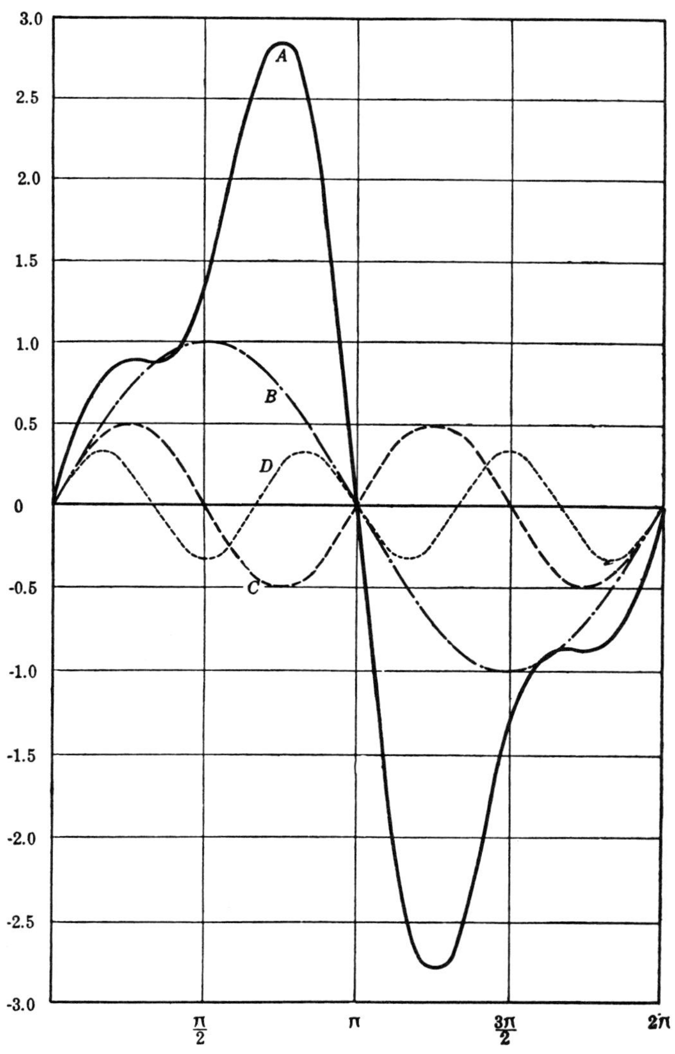

FIG. 52.—A composite curve (full line) built up from sine curves (dotted lines by Fourier's series.

Curve B: $y = \sin x$. Curve D: $y = \frac{1}{3} \sin 3x$.
Curve C: $y = \frac{1}{2} \sin 2x$. Curve A: $y = 2(\sin x - \frac{1}{2} \sin 2x + \frac{1}{3} \sin 3x)$.

by sines or cosines does not reveal any theoretical law underlying the phenomenon. It is of great practical value, however, to find empirical equations which permit accurate calculations.

The sine and cosine curves were discussed on page 66. By combining two or more of these curves and by changing the amplitudes, a composite curve can be built up, having almost any variety of curvature. An example of such a composite curve built up from different sine curves is shown in Fig. 52.

Now it is possible to work backwards and find a set of sine curves or cosine curves which will add together to give the curve corresponding to very complicated functions. The equations for the sine and cosine curves are known, so it becomes possible to find an equation representing a very complex curve. It is possible, theoretically, to find an equation for the profile of a man's face, but of course the labor involved would be great.

According to Fourier's theorem,

$$y = A_0 + a_1 \sin x + a_2 \sin 2x + \cdots b_1 \cos x + b_2 \cos 2x + \cdots$$

The constants A_0, a and b may be evaluated by integration, after multiplying through by dx. The details cannot be given here.

A great many applications of Fourier's theorem are found in physics, particularly in problems involving wave motion and heat conduction.

Simpson's Rule.—When the area under a curve is to be determined and the equation for the curve is unknown, Simpson's rule offers a good approximation. The rule depends on the fact that most curves can be expressed in terms of a parabolic equation of the form $y = a + bx + cx^2$; and on a rule of analytical geometry which states that the area of a parabola cut off between two ordinates is equal to two-thirds of the parallelogram of the same base and height.

If the area is divided up into strips as shown in Fig. 53 the area of a strip may be considered as approximately (1) a rectangle, (2) a trapezoid, (3) a parabola cut by two parallel ordinates. The third approximation is much the more accurate. The singly shaded area gives the error involved in (1).

According to the *trapezoidal rule* the area, S, is given by

$$S = (\tfrac{1}{2}y_1 + y_2 + y_3 + \cdots y_{n-1} + \tfrac{1}{2}y_n)\Delta x,$$

where y_1, y_2, y_3 + etc. are the heights of successive ordinates, and Δx is the constant spacing between them, along the X axis.

According to *Simpson's rule* the area is divided up into an even number, n, of vertical strips and the area, S, is given by;

$$S = \tfrac{1}{3}[(y_1 + y_n) + 4(y_2 + y_4 + y_6 + \cdots) + 2(y_3 + y_5 + y_7 + \cdots)]\Delta x.$$

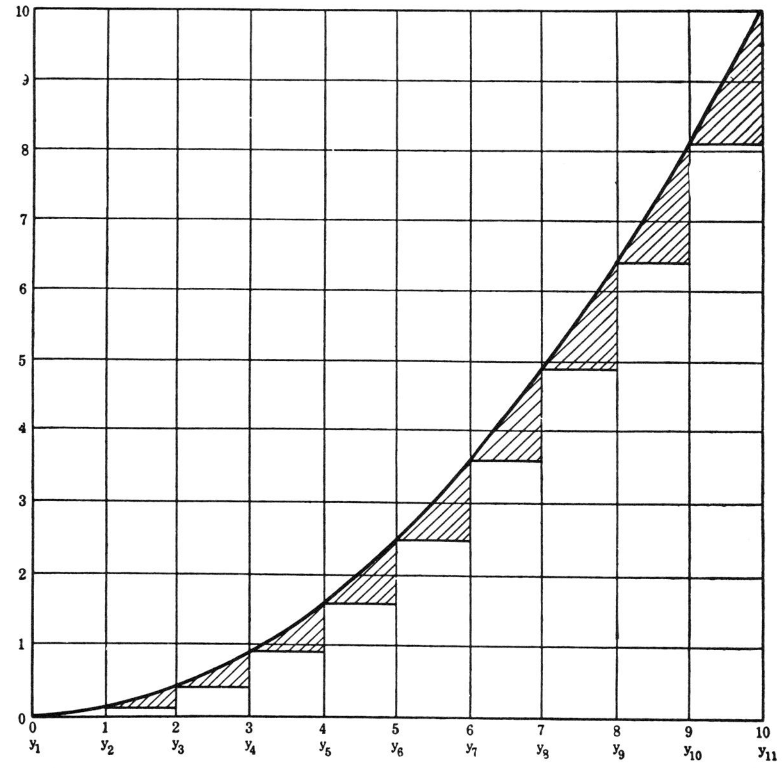

FIG. 53.—An application of Simpson's rule to the calculation of an area. The equation is $y = 0.1x^2$ and the area under the curve is 33.33. The shaded area represents the error involved by calculating the area as if it were composed of rectangles.

In words the area is equal to one-third of the sum of the following quantities, multiplied by the constant interval along the X axis between the ordinates.

(*a*) The first and last ordinates,

(*b*) Four times the sum of all the even ordinates,

(*c*) Twice the sum of all the odd ordinates except the first ordinate, y_1, and the last ordinate, y_n.

It is of course unnecessary to plot the curve, and the only restriction is that the distance between the ordinates must be uniform. The more frequently the ordinates are taken, *i.e.*, the smaller the value of Δx, the more accurate are the results.

As an example, it is well to take a function, the exact value of which is known. The area, S, under the curve $y = 0.1x^2$ is to be calculated between the ordinates at $x = 0$ and $x = 10$. $S = \sum_0^{10} y\Delta x = \sum_0^{10} 0.1x^2\Delta x$. This area can be obtained best by integration, and it is found to be

$$\int_0^{10} 0.1x^2dx = 0.1\left[\frac{x^3}{3}\right]_0^{10} = \frac{100}{3} = 33.33.$$

In the approximation methods, y must be calculated for the different values of x.

In the equation $y = 0.1x^2$, y has the following values

	y_1	y_2	y_3	y_4	y_5	y_6	y_7	y_8	y_9	y_{10}	y_{11}
x.......	0	1	2	3	4	5	6	7	8	9	10
y.......	0	0.1	0.4	0.9	1.6	2.5	3.6	4.9	6.4	8.1	10

By Simpson's rule,

$$\begin{aligned} S &= \tfrac{1}{3}[(y_1 + y_{11}) + 4(y_2 + y_4 + y_6 + y_8 + y_{10}) + 2(y_3 + y_5 + y_7 + y_9)] \\ &= \tfrac{1}{3}[(0 + 10) + 4(0.1 + 0.9 + 2.5 + 4.9 + 8.1) + \\ &\qquad 2(0.4 + 1.6 + 3.6 + 6.4)] = 33.3 \end{aligned}$$

By the trapezoidal rule, $S = 33.5$.

By plotting and counting small squares, S was estimated to be 33.24.

Simpson's rule may be applied to experimental data directly without plotting and it saves time when the equation connecting the two variables is not known.

Among other applications Simpson's rule has been used recently by White[1] to reduce the number of experimental observations necessary in determining the cooling correction of a calorimeter.

[1] White, *J. Am. Chem. Soc.*, **48**, 1148 (1926).

Exercises

1. What series is equivalent to e^x calculated by Maclaurin's theorem?

Ans.: $1 + x + \frac{x^2}{\underline{|2}} + \frac{x^3}{\underline{|3}} + \frac{x^4}{\underline{|4}} \cdots$

2. What series is equivalent to 5^x, calculated by Maclaurin's theorem?

3. What series is equivalent to $\log(x + y)$ as calculated by Taylor's theorem?

Ans.: $\log x + \frac{y}{x} - \frac{y^2}{2x^2} + \frac{y^3}{3x^3} \cdots$

4. What is the value of $\sin(x + y)$ in terms of a series, calculated by Taylor's theorem?

5. How may $(a + b)^n$ be expanded into a series by Taylor's theorem to give the regular binomial theorem (p. 201)?

6. What is the value of cos 40°, calculated by Maclaurin's theorem? The degrees must be changed to radians.

7. What is the value of $\int_0^1 \frac{dx}{1 + x}$ calculated by Simpson's rule? What is the true value?

Ans.: 0.693.

8. What is the value of $\int_1^{11} \frac{dx}{x}$ by Simpson's rule, taking $\Delta x = 2$; and taking $\Delta x = 1$?

9. A curve is drawn at random and the area under it between specified limits is determined by Simpson's rule, and by the methods of Chap. XXI.

CHAPTER XX

PROBABILITY

It has been said that the theory of probability is the logic of the physical sciences, and it is true that a great many of the phenomena of physical chemistry are expressions of the laws of probability.

For example, the speed of certain chemical reactions is determined by the probability of collisions between molecules. The velocities of molecules in a gas are distributed in accordance with probability laws. The emission of radiation from a hot body is such that a curve resembling a probability curve results when the intensity of light is plotted against the wave length. The chance that a molecule can acquire sufficient energy to break away from the other molecules at the surface of a liquid, depends on the temperature in such a way that the logarithm of the vapor pressure is proportional to the reciprocal of the absolute temperature. When the electrons are displaced in a mass of molecules (by an electrical discharge, for example) the return of the electrons to their normal positions gives rise to the emission of light, and the brightest lines of the spectrum are the ones corresponding to the most probable displacements. The number of atoms of a radio-active element, decomposing in a minute, depends on the probability that the different parts of the nucleus will arrange themselves in unstable configurations. Fortunately, in pure mathematics, a science of probability has been developed which is of value in interpreting these phenomena.

Not only do probability considerations enable us to push the ultimate explanation of natural phenomena one step farther back, but they enable us to carry out many practical calculations, such as the determination of errors in a series of experimental measurements. Statistical mechanics is the name given to that important branch of physical chemistry which deals with probability calculations.

Entropy is another conception of physical chemistry which has been aided by the application of probability theorems. It is known that chemical reactions occur spontaneously in such a way as to increase the entropy of the reacting systems. It is known also that systems tend to arrange themselves in such a way as to assume the most probable configuration; and the most probable arrangement is the arrangement of greatest disorder. It is the general behavior of any system, that things arranged in any specified order tend to lose this order as the system undergoes changes. There is then a direct relationship between entropy and probability and it has been possible to express the relation quantitatively and mathematically.

Permutations and Combinations

Permutations are the different orders in which things can be arranged.

For example if the three letters a, b, and c are taken two at a time, they may be written ab, ac, ba, bc, ca, and cb. It is evident by inspection that there are six and only six different orders; that is the number of permutations is six.

If there are more letters, $a, b, c, d, \ldots n$, etc., to be arranged in permutations, the first position may be occupied by any one of the n different letters; that is, it may be filled in n different ways.

The second place can be filled by any of the n letters except the one which occupies the first position, or, in other words, the second place can be filled in $(n - 1)$ different ways. The whole number of permutations of two letters then is $n(n - 1)$. This formula checks with the first example given where there were three letters and the number of permutations, P, was six. By calculation, $P = n(n - 1) = 3(3 - 1) = 6$.

In the general case it is desired to find the number of permutations of n different things taken, r, at a time. Proceeding as before, the first position may be filled in n different ways, the second in $(n - 1)$ different ways, the third in $(n - 2)$ or $(n - (3 - 1))$ different ways and the fourth in $(n - 3)$ or in $(n - (4 - 1))$ ways. Continuing in the same way the rth position can be filled in $(n - (r - 1))$ or $(n - r + 1)$ ways.

The total number of permutations, P, then is the number of different possibilities of each position, multiplied by or combined with each of the other possibilities, and

$$P_r^n = n(n-1)(n-2)(n-3)\ .\ .\ .\ (n-r+1) \tag{1}$$

For the special case where $r = n$ the last term drops out and

$$P_r^n = |\underline{n}. \tag{2}$$

Combinations are the different groupings which can be formed from the available units, without regard to the order in which they are placed. In the case of the three letters a, b, and c, taken two at a time there were six permutations, but there are only three combinations, ab, ac, and bc, because $ba = ab$, $ca = ac$, and $cb = bc$ when the restriction of order is removed.

The general case of *the number of combinations of n things taken r at a time is*

$$C_r^n = \frac{n(n-1)(n-2)(n-3)\ \cdot\cdot\cdot\ (n-r+1)}{|\underline{r}} \tag{3}$$

This relation is derived from the two preceding theorems by simply dividing the number of permutations of n things taken r at a time, by the number of permutations possible among the r things themselves.

To illustrate, in the case of the four letters a, b, c, and d, there are four different ways in which the letters can be combined if three are taken at a time. It can be seen that the only possible combinations are abc, abd, bcd, and acd, any other arrangement being equivalent to one of these four already given. This fact is in agreement with the formula, for

$$C_r^n = \frac{4(4-1)(4-3+1)}{1 \times 2 \times 3} = 4$$

Now equation (3) may be arranged in a more convenient form for calculation, by multiplying both numerator and denominator by $|\underline{(n-r)}$, giving

$$C^n = \frac{|\underline{n}}{|\underline{r}\ |\underline{(n-r)}} \tag{4}$$

Taking the same problem as before with four letters a, b, c, and d arranged in groups of three,

$$C_r^n = \frac{1 \times 2 \times 3 \times 4}{(1 \times 2 \times 3)(1)} = 4$$

Sometimes it is necessary to know the total number of combinations, including combinations with from 1 to n individuals in each group. For example, in the preceding problem, how many combinations are possible taking not only three in a group, but also two and one and four in a group? It can be shown that *the sum of all the combinations, S_c is given by a series which may be reduced to the form*

$$S_c = 2^n - 1 \tag{5}$$

When calculating factorials, a considerable saving may be effected by the use of Stirling's approximation formula, according to which

$$\underline{|n} = n^n e^{-n} \sqrt{2\pi n} \left[1 + \frac{1}{12n} + \frac{1}{288n^2} + \cdots \right]$$

The material in the brackets is practically unity for all large values of n. When $n = 20$, the error is less than one-half of one per cent if the material in the brackets is taken equal to 1.

Probability Theorems

The probability of an event depends on the number of possibilities.

$$\text{Probability} = \frac{\text{number of ways the event occurs}}{\text{number of possible ways the event can occur}} \tag{6}$$

There are six sides to a die and there is one chance in six that a specified side, number 4 for example, will turn up. There are two sides to a coin and the chances are 1 in 2 or "fifty-fifty" that heads will turn up. Expressed mathematically, if an event can happen in a ways or b ways and in no other way, the probability that the event will happen in a ways is $\frac{a}{a + b}$. Likewise the probability that it will happen in b ways is $\frac{b}{a + b}$.

The a may correspond to "yes" and b to "no," or a may represent a "success" and b a "failure." If a marksman hits the bull's-eye once and misses it 9 times, his chance of hitting it is

$$\frac{a}{a + b} = \frac{1}{1 + 9} = \frac{1}{10}$$

and his chance of missing it is

$$\frac{9}{1 + 9} = \frac{9}{10}$$

It is to be noted that the probabilities of success and failure add up to 1. In the example just given

$$\frac{1}{10} + \frac{9}{10} = 1$$

Then if p = probability of success, 1 − p = probability of failure. (7)

It follows also that a probability of 1 is equivalent to certainty for then the number of ways the event occurs is the number of possible ways it can occur. Certainty of failure corresponds to zero probability.

The probability that two independent events will occur simultaneously is equal to the product of the probabilities of the separate events, or

$$p = q \times r \tag{8}$$

where q is the probability of the occurrence of the first event and r is the probability of the occurrence of the second event and p is the probability that the two events will occur simultaneously.

This is one of the most important theorems in physical chemistry. For example the chance of throwing a four is $\frac{1}{6}$ and the chance of throwing a double four with two dice is $\frac{1}{36}$. The theorem can be readily understood by an examination of the following table, where the large numbers refer to the first die and the small numbers refer to the second die.

Table I.—Possible Combinations of a Pair of Dice

First Die	Two Dice
1	$1,_1$ $1,_2$ $1,_3$ $1,_4$ $1,_5$ $1,_6$
2	$2,_1$ $2,_2$ $2,_3$ $2,_4$ $2,_5$ $2,_6$
3	$3,_1$ $3,_2$ $3,_3$ $3,_4$ $3,_5$ $3,_6$
4	$4,_1$ $4,_2$ $4,_3$ $4,_4$ $4,_5$ $4,_6$
5	$5,_1$ $5,_2$ $5,_3$ $5,_4$ $5,_5$ $5,_6$
6	$6,_1$ $6,_2$ $6,_3$ $6,_4$ $6,_5$ $6,_6$

When the first die turns up 1, the second die may turn up 1, 2, 3, 4, 5, or 6, for there is no connection between the two dice. This makes six permutations with No. 1 of the first die. Likewise there are six permutations with No. 2 and in all, there are 36 and only 36 possible permutations. Among these 36 possibilities there is one double 4, and the probability of throwing a double four (or any other specified pair) is $\frac{1}{36}$ as given by the formula.

If there are 10 molecules in a 2-liter space, and an imaginary wall divides the space into two equal parts, the chance of any one molecule being in the first liter partition is $\frac{1}{2}$. The chance of a second molecule being on the same side of the partition is also $\frac{1}{2}$. The chance that the two specified molecules will both be on the same side is $\frac{1}{2} \times \frac{1}{2}$, or $(\frac{1}{2})^2$, or $\frac{1}{4}$. The chance that all 10 molecules will be on the same side is $(\frac{1}{2})^{10} = 1/1{,}024$. In other words, the chances are over a thousand to one against all ten molecules being together in 1 liter space and leaving the other liter space empty. If there is a whole gram molecule of molecules, 6×10^{23}, the chance of all the molecules being in half of the container and leaving the other half empty is

$$\left(\frac{1}{2}\right)^{6\times 10^{23}} = \frac{1}{2^{6\times 10^{23}}}$$

The number is so extremely small that there is practically no chance of a spontaneous separation of molecules. Laboratory experience all goes to show, as would be expected, that the spontaneous changes are the ones which make the concentrations uniform throughout the vessel. This phenomenon is, in fact, an application of the second law of thermodynamics, and an illustration of the theorem that a condition of maximum entropy corresponds to a condition of maximum probability.

It should be noted in passing that whereas there is only one chance in 36 for a $4,_4$ or a $1,_1$, there are two chances for a four and a one, corresponding to $4,_1$ and $1,_4$.

The probability, p, that one of two (or more) mutually exclusive events will occur is the sum of the probabilities of their separate occurrence, q and r.

$$p = q + r \tag{9}$$

For example, what is the probability that a die will turn up 5 or an even number? The chance of a 5 is $\frac{1}{6}$ and the chance of an even number is $\frac{3}{6}$. The events are mutually exclusive, since if it is five it cannot be an even number, and if an even number turns up, it cannot be a five.

$$p = \frac{1}{6} + \frac{3}{6} = \frac{2}{3}$$

and there are two chances in 3 for either a five or an even number. By a preceding theorem (7) the chance of an odd number, other than 5, is $1 - \frac{2}{3} = \frac{1}{3}$. This can be checked easily for there are only two remaining odd numbers (Nos. 1 and 3) out of six possibilities and the probability is, therefore, $2/6$ or $1/3$.

A considerable amount of confusion may arise in the application of probability laws through a failure to distinguish between independent events and mutually exclusive events.

It can be shown that *if the probability of an event happening on a single trial is p, the probability P that it will occur r times in n trials is given by the equation*

$$P = \frac{n(n-1)(n-2) \cdots (n-r+1)p^r(1-p)^{n-r}}{\underline{|r}} \qquad (10)$$

This equation is the basis of most of the probability formulas.

One of the outstanding features of the laws of probability is their dependence on the number of trials. The probability of throwing heads is 0.5, but no one imagines that if he throws heads the first time, he will throw tails the second time and then heads the third and tails the fourth, and so on. As a matter of fact three or four heads may come along in succession and one cannot expect a head because tails came up last time. One can be sure, however, that if he throws a very large number of times, the average will tend to show one-half heads and one-half tails. In one classic experiment 2,048 heads were thrown out of 4,048 trials and with a still greater number of trials the ratio would have been more nearly $1/2$.

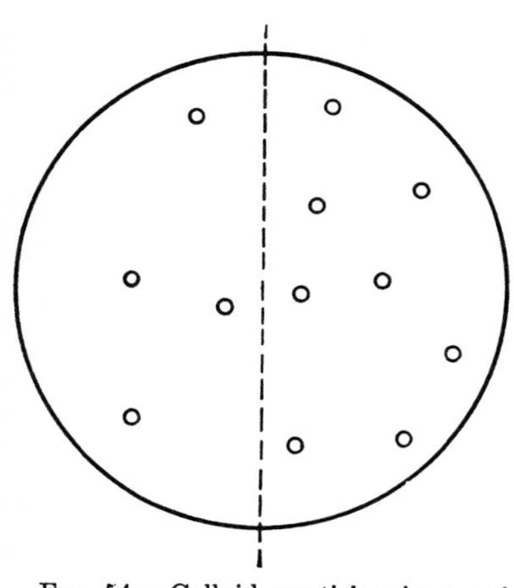

FIG. 54.—Colloid particles in a microscopic field illustrating theorems of probability. The general tendency is for particles to move from concentrated solutions to more dilute ones, but exceptions occur when the total number is small.

The probability laws must be considered as limiting laws, which are approached as the number of trials is increased. (11)

Fortunately, in applications to chemistry there are so very many molecules involved in any detectable phenomenon that the probability laws can be applied with absolute rigor. The failure of the general laws when the number of trials is small is well illustrated in Fig. 54, where 12 colloid particles are in the field of an ultra-microscope. There are eight on one side of the dividing line and four on the other. According to the second law of thermodynamics (p. 215) and on the basis of common sense, it is to be expected that the particles will diffuse from the concentrated side to the dilute side, giving 6 particles on each side, but actual experiment will show that, although the tendency is in this direction, sometimes the particles will wander over from the dilute to the concentrated side, and there may be at times 3 on one side and 9 on the other or even 2 and 10. These changes happen spontaneously and they are in direct violation of the second law of thermodynamics, but it must be remembered that this law, like many others in physical chemistry, is a statistical law and cannot be applied to a system of only twelve units. In a liter of gas (under normal conditions) there are 3×10^{21} units and the second law of thermo dynamics, or any probability theorem, may be applied with precision.

The Probability Curve

When a marksman shoots at a target most of the shots hit near the center, with scattering hits in the outer rings as shown in Fig. 55. If a line is drawn through the center to divide the target into + and −, or right and left, and the number of hits between successive rings is plotted in sequence, the graphs at the right of Fig. 55 are obtained. The target of a poorer marksman is shown below, with the corresponding flatter curve.

Two important deductions are to be drawn from these curves: first, that positive deviations occur as frequently as negative deviations, making the curve symmetrical; and second, that small deviations occur more frequently than large ones, making the curve peaked.

Now this curve is very frequently met with, in many different fields. Whenever an experimenter tries to make accurate measurements, his observations give a similar curve. The distribution of velocities among the molecules of a gas, or the statistical results of an investigation in economics are all recognized as probability phenomena by their resemblance to this curve.

As a practical result of this relationship, it is possible to calculate the peak of the curve; *i.e.*, the best value from a series of measurements. It amounts to locating the bull's-eye by plotting the hits on the target.

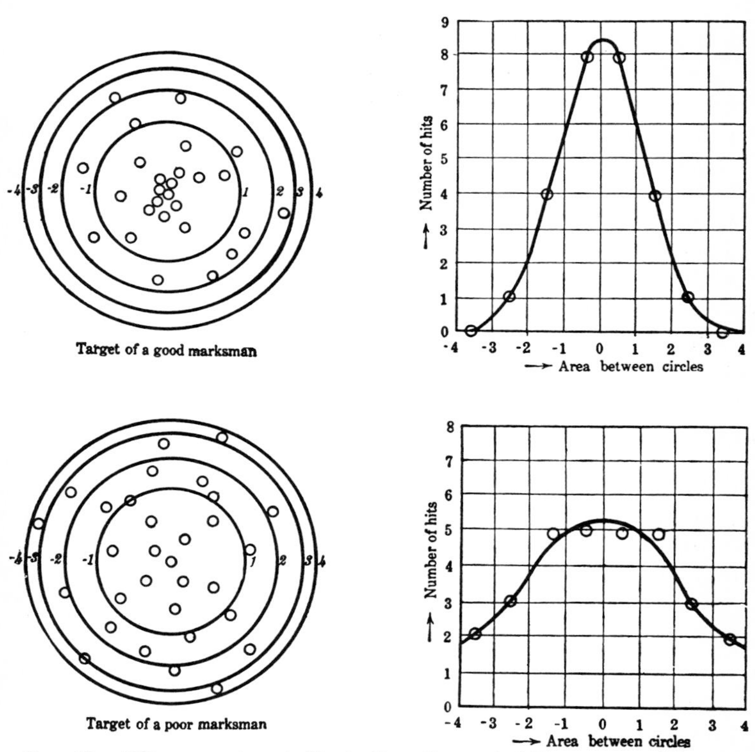

FIG. 55.—Hits on a target illustrating the probability curve. The areas between circles are all equal. The number of hits in each area is plotted at the right.

Curves of this type may be represented by the equation

$$y = ke^{-h^2x^2} \tag{12}$$

There are mathematical deductions for this probability equation but the strongest argument for it is the fact that it works. Remarkable checks have been obtained in a great many ways and accurate experimental measurements invariably approach the curve more closely as the number of measurements increases.

In this equation there are two variables, y being the frequency of occurrence of the deviation x; and two constants k and h. k determines the height at which the curve cuts the Y axis, for on the Y axis $x = 0$ and $y = ke^{-h^2 0^2} = k$. k is equal to the number of correct values.

For a fixed value of k, h determines the steepness of the curve. A high value of h gives a steep slope in the neighborhood of the central ordinate, and the smaller area under the curve corresponds to fewer errors, or greater precision. In the curves of Fig. 55 the lower curve corresponding to the poor marksman

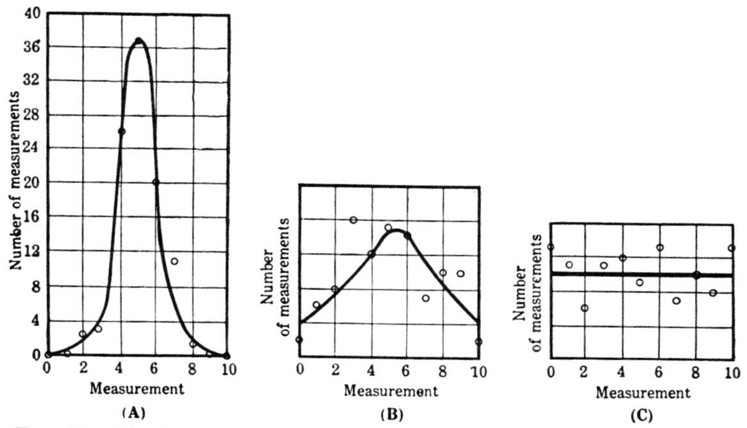

FIG. 56.—Graphs of experimental measurements illustrating the probability curve. The abscissas are experimental determinations of the electrical resistance of a wire. (*A*) corresponds to more favorable conditions than (*B*). (*C*) corresponds to pure chance.

would have a small k corresponding to a small number of bull's-eye hits, and a small h corresponding to a flat curve and many errors. If the marksman was blindfolded, there would be no peak and the curve would be a horizontal line, since a hit in the outer ring would be just as likely as a hit in the bull's-eye.

In Fig. 56, are shown the results of a hundred readings of a resistance (40.5 ohms), measured on a Wheatstone bridge with alternating current and a telephone. The dials are turned while the eyes are closed, until a minimum sound is registered in the telephone and then the dials are read. The measurement is not an easy one to make with precision. The first graph,

A, is plotted from experiment A, where conditions were favorable. Graph B corresponds to experiment B, where the data were taken under unfavorable conditions with considerable noise in the room. Data C and Graph C correspond to the dial turnings of pure chance while the telephone was disconnected.

TABLE II.—100 MEASUREMENTS OF AN ELECTRICAL RESISTANCE

Number of readings	Dial readings (+40 ohms)										
	0.0	0.1	0.2	0.3	0.4	0.5	0.6	0.7	0.8	0.9	1.0
A. Accurate	0	0	2	3	26	37	20	11	1	0	0
B. Inaccurate	2	6	8	16	12	15	14	7	10	10	2
C. Chance	13	11	6	11	12	9	13	7	10	8	13

The accurate measurements, A, have the highest peak and steepest curve corresponding to a large k and a large h. The irregularities of the curve would be reduced by taking many more readings. The apparent dips in the curve prominent in C may be due to the superimposition of a mechanical factor, perhaps the position of the dial handle.

It can be shown from the general expression $y = ke^{-h^2x^2}$ that

$$P = \frac{h}{\sqrt{\pi}}\int_{x_0}^{x} e^{-h^2x^2}dx \tag{13}$$

where P is the probability that an error will lie between any two limits x and x_0. This equation may be written also

$$P = \frac{2}{\sqrt{\pi}}\int_{0}^{hx} e^{-h^2x^2}d(hx) \tag{14}$$

where P is the probability that an error will be numerically less than x. Tables have been compiled for error computations, giving the value of P for various values of k.

ERRORS

Every physical or chemical measurement is subject to error, the magnitude of which depends on the character of the measurement, the refinement of the apparatus. and the care of the observer. In accurate work several measurements are made, and the average of all the values is considered to be the correct value.

The important question arises: How nearly correct is the arithmetical mean or the average of a series of measurements? There are different ways of answering this question. According to one procedure the deviations of the mean are plotted to give a probability curve and the constants h and k of the equation are determined. Simpler methods have been adopted and there are three standard ways of expressing the error of a series of measurements, the average error, the mean square error, and the probable error.

To illustrate the application of these three calculations the measurements of a bar of iron are given in Table III.

TABLE III.—CONSIDERATION OF ERRORS IN A PHYSICAL MEASUREMENT

Experiment number	Measurement, in centimeters	Deviation from average (v)	Positive deviation	Negative deviation	Square of deviation
1	1.05	0.00	....	......	0.0000
2	1.04	−0.01	....	−0.01	0.0001
3	1.02	−0.03	....	−0.03	0.0009
4	1.06	0.01	0.01	......	0.0001
5	1.06	0.01	0.01	......	0.0001
6	1.07	0.02	0.02	......	0.0004
7	1.03	−0.02	....	−0.02	0.0004
8	1.07	0.02	0.02	......	0.0004
Total.......	8 |8.40	8 |0.12	0.06	0.06	0.0024
Average.....	1.05	0.015			

The average error is the simplest error to calculate and it gives at once a measure of the magnitude of the errors.

The average error, a, of a single determination is the average deviation of a determination from the average result, regardless of sign.

$$a = \pm \frac{\Sigma v}{n} \tag{15}$$

where Σv is the sum of all the deviations from the average, and n is the number of measurements.

In the preceding example,

$$a = \pm\frac{\Sigma v}{n} = \pm\frac{0.12}{8} = \pm 0.015.$$

It can be shown that *the average error, A, affecting the averaged value of all n observations is given by the expression*

$$A = \pm\frac{\Sigma v}{n\sqrt{n}}. \tag{16}$$

In the preceding example,

$$A = \pm\frac{\Sigma v}{n\sqrt{n}} = \pm\frac{0.12}{8 \times 2.9} = \pm 0.005$$

The mean-square error, m, is the error whose square is the average of the squares of all the errors. Sometimes this error is called the root-mean-square error, and sometimes the mean error, but the latter name leads to confusion with the average error. It depends on the principle of least squares, according to which the best value is the one which makes the sum of the squares of the deviations a minimum. The sign of the error can be neglected since squares of either positive or negative numbers are positive. The *mathematical expression for the mean square error, m, of single observations is given by,*

$$m = \pm\sqrt{\frac{\Sigma(v^2)}{n - 1}} \tag{17}$$

and the *mean square error, M, of the average of all the observations is given by*

$$M = \pm\sqrt{\frac{\Sigma(v^2)}{n(n - 1)}}. \tag{18}$$

In the numerical example given above

$$m = \pm\sqrt{\frac{0.0024}{7}} = \pm 0.018; \text{ and } M = \pm\sqrt{\frac{0.0024}{8 \times 7}} = \pm 0.0065$$

The probable error, r, is the error which has been most commonly applied to a series of accurate measurements. It is an error such that the number of errors which are greater than r, is equal to the number of errors which are less than r.

The significance of the probable error is illustrated in Fig. 57, where the shaded area represents the errors less than the probable errors R and $-R$, and the unshaded area represents the errors which are greater than the probable error. The shaded and unshaded areas are equal. *The probable error, r, of a single measurement is given by the expression*

$$r = \pm 0.6745\sqrt{\frac{\Sigma(v^2)}{n - 1}} \tag{19}$$

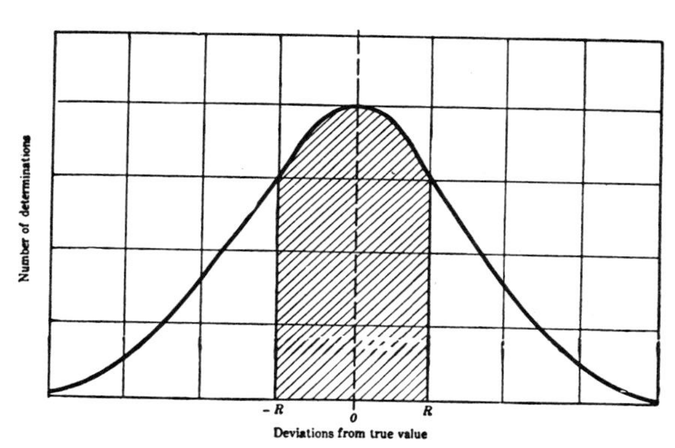

FIG. 57.—The graphical significance of the probable error, R. The value of R is such that the shaded area which represents errors less than the probable error is equal to the unshaded area which represents errors greater than the probable error.

and the probable error, R, of the *average of all the measurements is given by the expression*

$$R = \pm 0.6745\sqrt{\frac{\Sigma(v^2)}{n(n - 1)}} \tag{20}$$

In the numerical example given before

$$r = \pm 0.6745\sqrt{\frac{0.0024}{7}} = \pm 0.012 \text{ and } R = \pm 0.6745\sqrt{\frac{0.0024}{8 \times 7}} = \pm 0.0044$$

The labor involved in squaring all the deviations in a large number of observations has led to the invention of an approximation formula which is easier to use and which is nearly as accurate. The approximate value of the probable error of the average of a series of observations is given by the following expression, where the sign of the deviation, v, is neglected,

$$R = \pm 0.8453\frac{\Sigma(v)}{n\sqrt{n - 1}} = \pm K\Sigma v. \tag{21}$$

The values of $\frac{\pm 0.8453}{n\sqrt{n-1}}$ can be calculated for different values of n as shown in Table IV and used directly to determine the probable error of the averaged result.

TABLE IV.—NUMERICAL VALUES OF $K = \frac{\pm 0.8453}{n\sqrt{n-1}}$

When n =	1	2	3	4	5	6	7	8	9	10
K =		0.423	0.199	0.122	0.085	0.063	0.049	0.040	0.033	0.028
When n =	11	12	13	14	15	16	17	18	19	20
K =	0.024	0.021	0.019	0.017	0.015	0.014	0.012	0.011	0.011	0.010

Applying the formula to the preceding numerical example where $n = 8$, $R = \pm K\Sigma v = \pm 0.04 \times 0.12 = 0.0048$.

In arithmetical operations involving probable errors, the following formulas apply:

A is a quantity with a probable error $\pm a$.

B is a quantity with a probable error $\pm b$.

The probable error, R, of the operation is as follows,

$A + B$, or $A - B$,	$R = \pm\sqrt{a^2 + b^2}$
$A \times B$,	$R = \pm\sqrt{(Ab)^2 + (Ba)^2}$
$B \div A$,	$R = \pm\sqrt{\frac{(Ba/A)^2 + b^2}{A^2}}$

For very rough approximations, these formulas may be reduced to the following general rules:

In addition or subtraction the largest numerical error in the quantities is retained as in the following example,

$$\begin{array}{r} 5.4 \ \pm 0.3 \\ 3.50 \pm 0.05 \\ \hline 8.90 \pm 0.3 \end{array}$$

In multiplication or division the percentage error of the least accurate factor is retained as in the following example,

$(25.0 \pm 0.5) \times (2.000 \pm 0.001) = 50.0 \pm 2$ per cent error $= 50.0 \pm 0.01$
(2 per cent error) (0.05 per cent error)

THE USE OF CALCULATED ERRORS

The theory of probability is of little significance when the constant errors are greater than the accidental errors. Inconsistent results indicate inaccurate work, but consistent results do not necessarily assure accuracy, and the probable error shows only how uniformly the experimenter has made his measurements. In the opinion of some, most of the measurements of physical chemistry do not warrant the calculation of the probable error. Very consistent measurements of the electrical conductance of a solution might still be quite inaccurate on account of a failure of the thermostat; and as a rule it is more important to improve the apparatus than to take a large number of readings.

The calculation of the average of a series of measurements, together with the average deviation from this mean, regardless of sign, is the simplest way of recording errors, and it seems to be gaining in favor among chemists. The mean-square error has been used mostly by German and French writers while the probable error has been used largely by English and American writers.

The three methods for indicating the precision (not necessarily the accuracy) of the work are connected by simple constants as follows:

$$\text{Probable error } (R) = 0.6745 \text{ mean-square error } (M)$$
$$\text{Probable error } (R) = 0.8453 \text{ average error } (A)$$

Exercises

1. How many permutations are possible with seven different colors, taken two at a time? How many combinations? *Ans.:* 42 permutations.

2. Twenty different metals are to be combined in alloys, regardless of relative compositions. How many different binary alloys (containing two metals) are possible?

How many ternary alloys are possible?

3. In how many throws of dice is there an even chance of throwing double sixes at least once? *Ans.:* 25.

$$\left(\frac{35}{36}\right)^n = \frac{1}{2}. \quad n = \frac{\log 2}{\log 36 - \log 35} = 25$$

4. Four coins are tossed in the air at once. What is the chance that all four will come up heads? What is the chance of tossing heads three times in succession? How many times is it expected that this run of three will occur in 10,000 tosses?

5. One hundred colloid particles are suspended in 2 cc. of solution. What is the chance of drawing out 60 particles when 1 cc. is removed? Out of a thousand trials, how many times can one expect to draw out 45 particles? Ans.: One chance in 93 of withdrawing 60.

6. Ten molecules of hydrochloric acid (HCl) are decomposed by electrolysis and then recombined by the action of light. How many hydrogen atoms will be paired with their original chlorine atoms?

7. In how many different ways can a hand of 13 cards be dealt from a pack of 52 cards? *Ans.:* 635,013,559,600.

8. What is the chance of dealing any specified hand (13 spades, for example) from a pack of 52 cards?

9. The following readings were taken on a polarimeter:

20.25	20.31
20.27	20.26
20.31	20.21
20.22	20.28

What is the average reading and what is the average error; the mean square error; and the probable error? *Ans.:* Average error, A, $=0.01$.

10. The following values were obtained for the heat of vaporization of benzene. What is to be taken as the best value for this quantity? What is the average error; the probable error; and the mean-square error?

94.40, 94.37, 94.30, 94.22
94.48, 94.24, 94.38, 94.44

CHAPTER XXI

GRAPHICAL METHODS IN PHYSICAL CHEMISTRY

Graphical representation is perhaps the simplest means for showing the influence of one variable on another. It is the easiest way of describing a phenomenon, but as experimental measurements of a science become more accurate, the graphical methods must be abandoned for the greater precision of mathematical computation. Graphical computations are not common in astronomical calculations nor in certain branches of physics, but in many measurements of physical chemistry they are still adequate. As new improvements are made in apparatus and in the execution of experiments it is likely that parts of physical chemistry will outgrow the graphical methods. In sciences which are still further removed from the exact science of mathematics, as for example in economics, the graphical methods are used still more than in chemistry.

Drawing Curves

One of the commonest procedures in physical chemistry is the plotting of experimental data—usually on rectangular coordinates. The results are easily described in this way and the reader can grasp the situation with the minimum of effort. Another advantage of the graphical representation lies in the convenient way of determining averages. Each individual measurement must be recorded as a point on the coordinate paper but a smooth curve is drawn through the various points.

In Fig. 58 the rather discordant points represent the progress of a chemical reaction with time. The curve is drawn in such a way as to make positive errors counterbalance negative ones. As a general but not infallible rule, natural phenomena are represented by smooth curves (or rather the discontinuous jumps corresponding to individual molecules or electrons or quanta are so small as to appear continuous) and the full line represents the true relation more correctly than does an unaveraged line joining points which inexperienced students are inclined to draw.

In drawing a straight line through a number of points an opaque ruler is not satisfactory because the points below the line are obscured. A straight line scratched on the bottom of a strip of celluloid, or a stretched thread is helpful in locating the best straight line. After its location is established, the line may be drawn in with a ruler.

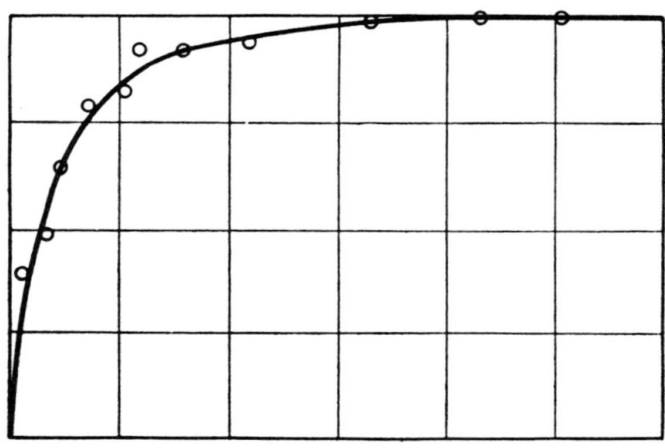

FIG. 58.—Experimental data are represented as points, and a smooth curve is drawn so as to give a graphical average.

In drawing curved lines a spline or a French curve is necessary. The former is a flexible ruler and the latter is a piece of celluloid cut with various curved edges so that part of the edge is sure to fit the experimental data. A spline is indispensable for drawing lines which deviate but slightly from straight lines. A spline costs $1 or more but an inexpensive one may be made by inserting a length of wire-solder into a snugly fitting rubber tube. Usually it is necessary to draw several trial curves in light pencil through the points before a satisfactory one is obtained.

The deviations of the experimental points from the smooth curve give an idea of the precision of the results. If they do not fall consistently near the smoothed out curve the measurements may be considered comparatively inaccurate. If they fall entirely on the line the measurements are shown to be consistent but there may still be a constant error in the measurements. According to convention the size of the "points," *i.e.*, the diameter of the circles or the lengths of the crosses, should show the accuracy of the measurement. For example, the circles of Fig. 58 cover a distance along the ordinate corresponding to 2 mm. of pressure and indicate that the measurements were not accurate to more than 2 mm.

Finding Empirical Equations

A frequent problem confronting the experimenter is that of finding a mathematical equation which will express his results. The equation may be fitted either to the original data or to the smoothed out curve. The former procedure is the better; the latter is to be adopted only in case the measurements are too inaccurate to warrant more than a common-sense graphical average. The plotting of the curve is always helpful in deciding what type of equation is most likely to fit the data.

In some cases the law governing the phenomenon is known and a theoretical equation can be used at once. In most cases the law is unknown or there are several laws involved, and then it is necessary to find an empirical equation. A good example of an empirical and a theoretical equation was given on page 63. Frequently an empirical equation leads to a theoretical equation and points the way to the natural law which governs the phenomenon.

In starting a search for an empirical equation it is necessary to determine the type of the equation and then to evaluate the constants. First, the equation of the straight line is tried because it is the simplest to determine and the easiest to use in subsequent calculations. If the line is not straight but nearly so, a parabolic equation $y = a + bx + cx^2$, or $y = a + bx + cx^2 + dx^3$ is tried next. As a third choice an exponential equation is tested by plotting one variable against the logarithm of the other. If the line plotted on semilogarithm paper is not straight it may be possible as a fourth choice to apply the equation $y = a + bx + cx^2$ to the semilogarithm graph. The semi-log equation is so common in physical chemical phenomena that in many cases it should be tried first. A series of measurements may be expressed by any one of a great number of equations, but the simplest one is usually chosen.

Solving Simultaneous Equations.—After choosing the type of equation, the next thing is to evaluate the constants, and there are two general ways in which this can be done. According to one procedure typical points are chosen and the specific values of x and y are introduced. The resulting equations are then solved simultaneously for the unknown constants.

For example, if $y = 4$ when $x = 1$, and $y = 6$ when $x = 2$ and the equation is $y = a + bx$,

$$4 = a + b \times (1)$$
$$6 = a + b \times (2)$$

or

$$-2 = -b \text{ (subtracting)}$$
$$b = 2$$
$$a = 4 - 2 = 2 \text{ (substituting } b = 2)$$

Then

$$y = a + bx = 2 + 2x$$

In a more general case, $y = a + bx + cx^2$, and x_1y_1, x_2y_2 and x_3y_3 are specific points.

Then

$$y_1 = a + bx_1 + cx_1^2$$
$$y_2 = a + bx_2 + cx_2^2$$
$$y_3 = a + bx_3 + cx_3^2$$

There are three unknowns and the three equations are sufficient to evaluate the constants, by solving the equations simultaneously, taken two at a time (p. 267). If the equation is of the form

$$y = a + bx + cx^2 + dx^3$$

there are four unknowns and four equations are necessary.

The points chosen may be experimental points or points taken from the smooth curve. The trouble with this method lies in the fact that there are several measurements involved and only three or four are used in the calculation. The other measurements should be given just as much weight as the ones chosen.

Reduction to a Straight Line.—In a second and better method for evaluating the constants of an empirical equation, some function of the variables is chosen which will yield a straight line. The equation of the straight line is readily obtained (Chap. IV) and from it the unknown constants may be calculated.

A few special examples follow:

(1) $y = a + bx^2$.—In this equation one variable depends on the square of the other variable. It represents a parabolic curve.

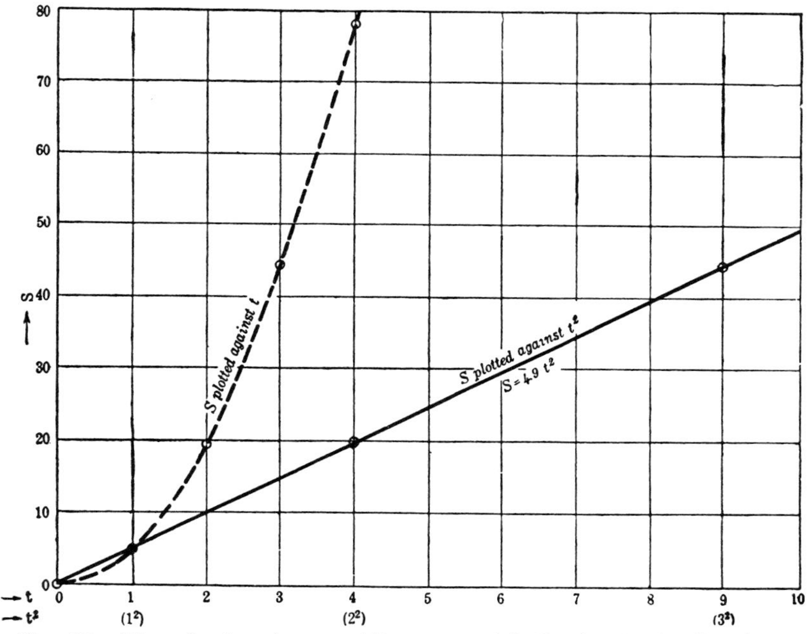

Fig. 59.—The reduction of a curved line to a straight line by plotting S against t^2 instead of against t.

If x^2 is plotted against y, a straight line will result, for if x^2 is set equal to another variable, z for example, the equation becomes $y = a + bz$ which is the equation of the straight line. The constants a and b obtained for this straight line are identical with the constants of the original equation.

The method may be illustrated with the case of a falling stone. It is known that the distance, S, travelled is proportional to the square of the time, t, but it is desired to evaluate the proportionality factor, b, in the equation $S = bt^2$. The experimental data are given below.

TABLE I.—DISTANCE OF A FALLING STONE AT DIFFERENT TIMES

t (seconds)	0	1	2	3	4	5
S (meters)	0	4.9	19.6	44.1	78.4	122.5
z or t^2 (seconds)	0	1	4	9	16	25

On plotting S against t the curved line of Fig. 59 is obtained but on plotting S against z (or t^2) the straight line is obtained. The equation for the straight line $y = bz$, is readily found to be $y = 4.9z$. Since $b = 4.9$ the original equation is $y = 4.9t^2$.

(2) $y = a + bx^n$ where n is known. The constants a and b can be evaluated as in the preceding case by finding a and b for the straight line produced by plotting values of x^n along the X axis and the corresponding values of y along the vertical axis.

(3) $y = x/(a + bx)$. This is an hyperbolic curve. It may be written in the form $\frac{1}{y} = b + \frac{a}{x}$, and it is then evident that a straight line results when $1/y$ is plotted against $1/x$. The values of a and b may be readily obtained from the straight line and substituted into the original equation.

(4) $y = a + bx + cx^2$.—This equation is the general equation for a parabola and it has been discussed before (pp. 48 and 202). Next to the straight line and the semi-logarithm graph it is perhaps the most useful empirical equation in physical chemistry. If $\frac{y_1 - y}{x_1 - x}$ is plotted against x (where x_1, y_1 is any specified point on the curve) a straight line is produced, and from this straight line the constants of the original equation may be determined.

The reason for this statement is evident from the following,

$$y = a + bx + cx^2 \text{ (general equation)}$$
$$y_1 = a + bx_1 + cx_1^2 \text{ (equation at point } x_1, y_1\text{)}$$
$$y_1 - y = b(x_1 - x) + c(x_1^2 - x^2)$$

or

$$\frac{y_1 - y}{x_1 - x} = b + c\frac{(x_1^2 - x^2)}{(x_1 - x)} = b + c(x_1 + x) = (b + cx_1) + cx.$$

Since $(b + cx_1)$ and c are constants the equation is the equation of a straight line, $(b + cx_1)$ being the intercept on the y axis and c being the slope of the straight line. From these two quantities, c and b can be evaluated and a can then be determined, since it remains the only unknown in the equation $y_1 = a + bx_1 + cx_1^2$. The numerical values of the three constants, a, b, and c are then put into the original equation.

The following example illustrates the method where S = the specific heat of ethyl alcohol and t = the temperature.

TABLE II.—THE SPECIFIC HEAT OF ETHYL ALCOHOL

t	S	$(50 - t)$	$(0.6633 - S)$	$\frac{0.6633 - S}{50 - t}$
0	0.50680	50	0.15650	0.003130
10	0.53594	40	0.12736	0.003184
20	0.56617	30	0.09713	0.003238
30	0.58746	20	0.06584	0.003292
40	0.62984	10	0.03346	0.003346
50	0.66330	0	0.00000	

When $(0.6633 - S)/(50 - t)$ is plotted against t, the straight line shown in Fig. 60 is obtained. It is found to have the equation

$$\frac{0.6633 - S}{50 - t} = 0.000{,}0054t + 0.003{,}13$$

$$S = 0.6633 - (50 - t)(0.000{,}054t) - (50 - t)(0.003{,}13)$$
$$S = 0.6633 - 0.000{,}270t + 0.000{,}0054t^2 - 0.1565 + 0.003{,}13t$$
$$S = 0.5068 + 0.002{,}86t + 0.000{,}0054t^2$$

(5) $y = ae^{bx}$.—This is the exponential or semilogarithmic curve so often referred to on preceding pages. Taking logarithms of both sides it becomes

$$\log y = \log a + bx \log e = \log a + 0.4343bx$$

Log y is seen to be a straight-line function of x. By plotting log y against x and finding the equation for the straight line it is possible to evaluate the constants log a and $0.4343b$. These

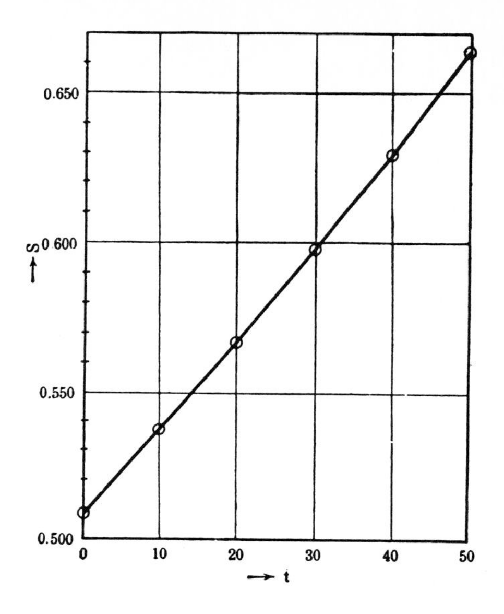

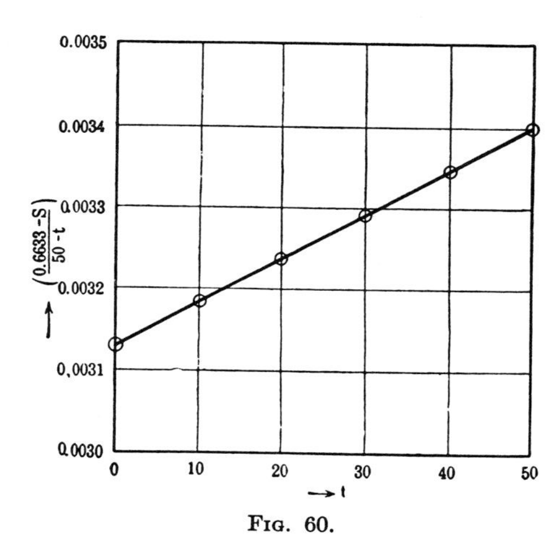

FIG. 60.

In the upper graph the specific heat, S, of ethyl alcohol is plotted against the temperature, t, giving a curve of the type $S = a + bt + ct^2$.

In the lower graph the constants a, b and c are evaluated by plotting $\frac{S_1 - S}{t_1 - t}$ against t, to give a straight line.

constants, or rather the corresponding constants, a and b, are then put into the logarithmic equation or into the original exponential equation. Problems of this kind have been worked out already on pages 61 and 137.

(6) $y = ax^b$.—This equation represents a parabolic, or hyperbolic curve. Taking logarithms of both sides of the equation,

$$\log_{10} y = \log_{10} a + b \log_{10} x$$

$\log_{10} y$ is a straight line function of $\log_{10} x$ and the value of $\log_{10} a$, and b may be found by simply finding the equation for the straight line which results when $\log_{10} y$ is plotted against $\log_{10} x$. The numerical values of a and b are then put into the original equation. An example was given on page 65.

(7) $y = a + bx + cx^2 + dx^3 \cdot \cdot \cdot mx^n$.—When the simpler equation $y = a + bx + cx^2$ fails the introduction of a third term is sometimes necessary, although the labor of computation is increased. The constants may be evaluated by solving a sufficient number of equations simultaneously, or in other ways.

The Method of Averages.—When the best straight line is drawn through a series of discordant data, the experimenter averages all the points in a common sense way, and the equation of his line corresponds to selected points. In a great many cases this procedure is sufficiently accurate and the data do not warrant any further precision. It is true, however, that every experimental point should be given equal weight, and it is possible that the experimenter's judgment as to the best straight line, may be at fault.

Methods have been worked out which eliminate this chance for personal error, the method of averages; and the method of least squares, which is discussed in the following section. Both methods give constants which are based on *all* the points and they may be computed mechanically without plotting the curve and without exercising any judgment in plotting. The method of averages is the easier to calculate, but the method of least squares is used where the highest precision is demanded.

The method of averages gives the equation for the line which makes all the positive and negative errors cancel. When the straight line passes through the origin and ($y = bx$), it is only necessary to add up all the values of y and divide by the sum of all the values of x.

$$b = \frac{\Sigma y}{\Sigma x}$$

This relation follows from the condition that the average line tends to make the algebraic sum of all the deviations from the line along the Y axis zero, by balancing positive errors against negative ones.

$$\Sigma(y - y_{calc}) \text{ or } \Sigma(y - bx) = 0.$$

Rearranging

$$\Sigma y - b\Sigma x = 0 \text{ and } \Sigma y = b\Sigma x \text{ and } b = \frac{\Sigma y}{\Sigma x}$$

A specific example is shown in Table III, where the values of y_{calc} have been calculated from the equation $y = bx$, after evaluating b.

TABLE III.—FINDING AN EQUATION ($y = bx$) BY THE METHOD OF AVERAGES

x	y	$y_{calc} = \frac{\Sigma y}{\Sigma x} x$	$y - y_{calc}$
0	0	0	0
1	21	20.1	+0.9
2.5	53	50.2	+2.8
3	58	60.3	−2.3
5	99	100.5	−1.5
11.5	231		−0.1
$\Sigma x = 11.5$	$\Sigma y = 231$		

$$b = \frac{\Sigma y}{\Sigma x} = \frac{231}{11.5} = 20.1$$

The best line obtained by the method of averages is $y = 20.1x$.

When the straight line does not pass through the origin there are two constants to evaluate (a as well as b) and it is necessary to divide the data into two groups as nearly equal as possible.

Then $\Sigma y = \Sigma a + b\Sigma x$ and $\Sigma' y = \Sigma' a + b\Sigma' x$. The two equations may be solved simultaneously to evaluate the two unknowns, a and b.

A specific example follows

TABLE IV.—FINDING AN EQUATION ($y = a + bx$) BY THE METHOD OF AVERAGES

	x	y	y_{calc}	$y - y_{calc}$
Σ	1	14	15.44	−1.44
	2	21	20.33	+0.67
	3	26	25.22	+0.78
	6	61		
Σ'	4	32	30.11	+1.89
	5	34	35.00	−1.00
	6	39	39.89	−0.89
	15	105		0.01

Dividing into two groups of 3 and 3,

$\Sigma x = 6$	$\Sigma y = 61$	$\Sigma y = \Sigma a + b\Sigma x$	$61 = 3a + 6b$
$\Sigma' x = 15$	$\Sigma' y = 105$	$\Sigma' y = \Sigma' a + b\Sigma' x$	$105 = 3a + 15b$

Solving these two equations simultaneously

$$b = 4.89 \text{ and } a = 10.55.$$

The equation then is $y = 4.89x + 10.55$ and it is seen to express all the data fairly well. The positive and negative deviations balance each other.

Although the method of averages has been applied here only to simple cases it will be realized that the method is perfectly general. It can be used to advantage in the problems of the preceding section for finding the equations of the straight lines which result from the reduction of the other general formula. For example, in the case of the equation $y = a + bx + cx^2$, after tabulating t and $\frac{0.6633 - S}{50 - t}$ in Table II, the best way to find the straight line relation is by the method just illustrated in Table IV. Having evaluated the constants for this straight line (without graphing), they may be used directly in the original equation relating y and x.

The Method of Least Squares.—This method is based on the theorem that the best line is the line which makes the sum of the squares of the deviations a minimum. According to this criterion $\Sigma(y - y_{calc})^2$ is to be a minimum and $\frac{d\Sigma(y - y_{calc})^2}{db} = 0$, where b is the coefficient of x. In the simplest case where $y = bx$,

$$\frac{d\Sigma(y - y_{calc})^2}{db} = \frac{d\Sigma(y - bx)^2}{db}$$

$$\frac{d\Sigma(y - bx)^2}{db} = 2\Sigma(-x)(y - bx) = 2\Sigma(-xy) + 2\Sigma bx^2$$

Setting equal to zero for the conditions of a minimum (p. 102)

$$-\Sigma xy + b\Sigma x^2 = 0$$

$$b = \frac{\Sigma xy}{\Sigma x^2}$$

Applying this equation to the data of Table III, Table V is obtained.

TABLE V.—FINDING AN EQUATION, $y = bx$, BY THE METHOD OF LEAST SQUARES

x	y	xy	x^2	y_{calc}	$(y - y_{calc})^2$
0	0	0	0	0	
1	21	21	1	20	1
2.5	53	132	6.2	50	9
3	58	174	9	60	4
5	99	495	25	100	1
		822	41.2		15

$$b = \frac{\Sigma xy}{\Sigma x^2} = \frac{822}{41.2} = 20.0$$

The best line by the method of least squares is $y = 20x$, and other values of b will make the sum of the squares of the deviations larger than 15.

When the equation is of the form $y = a + bx$, the experimental data are divided into two groups, as described in the method of averages, and the two sets of data give equations which can be solved simultaneously for the two constants, a and b. As pointed out before, most of the empirical equations can be reduced to functions which will give a straight line and the constants of this equation can be determined accurately by the method of least squares. The constants for this new equation may then be applied to the original equation.

A somewhat different application of the method of least squares to a parabolic equation of the type $y = a + bx + cx^2$ is described by Leland.[1] The constants are all summed up, and "normal equations" are obtained by cross multiplication of constants. The normal equations are then solved simultaneously. It is not necessary to first find functions which will give an approximately straight line. For accurate results in mechanical computation the method is probably the best.

Abruptly Changing Curves

If a curve appears to have abrupt changes in slope it is usually better to divide it up into sections and employ different equations for the different parts. The solubility-temperature curve of sodium sulphate is an example. Abrupt change is caused by a change of chemical composition, for the hydrated salt $Na_2SO_4 \cdot 10H_2O$ is stable below 32.38° and the anhydrous form Na_2SO_4 is stable above this transition temperature. It is to be expected that these two different substances will behave differently and will require two different equations.

Even when the break is not abrupt it is often more convenient to apply different equations over different parts of the curve.

Finding Tangents

After obtaining experimental measurements, it is frequently necessary to find the derivative of one variable with respect to the other. For example, the derivative of the heat absorbed with respect to temperature is defined as heat capacity; the derivative of chemical change with time permits a calculation of the velocity constant; and the derivative of the molal volume of a solution with respect to concentration (mol fraction) gives the partial molal volume.

[1] Leland, "Practical Least Squares," pp. 29, 33, and 143. McGraw-Hill Book Company, Inc., New York (1921).

If the relation between the two variables can be expressed by an equation, either theoretical or empirical, it is a simple matter to find the derivative by the rules of differential calculus.

In many cases the relation between the variables is expressed only as a curve, and the easiest way of finding the derivative is by drawing a tangent to the curve and finding the equation of the tangent. For approximate work it is a simple matter to draw a tangent to the curve, but for precision work or for curves of gradual slope the drawing of the true tangent is a difficult problem.

A mechanical device, consisting of a mirror mounted on a square, is a considerable help in finding, graphically, the slope of a curve. When the mirror is exactly at right angles to the line, the line and its image appear to be one continuous unbroken line, while at all other angles the line appears to deviate at the edge of the mirror. When the mirror is in this position a line at right angles to the mirror is parallel to the true tangent and must give the slope of the curve at the point where the mirror intersects the line.

An accurate mirror (a microscope slide silvered on the front, for example) is glued to a steel square or a drawing triangle, and when the line and its image appear unbroken, a line is drawn perpendicular to the mirror. The slope of this line, which is also the slope of the tangent, is obtained from the graph paper. It is well to test the mounting of the mirror by drawing the line a second time after turning the mirror through 180°. If the mirror is mounted at exactly right angles the two lines, so drawn, will coincide. This simple device, called a tangentimeter, is described by Latishaw[1] and used with marked success by Haasche and Patrick[2] in finding the reaction rate of a third-order reaction.

Two ingenious methods of plotting enable one to find the derivative with greater precision. They are applicable in the study of solutions. In the method of intercepts, due to Roozeboon, the intersection of the tangent with the ordinate at 0 or 100 per cent, gives the desired quantity when the compositions of the solution are plotted along the X axis.

[1] Latishaw, *J. Am. Chem. Soc.*, **47,** 793 (1925).

[2] Haasche and Patrick *J. Am. Chem. Soc.*, **47,** 1207 (1925).

In a second method, due to G. N. Lewis, an approximate value of the desired function is calculated arithmetically and deviations from this calculated quantity are plotted as logarithms in such a way as to increase the accuracy greatly. These methods are described by Lewis and Randall.[1]

Graphical Integration

It is frequently necessary to know the area enclosed under a curve when the equation of the curve is not known, or when the expression cannot be readily integrated. Under these conditions, the area can be approximated in several ways.

Counting Squares.—The most obvious way of determining the area is to plot accurately on small squared graph paper and count all the squares, estimating the fractions of squares along the boundary. It is well to block out large rectangles in the area and write down in each its area as obtained by simple multiplication. The remaining small squares are then counted and checked off. In estimating fractions of squares it is a good plan to first count and check off the ones that are approximately half squares, and one that is a little more than half can be balanced against one which is less than half by the same amount. The fractions of squares which have not been checked off are then added up as tenths of squares.

If the squares are small, compared with the whole area, the results are surprisingly accurate, and the labor is not great.

As a check, a rectangle may be drawn covering the whole area and the area outside the curve is determined. The area inside the curve plus the area outside the curve must add up to give the area of the rectangle.

Some graphs are easier to calculate than others. The types which require the finding of the area out to an infinite distance along one of the axes are particularly unsatisfactory. Sometimes this difficulty may be avoided by changing the function to be plotted. Sometimes it is advantageous to plot the difference between the experimental function and some arbitrary function,

[1] Lewis and Randall, "Thermodynamics and the Free Energy of Chemical Substances," pp. 36–41. McGraw-Hill Book Company, Inc., New York (1923).

chosen so as to simplify the extrapolation to infinity. For example, in the study of solutions, Lewis uses a function which approaches zero as infinite dilution is approached. The art of these special graphical methods is described by Lewis and Randall.[1]

Weighing.—If the paper is reasonably uniform in thickness, a very satisfactory method consists in cutting the figure out with scissors or a sharp knife, and weighing it. A rectangle of similar area is also cut out and weighed, and the area of the rectangle is easily obtained by multiplication of the sides. The following simple proportion may be used, where X is the unknown area under the curve.

Area X (area graph): area rectangle:: wt. X: weight rectangle.

Planimeter.—The planimeter is a mechanical device consisting of a lever arm and a wheel, running along the curve. The area covered is read off on the machine. With a good planimeter, excellent results may be obtained on large drawings with little effort. The planimeter and other mechanical devices are described by Lipka.[2]

Exercises

1. A nearly straight line is drawn at random on graph paper with a spline and its equation is determined.

2. A curve is drawn at random with a French curve and its equation is determined.

3. What is the equation for the following data?

x	1	2	3	4	5	6	7
y	1	4	9	16	25	36	49

Ans.: $y = x^2$.

4. What is the equation for the following data?

Pressure	1	2	3	4	5	10	20	50
Volume of gas	22.4	11.2	7.5	5.6	4.5	2.2	1.1	0.4

5. What is the best straight line for the following data?

x	1	2	3	4	5	6	7	8	9	10
y	12.2	14.3	15.9	18.7	20.4	21.7	24.5	26.0	28.2	30.1

[1] Lewis and Randall, "Thermodynamics and the Free Energy of Chemical Substances," pp. 266, 269–275, *et seq.* McGraw-Hill Book Company, Inc., New York (1923).

[2] Lipka, "Graphical and Mechanical Computation," pp. 246–256. John Wiley and Sons, New York (1918).

6. What is the equation for the following experimental data?

Temperature................	32.41	33.31	38.99	39.85	45.61
Specific heat benzene.........	0.4132	0.4152	0.4191	0.4205	0.4296
Temperature (continued)............		46.41	52.25	52.98	58.82
Specific heat benzene (continued)		0.4315	0.4362	0.4414	0.4522

7. What is the equation for the solubility of $AgNO_3$ from the following data:?

Temperature (t).....................	0	10	20	30	40	50
Solubility (grams per 100 grams H_2O) (s).	115	160	215	270	335	400

Ans.: $s = 112.5 + 4.5t + 0.025t^2$.

8. A clover leaf or other leaf is laid flat on graph paper and a line traced around its edge. What is the area of the leaf in square centimeters as determined by (*a*) counting squares, (*b*) cutting out and weighing?

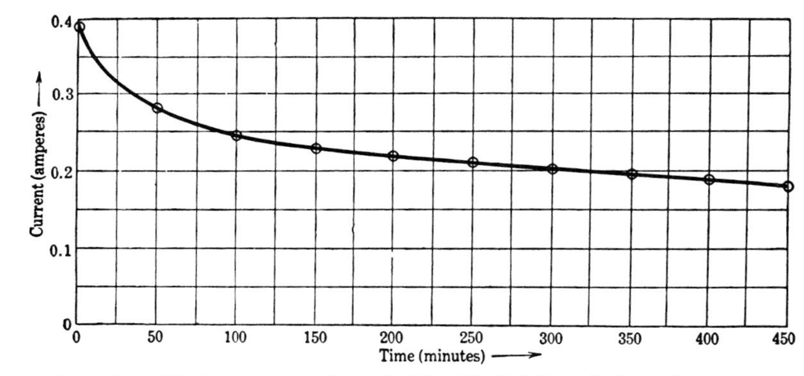

FIG. 61.—Discharge curve of an electrical flash-light cell through a constant resistance. The area under the curve gives the quantity of electricity consumed. The area is obtained by counting squares or by Simpson's rule in Prob. 10.

9. In Fig. 62, the total area represents radiation passed through a certain cell and the blackened area represents the radiation absorbed. What fraction of the total radiation was absorbed?

Ans.: 540/11,450 = 4.71 per cent.

10. When a flashlight cell was discharged through a constant resistance of 4 ohms the following data were obtained:

Time (minutes)........	0	50	100	150	200	250	300	350	400	450
Current (amperes)......	0.39	0.28	0.25	0.23	0.22	0.21	0.20	0.20	0.19	0.18
Voltage (volts).........	1.55	1.13	1.01	0.94	0.89	0.85	0.82	0.80	0.77	0.74

The data are plotted in Fig. 61. What was the ampere-minute capacity (the area under the curve) as determined by (*a*) graphing and counting squares and (*b*) by Simpson's rule?

(*c*) How many grams of zinc were consumed? (By Faraday's law 32.5 g of zinc are equivalent to 26.8 ampere-hours or 96,500 coulombs.)

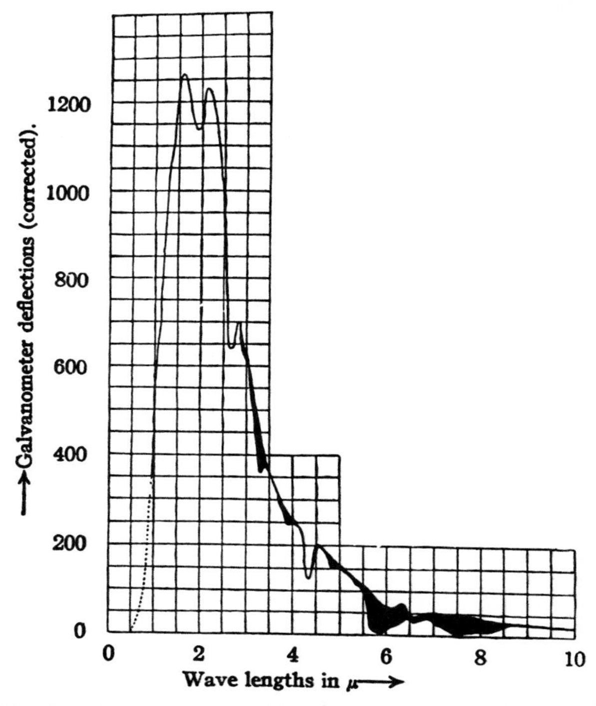

FIG. 62.—The distribution of radiation in the Nernst filament at 1 amp. The shaded areas represent absorption by 8 cm. of nitrogen pentoxide at 51-mm. pressure.[1] (Prob. 9.)

(*d*) How many watt-hours of energy were consumed. (1 watt-hour = 1 volt-ampere-hour.) The voltage may be calculated by multiplying the current by the external resistance of 4 ohms. (*e*) How many joules were consumed? (1 joule = 1 volt-ampere-second.) (*f*) How many calories were consumed? (P. 271.)

[1] DANIELS, *J. Am. Chem. Soc.*, **48**, 609 (1926).

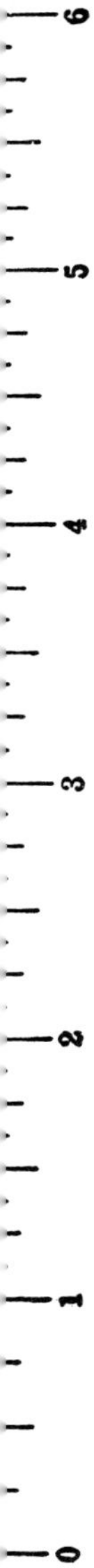

SCALE OF INCHES
FIG. 64.

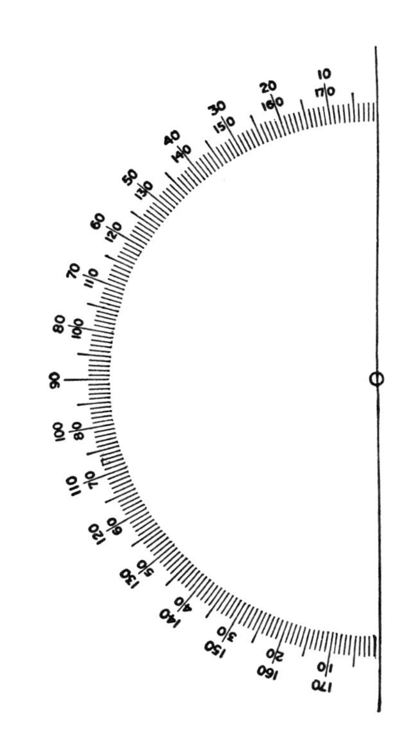

FIG. 63.—The protractor.
(The paper under the graduated semicircle may be cut out with a sharp knife, for measuring angles.)

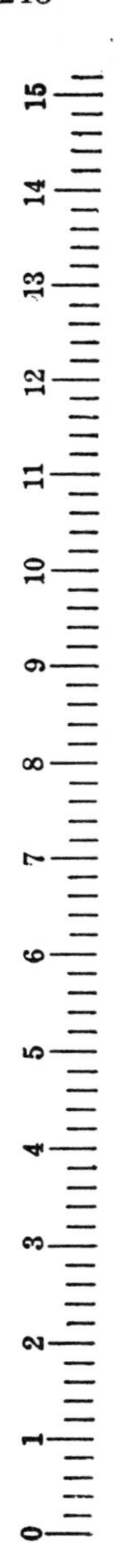

CENTIMETER SCALE

FIG. 65.

APPENDIX I

PROBLEMS IN PHYSICAL CHEMISTRY

In addition to the problems already given in the text the following examples have been taken at random from standard text books on physical chemistry and from recent journal articles. References are given for the benefit of those who are interested in the chemical applications.

1. What is the equilibrium constant, K, of the reaction between potassium cyanide and acetic acid?

$$K = \sqrt{\frac{18 \times 10^{-6}}{7 \times 10^{-10}}} = ?$$

(Millard, "Physical Chemistry for Colleges," 1926, p. 270. McGraw-Hill Book Company, Inc., New York.)

2. The valence factor, w, in the Debye-Hückel theory of electrolytic dissociation is given for a uni-bivalent electrolyte by the following equation:

$$w = \left(\frac{2 \times 1^2 + 1 \times 2^2}{2 + 1}\right)^{3/2}$$

What is the numerical value of w?

(Taylor, "Elementary Physical Chemistry," 1927, p. 447. D. van Nostrand and Co., New York.)

3. The ionization constant of a weak acid is given by

$$K = \frac{x^2}{(1 - x)v}$$

What is the value of x in terms of K and v?

(Getman, "Outlines of Theoretical Chemistry," 1927, p. 503. John Wiley and Sons, New York.)

4. The dissociation constant of hydroiodic acid is given by the expression

$$K = \frac{(a - x)(b - x)}{4x^2}$$

What is the value of x in terms of a, b and K?

(Getman, "Outlines of Theoretical Chemistry," p. 370.)

5. The degree of dissociation, α, of nitrogen tetroxide may be calculated from the following equations,

$$\log K = \log T - \frac{2866.2}{T} + 9.13242$$

$$K = \frac{4(760)\alpha^2}{1 - \alpha^2}$$

What is the value of α when $T = 273 + 41 = 314°$; when $T = 333.9°$?

(McCollum, *J. Am. Chem. Soc.*, **49**, 36, (1927).

6. When 0.8 cc. of nitric oxide was oxidized per second in a certain gas chamber a current of 1.7×10^{-13} amp. was produced. What is the ratio of number of molecules reacting to the number of electrons reaching the electrodes (p. 271)? The temperature was 25° C.

(Brewer and Daniels, *Trans. Am. Electrochem. Soc.*, 266 (1923).

7. In the Debye-Hückel theory of electrolytic dissociation the following equation occurs:

$$j = w\frac{e^2}{6DkT}\sqrt{\frac{4\pi e^2}{DkT}n\Sigma\nu_i}$$

e = charge on the ion = 4.77×10^{-10} electrostatic units.
D = dielectric constant = 88.23 when $T = 273$.
k = gas constant per molecule = 1.371×10^{-16} ergs.
n = number of ions per cc. = $6.06 \times 10^{20}m$, where m is the molal concentration.

Substituting these values how may it be shown that $j = 0.263w\sqrt{\nu m}$ where ν is the total number of ions?

(Taylor, "Elementary Physical Chemistry," p. 446.)

8. According to the equation of Ramsay and Shields

$$\gamma(Mv)^{2/3} = k(t_c - t - 6)$$

γ = surface tension; M = molecular weight, v = specific volume (the reciprocal of the density); k = a constant, 2.12, for unassociated liquids; t = temperature and t_c = the critical temperature Centigrade.

(*a*) What is the critical temperature of carbon disulphide if at 19.4° C. the density is 1.264, the surface tension is 33.58 and the molecular weight is 76?

(*b*) If the density and surface tension are determined at *two* different temperatures it is possible to calculate a value for M, the molecular weight in the liquid state. What is the equation which gives M in terms of γ_1, d_1, t_1, and γ_2, d_2, t_2?

(Getman, "Outlines of Theoretical Chemistry," p. 88.)

9. At 25° the solubility of iodine in CCl_4 is 30.33 g. per liter and in water it is 0.00132 mols per liter. One liter of CCl_4 containing 25 g. of iodine is shaken with 2 liters of water until equilibrium is established. How many grams of iodine are dissolved in the water layer? The solute distributes itself in dilute solutions in the ratios of the solubility.

If the CCl_4 layer is shaken with 5 successive portions of 2 liters each, how many grams of iodine will be left in the water layer?

After ten such extractions how many grams will be left?

(Washburn, "Principles of Physical Chemistry," p. 184 (1921). McGraw-Hill Book Company, Inc., New York.)

10. Hostetter and Roberts give a good example of finding the points of inflection of curves by plotting $\Delta y/\Delta x$ against x and finding the location of the maxima.

(*J. Am. Chem. Soc.*, **41**, 1341, (1919).

11. In the adsorption isotherm

$$\left(\frac{x}{m}\right)^n = kc$$

x is the weight of material adsorbed by the weight, m, of adsorbing material, from a solution whose concentration is c. k and n are constants.

n and k may be evaluated by plotting log (x/m) against log C. Why?

What are the numerical values of k and c calculated from the following data of Freundlich?

ADSORPTION OF ACETIC ACID BY CHARCOAL AT 25° C.

Concentration (mols per liter)	x/m	Concentration	x/m
0.018	0.467	0.268	1.55
0.031	0.624	0.471	2.04
0.062	0.801	0.882	2.48
0.126	1.11	2.79	3.76

(Getman, "Outlines of Theoretical Chemistry," p. 321.)

12. In studying the influence of ionic concentration, c, on the vapor pressure, P, the following data were obtained.

c	0	0.01	0.02	0.04	0.06	0.08
P/c (in arbitrary units).......	40	130	190	260	305	380

These data may be represented by an empirical equation of the type

$$\log_{10} \frac{P}{c} = ac^b + k$$

What are the numerical values of the constants a, b, and k?

(Washburn, "Principles of Physical Chemistry," p. 273.)

13. In the equation of Young

$$T' = T + \frac{144.86}{T^{0.0148T}}$$

T is the boiling point of a straight chain hydrocarbon and T' is the boiling point of the next higher member of the homologous series.

If the boiling point of heptane is 101° C. what is the boiling point of octane?

T, as usual, indicates temperature on the absolute scale.

14. How may the following equation be solved for Q when k is known?

$$\log k = 10.0203 + \log Q - \frac{Q}{4.571}$$

Ans.: graphical interpolation.

(Dushman, *J. Am. Chem. Soc.*, **43**, 397, 1921.)

15. According to the first law of thermodynamics $U = Q - W$ and according to the second law

$$dW = Q\frac{dT}{T}$$

How may these two equations be combined with the elimination of Q? The resulting equation is known as the Gibbs-Helmholtz equation.

16. The heat capacity of a substance varies with temperature as follows:

$$C_p = a + bT + cT^2 + \cdot \cdot$$

The difference between the heat capacities of the products and the reactants in a chemical reaction is given by the expression

$$\Delta C_p = a_0 + b_0 T + c_0 T^2 + \cdot \cdot \cdot$$

where a_0 is the difference of all the a's, b_0 of all the b's and so on.

According to an important theorem of thermochemistry

$$\left(\frac{\partial \Delta H}{\partial T}\right)_p = \Delta C_p$$

where ΔH is the increase in heat content of the reaction.

Substituting

$$\partial \Delta H = a_0 \partial T + b_0 T \partial T + c_0 T^2 \partial T$$

and integrating

$$\Delta H = a_0 T + \tfrac{1}{2} b_0 T^2 + \tfrac{1}{3} c_0 T^3 + \cdot \cdot \cdot + \Delta H_0$$

where ΔH_0 is an integration constant which may be evaluated from the data of an experiment. The steps in integration are to be reviewed.

17. How can it be shown that the two following equations are equivalent?

$$k = Se^{-\frac{Q}{RT}}$$

$$\frac{d \log_e k}{dT} = \frac{Q}{RT^2}$$

k = specific reaction rate of a chemical reaction.
T = absolute temperature.
Q = critical increment.
S and R are constants.

18. The rate of cooling of a hot body is given by the two following equations $\theta = be^{-at}$ or,

$$a = \frac{2.303}{t_1 - t_2} \log_{10} \frac{\theta_2}{\theta_1}$$

where θ_1 = temperature at time t_1.
and θ_2 = temperature at time t_2.
How can the first equation be converted into the second?

19. The pressure, P, of the atmosphere at an altitude h kilometers is given by the expression $P = P_0 e^{-kh}$, P_0 being the pressure at sea level (760 mm.). What is the value of k calculated from the following data?

at 10 kilometers p = 199.3 mm.
at 20.2 kilometers p = 42.2 mm.
at 50.0 kilometers p = 0.32 mm.
An average of the three results may be taken.

20. The weight of water imbibed by gelatine in t minutes is w_t and the maximum weight of water which can be taken up under the conditions is w_∞. Then

$$\frac{dw}{dt} = k(w_\infty - w_t)$$

How may k be evaluated through integration of this equation?

(Getman, "Outlines of Theoretical Chemistry," 1927, p. 313.)

21. Why are the following expressions equivalent?

$$\frac{d \log_e P}{dT}; \frac{dP}{PdT}; \text{ "percentage change in } P \text{ per degree."}$$

P = vapor pressure.

T = absolute temperature.

22. According to the Clapeyron equation for vapor pressure P is related to absolute temperature by the equation,

$$\frac{dP}{dT} = \frac{Q}{TV}$$

Q = molar heat of vaporization.

V = molar volume of vapor.

Assuming $PV = RT$, how does it follow that when log P is plotted against log T, the slope of the line is directly proportional to Q/T. Q/T is the molal entropy of vaporization.

(Wilson and Bahlke, *J. Ind. Eng. Chem.*, **16,** 115 (1924) give numerical examples of this equation.)

23. It can be shown from thermodynamics that

$$\left(\frac{\partial \Delta F}{\partial T}\right)_P = \frac{\Delta F - \Delta H}{T}$$

How does it follow that

$$\frac{d(\Delta F/T)}{dT} = -\frac{\Delta H}{T^2}$$

and that

$$\frac{d(\Delta F/T)}{d(1/T)} = \Delta H$$

T = absolute temperature.

ΔF = increase in free energy.

ΔH = increase in heat constant.

(Lewis and Randall, "Thermodynamics and the Free Energy of Chemical Substances," 1923, p. 173. McGraw-Hill Book Company, Inc., New York.)

24. Given the equation for the influence of temperature on the rate of a chemical reaction

$$\frac{d \log_e k}{dT} = \frac{A}{RT^2}$$

How may this equation be changed to correspond to the equation for a straight line?

$$\log_e k = \frac{-A}{RT} + C$$

If $\log_{10} k$ is plotted against $1/T$ how may the constants A and C be evaluated? A is the critical increment, the excess energy required to put the molecule into a reactive condition. The calculation may be checked with a numerical example given in Taylor's "Elementary Physical Chemistry," on page 167.

25. Given

$$\frac{dlnK}{dT} = \frac{Q}{RT^2} \text{ and } K = \frac{k_1}{k_2}$$

How can this equation be broken up into two equations

$$\frac{dlnk_1}{dT} = \frac{A_1}{RT^2} + B_1 \text{ and } \frac{dlnk_2}{dT} = \frac{A_2}{RT^2} + B_2$$

B is independent of T. A_1 and A_2 are such quantities that their difference is the heat of reaction, Q. ln is frequently used for $\log_e$.

26. The equilibrium constant K, temperature T and heat of reaction Q, are related by the following equation

$$\frac{dlnK}{dT} = \frac{Q}{RT^2}$$

How may this reaction be transformed to give

$$\frac{dlnK}{d1/T} = \frac{Q}{R}.$$

If $\log_{10} K$ is plotted against $1/T$ all points fall on a straight line. The slope of this line multiplied by $2.303R$ gives Q the heat of reaction.

27. Ethyl alcohol boils at 76.1° under a pressure of 700 mm. and at 72.4° under a pressure of 600 mm. What is the heat of vaporization calculated by the equation

$$\log_{10} \frac{P_2}{P_1} = \frac{Q}{2.303R} \frac{(T_2 - T_1)}{T_2T_1}$$

Using the calculated value of Q and one of the pressures, what is the boiling point at 760 mm.?

28. In the derivation of the reaction isochore of van't Hoff

(a) If $W = RT \log_e K$, $dW = ?$

(b) If $\frac{d[\log_e (RT)^{\Sigma n - \Sigma n'} K]}{dT} = \frac{Q}{RT^2}$, $\frac{d \log_e K}{dT} = ?$

(Getman, "Outlines of Theoretical Chemistry," p. 377.)

29. How many calories of heat are required to raise a mol of steam from

$$T = 373 \text{ to } T = 423?$$

$$\text{Heat absorbed} = \int_{373}^{423} (8.81 - 0.0019T + 0.000,0022T^2)dT$$

(Millard, "Physical Chemistry for Colleges," 224.)

30. The increase in entropy, S, of a gas when heated at constant pressure is given by the equation

$$dS = \frac{C_V dT + PdV}{T}$$

How can it be shown that if $C_V = \frac{3}{2}R$ for monatomic gas

$$S = \frac{3}{2}RlnT + RlnV + C$$

The integration constant C may be evaluated with experimental data on the entropy of monatomic helium.

(Lewis and Randall, "Thermodynamics and the Free Energy of Chemical Substances," 1923, p. 446. McGraw-Hill Book Company, Inc., New York.)

31. For the infra red absorption of chloroform at a wave length of 3.3μ

$$\frac{I}{I_0} = \frac{1}{3.50s}(1 - e^{-3.50s})$$

What are the numerical values of I/I_0 when $s = 1$; $s = 5$; and $s = 10$? I/I_0 is the percentage transmission of light and s is the thickness.

(Bennett and Daniels, *J. Am. Chem. Soc.*, **49**, 54, 1927.)

32. In a termolecular reaction with special initial concentrations the following equation occurs

$$\frac{dx}{dt} = k\,(a - x)^2$$

What is the value of k obtained by integration?

(Taylor, "Elementary Physical Chemistry," p. 173.)

33. The velocity of a termolecular reaction is represented by the differential equation

$$\frac{dx}{dt} = k(a - x)(b - x)(c - x)$$

What is the integrated expression?

(Getman, "Outlines of Theoretical Chemistry," p. 432.)

34. The fundamental theorem of the kinetic theory of gases is $pv = \frac{1}{3}mnu^2$.

Why does it follow that $\frac{d(PV)}{dT}$ = a constant and that $\frac{d^2(PV)}{dT^2} = 0$?

(Washburn, "Principles of Physical Chemistry," p. 29.)

35. What statements can be made concerning the behavior of perfect gases deduced from the differential equation

$$\frac{d^2(PV)}{dT^2} = 0$$

T = absolute temperature, P = pressure, V = volume.

(Washburn, "Principles of Physical Chemistry," p. 33.)

36. According to van der Waal's equation

$$P = \frac{RT}{V - b} - \frac{a}{V^2} \text{ where } a \text{ and } b \text{ are constants.}$$

What is the value of $\int VdP$ where dP is obtained by differentiating the van der Waal equation?

37. According to van der Waal's equation for a gas

$$\left(P + \frac{a}{V^2}\right)(V - b) = RT$$

Where are the maximum and minima in this equation located?

38. The equilibrium constant, K, for the decomposition of nitrogen tetroxide is given by the equation

$$\log K = \log T - \frac{2866.2}{T} + 9.13242. \quad \frac{d \log K}{dT} = ?$$

also

$$\frac{d \log K}{dT} = \frac{\Delta H}{(1.99)(2.303)T^2}. \quad \Delta H = ?$$

also

$$K = \frac{4(760)\alpha^2}{1 - \alpha^2}. \quad \frac{d \log K}{d\alpha} = ?$$

$$\frac{d\alpha}{dT} = \frac{\dfrac{d \log K}{dT}}{\dfrac{d \log K}{d\alpha}} = ?$$

What is the numerical value of K and of α when $T = 300°$? (McCollum, *J. Am. Chem. Soc.*, **49**, 36, 1927.)

39. The following equation gives the equilibrium constant for the decomposition of nitrogen tetroxide, according to Bodenstein.

$$\log K = \frac{2891}{T} + 1.75 \log T + 0.0046T - 8.92 \times 10^{-6}T^2 + 6.8148$$

$$\frac{d \log K}{dT} = ?$$

The equilibrium constant expressed in millimeters is given as a function of the degree of dissociation, α, by the following equation

$$K = \frac{(760)(4)\alpha^2}{1 - \alpha^2}, \quad \frac{d \log K}{d\alpha} = ?$$

$$\frac{d\alpha}{dT} = \frac{\dfrac{d \log K}{dT}}{\dfrac{d \log K}{d\alpha}} = ?$$

(Bodenstein, *Z. physik. Chem.*, **100**, 78, 1922.)

40. An approximate form of Maxwell's distribution law is

$$dn = \frac{2n}{a^2} e^{-c^2/a^2} c\,dc$$

m = mass of molecule.
n = the number of molecules.
c = translational velocity of the molecules.
a = the most probable velocity.

How may this equation be changed to read

$$dn = \frac{n}{RT} e^{-E/RT} dE$$

by substituting $a^2 = 2RT/mN$ and $E = \frac{1}{2}Nmc^2$?

This expression is integrated between ∞ and E to determine the number of molecules, n_E, which have a velocity in excess of that which corresponds to the kinetic energy E per N molecules.

(Taylor, "Elementary Physical Chemistry," p. 88.)

41. What fraction of the total number of collisions in nitrous oxide is effective in causing decomposition at 1,000° K., calculated from the following data?

Total collisions = 2.09×10^{28} per cubic centimeter per second.

Effective collisions = $\frac{1}{2}(3.58) \times 10^{16}$ per cubic centimeter per second.

If this calculated fraction is equal to $e^{-E/RT}$, what is the value of E if $R = 2$ calories and $T = 1{,}000$?

(Taylor, "Elementary Physical Chemistry," p. 169.)

42. In the adiabatic expansion of a perfect gas the following equation occurs

$$d \log_e P + \frac{Cp}{Cv} d \log_e V = 0$$

What does this equation give on integration?

43. The influence of temperature T on a chemical equilibrium is given by the following equation:

$$\frac{d \log_e K_p}{dT} = \frac{\Delta H}{RT^2}$$

ΔH = heat of reaction. R = gas constant.

If ΔH is not a constant but a function of temperature given by the equation

$$\Delta H = \Delta H_0 + aT + bT^2 + cT^3 + \cdots$$

what is the value of $\log_e K_p$?

44. The relationship between heat capacities is given by the equation

$$C_P - C_V = \frac{-T(\partial V/\partial T)_P^2}{(\partial V/\partial P)_T}$$

How may this equation be changed to read $C_P - C_V = \alpha^2 VT/\beta$, where α is the coefficient of thermal expansion and β is the coefficient of compressibility? This equation is important because it enables one to calculate heat capacity at constant volume from measurements made at constant pressure. The constant pressure measurements are very much easier to make, experimentally, than the measurements at constant volume.

(Lewis and Randall, "Thermodynamics and the Free Energy of Chemical Substances," p. 136.)

45. An important relation in chemical thermodynamics is given by the equation $\Delta F = \Delta H - T\Delta S$, where F is free energy (of G. N. Lewis), H is heat content and S is entropy. Δ signifies a finite increase in one of these properties. How can it be shown that

$$\frac{\partial(\Delta F)}{\partial T} = \frac{\partial(\Delta H)}{\partial T} - \Delta S - T\frac{\partial(\Delta S)}{\partial T}?$$

The first and third terms of the right hand side of this equation are each equal to ΔC_P so that the equation can be written $(\partial \Delta F/\partial T)_P = -\Delta S$.

What is the meaning in words of this equation?

(Lewis and Randall, "Thermodynamics and the Free Energy of Chemical Substances," p. 172.)

46. The following thermodynamical equations are to be expressed in words:

$$(\partial E)_P = C_P - P\left(\frac{\partial V}{\partial T}\right)_P$$

$$(\partial E)_T = T\left(\frac{\partial V}{\partial T}\right)_P + P\left(\frac{\partial V}{\partial P}\right)_T$$

$$(\partial E)_V = C_P\left(\frac{\partial V}{\partial P}\right)_T + T\left(\frac{\partial V}{\partial T}\right)_P^2$$

$$(\partial V)_E = -C_P\left(\frac{\partial V}{\partial P}\right)_T - T\left(\frac{\partial V}{\partial T}\right)_P^2$$

E = internal energy, T = absolute temperature, V = volume,
P = pressure, C_P = heat capacity at constant pressure.

(Lewis and Randall, "Thermodynamics and the Free Energy of Chemical Substances," p. 164.)

47. Given the following relations between osmotic pressure, P, vapor pressure, p, and mol fraction x and volume V and v of the liquid and vapor,

$$\left(\frac{\partial p}{\partial x}\right) = k. \quad \left(\frac{\partial p}{\partial P}\right) = -\frac{V}{v}. \quad pv = RT.$$

How can it be shown that

$$dP = -\frac{RT}{V}\, d \log_e x$$

(Washburn, "Principles of Physical Chemistry," p. 184.)

48. Given the following equation which is concerned with the change in the boiling point elevation with the change in concentration

$$d(\Delta T)_B = \frac{-R(T_{B_0} + \Delta T_B)_2}{L_v}\, d \log_e x$$

Using Maclaurin's theorem how can the integral of this equation be expressed as a power series in x?

If x is small, what simplification can be effected?

(Washburn, "Principles of Physical Chemistry," p. 200.)

49. What is the value of $\log_e x$ obtained by integrating the equation

$$d(\Delta T_F) = \frac{-R(T_{F_0} - \Delta T_F)^2}{L_{F_0} - \Delta C_P(\Delta T_F)}\, d \log_e x$$

ΔT_F = freezing point depression produced by a solute.
ΔC_P = difference in the molal heat capacity of the solvent as a liquid and a solid.
T_{F_0} = melting point.
L_{F_0} = molal heat of fusion.

(Washburn, "Principles of Physical Chemistry," p. 203.)

50. How may the equation

$$P = \frac{-RT}{V_0} \log_e \frac{p}{p_0}$$

be changed into the equation

$$P = \frac{RT}{V_o}\left[\frac{(p_0 - p)}{p_0} + \frac{1}{2}\frac{(p_0 - p)^2}{p_0} + \frac{1}{3}\frac{(p_0 - p)^3}{p_0} + \cdots\right]$$

(Washburn, "Principles of Physical Chemistry," p. 163.)

51. The following equation, which is used frequently in derivations in physical chemistry is to be developed and checked with definite numerical values of x,

$$e^{-x} = 1 - e^x$$

52. The relation between osmotic pressure, and osmotic pressure at equilibrium is given by the equation

$$dP = \frac{RT}{V_o} \times \frac{dx}{1 - x}$$

How may this be integrated to give

$$P = \frac{RT}{V_o}[-\log_e (1 - x)]$$

and then expanded to give

$$P = \frac{RT}{V_0}\left(x + \frac{1}{2}x^2 + \frac{1}{3}x^3 + \cdots\right)$$

(Findlay, "Osmotic Pressure," Longmans, Green and Co., London, 1919, p. 61.)

53. The following data are taken from Lewis and Randall's "Thermodynamics and the Free Energy of Chemical Substances." Problem 1, page 202 (1923).

p...	0	100	200	300	400	500	600	700	800	900	1,000
α...	(15.47)	15.47	15.47	15.60	15.85	15.97	16.10	16.14	16.17	16.15	16.13

The area under the curve or $\int_{15.47}^{16.13} p d\alpha$ may be calculated with the help of Simpson's rule.

The following ten exercises are taken from a recent number of the *Journal of the American Chemical Society*. This Journal has been given as an argument for the fact that chemists must understand mathematics.

(Taylor, "Elementary Theoretical Chemistry," preface (1927).)

54. Data on the vapor pressure of oxygen and nitrogen which can be fitted with an equation of the type

$$X = a + bx + cx^2$$

are given by Dodge and Dunbar, *J. Am. Chem. Soc.*, **49**, 604, 1927.

The constants of the equation are determined by the method of averages.

55. According to Roault's law

$$P = p_a x_1 + p_b(1 - x_1)$$

P = total pressure. p_a = vapor pressure of pure A.
p_b = vapor pressure of pure B.
x_1 = mol fraction A in liquid.

Assuming x_1 is constant,

$$\frac{dP}{dt} = ?$$

(Dodge and Dunbar, *J. Am. Chem. Soc.*, **49,** 603, 1927.)

56. Data for calibrating a thermocouple at low temperatures are given in *J. Am. Chem. Soc.*, **49,** 615, 1927 by Dodge Davis.

The data may be expressed by either of the two following equations:

$$E = -131.8046t + 18331.8(1 - e^{-0.003t})$$
$$E = 0.01282t^{2.369} - 74.564t$$

The constants of the equation are to be evaluated from the experimental data given in Table I of the reference.

57. Data on the vapor pressure of oxygen at different temperatures may be fitted by the equation

$$\log P = \left(\frac{A}{T}\right) + B + CT + DT^2$$

The numerical values of A, B, C, and D are to be evaluated and checked with the values obtained by Dodge and Davis. (*J. Am. Chem. Soc.*, **49,** 616, 1927.)

58. A good method is described by Randall and Scott for interpolating conductance data with the help of the equation

$$\log m = a \log k + \log b$$

(*J. Am. Chem. Soc.*, **49,** 645, 1927.)

59. In determinations of the activity of electrolytes from freezing point data the following expression must be evaluated graphically

$$\int_o^m \frac{j}{(m^{1/2})} dm^{1/2}$$

Graphs to which this calculation may be applied may be found in an article by Randall and Scott. (*J. Am. Chem. Soc.*, **49,** p. 651–3, 1927.)

60. Data for the dissociation constant, K, of bromine may be expressed by the equation

$$\log K_p = \frac{-9{,}000}{T} + 3.695$$

(DeVries and Rodebush, *J. Am. Chem. Soc.*, **49,** 665, 1927.)

61. Data on the compressibility of nitrogen and hydrogen can be expressed as an empirical equation

$$\frac{PV}{P_0V_0} = A + BP + CP^2 + DP^3$$

The method of least squares is used in evaluating the constants from the experimental data.

(Bartlett, *J. Am. Chem. Soc.*, **49**, 699, 1927.)

62. The equation

$$\left(\frac{r - 2x}{r - 2x + 2}\right)^2 \left(\frac{1 - x}{r - 2x + 2}\right) = \frac{Kx}{r - 2x + 2}$$

occurs in a research on the activity of molten lead chloride.

It is solved by assigning a value to K and a series of values to x and solving each for r.

(Hildebrand and Ruhle, *J. Am. Chem. Soc.*, 727, 1927.)

63. How is the equation $R_z = R_1 Z^{-\frac{2}{n-1}}$ obtained from the following equations?

$$\left(\frac{d\phi}{dR}\right)_{R_z} = a\frac{(Z^2e^2)}{R_z^2} - \frac{nB}{R_z^{n+1}} = 0$$

$$\phi = -\frac{ae^2Z^2}{R} + \frac{B}{R^n}$$

and

$$R_1^{n-1} = \frac{(nB)}{(ae^2)}$$

(Pauling, *J. Am. Chem. Soc.*, **49**, 770, 1927.)

A wealth of excellent mathematical applications may be found in Tolman's, "Statistical Mechanics with Applications to Physics and Chemistry." American Chemical Society Monograph. Chemical Catalog Co., New York (1927).

APPENDIX II

DEFINITIONS OF ADVANCED TERMS

The following terms, for which brief definitions are given, cannot be discussed in the pages of this book. For definitions of material discussed in the book the index should be consulted.

Angular Velocity: The velocity of a particle expressed in the angle covered per unit of time.

Bernoulli's Equation: A differential equation of the type $\frac{dy}{dx} + Py = Qy^n$, which may be transformed into a linear differential equation for solution.

Cardioid: A heart shaped curve corresponding to the equation $x^2 + y^2 + ax = a\sqrt{x^2 + y^2}$. A cardioid condenser is used in microscopy.

Catenary: A U-shaped curve corresponding to the equation

$$y = \frac{a}{2}(e^{\frac{x}{a}} + e^{-\frac{x}{a}})$$

Cissoid: A pointed curve corresponding to the equation

$$y^2(2a - x) = x^3.$$

Curvature: Rate of change of the slope of a curve with distance.

Cusp: A point where two branches of a curve come together with a common tangent and stop abruptly.

Determinant: A conventional arrangement of numbers in horizontal rows and vertical columns indicating diagonal multiplication. It is useful in solving simultaneous equations.

Double Integration: $\int\int f(x, y)dx\ dy$. Sometimes called a surface integral. An expression which is to be integrated for x, holding y constant; and for y holding x constant. The order is immaterial. It is useful in finding areas by integration.

Elliptic Integrals: $K = \int_0^{\pi/2} \frac{dz}{\sqrt{1 - k^2 \sin^2 z}}$

$$E = \int_0^{\pi} \sqrt{1 - k^2 \sin^2 z}dz$$

Certain physical problems involving harmonic motions can be expressed in terms of these integrals. Tables giving numerical values of the integrals have been compiled.

Explicit Function: When an equation in x and y is solved for y, y is an explicit function of x, as $y = x^2/2$. (Comparison with implicit function.)

Gamma Function: $\Gamma_{(n)} = \int_0^\infty e^{-x}x^{n-1}dx$. Certain physical problems can be expressed in terms of $\Gamma_{(n)}$ and solved by reference to tables which give numerical values of $\Gamma_{(n)}$.

Green's Theorem: An analytical expression of the fact that the accumulation of any substance within a given region is the excess of what passes inward through the boundaries, over that which passes outward.

Horner's Solution of Equation: A practical method based on trial division, for evaluating the roots of an equation to any desired degree of accuracy.

Hyperbolic Cosine: $\cosh x = \frac{1}{2}(e^x + e^{-x}) = 1 + \frac{x^2}{\underline{/2}} + \frac{x^4}{\underline{/4}} + \cdots$ Useful in certain integrations. $\cosh^{-1}$ = arc hyperbolic cosine.

Hyperbolic Sine: $\sinh x = \frac{1}{2}(e^x - e^{-x}) = x + \frac{x^3}{\underline{/3}} + \frac{x^5}{\underline{/5}} \cdots$ Useful in certain integrations. $\sinh^{-1}$ = arc hyperbolic sine.

Hyperbolic Tangent: $\tanh x = \frac{e^x - e^{-x}}{e^x + e^{-x}}$. $\tanh^{-1}$ = arc hyperbolic tangent.

Identical Equation: An equation which is always true no matter what values of the variable are taken, as for example $(x + 2)^2 = x^2 + 4x + 4$.

Implicit Function: When an equation in x and y is *not* solved for y, y is an implicit function of x, as $x^2 + 2y = 0$. (Comparison with explicit function.)

Inferior Limit: Lower limit in definite integration (p. 123).

Inverse Sine: Arc sine (p. 145).

Isobar: A condition of constant pressure

Isochore: A condition of constant volume.

Isentropic: A condition of constant entropy.

Isotherm: A condition of constant temperature.

Lagrange's Theorem: If $z = y + xF(z)$ where x and y may vary independently,

$$f(z) = f(y) + \frac{df(y)}{dy}F(y)\frac{x}{1} + \frac{d}{dy}\left[\frac{df(y)}{dy}(F(y))^2\right]\frac{x^2}{\underline{/2}} + \cdots$$

Matrix: A rectangular table composed of mn numbers arranged in m rows with n numbers in each row.

Matrix Multiplication: A process of multiplying matrices. Useful in interpreting certain "forbidden" combinations of quantum numbers in calculations of spectroscopy.

Maxwell's Distribution Law: An expression showing how the velocities of molecules are distributed in a mass of molecules. Essentially a probability curve.

Newton's Approximation Formula: A method, based on Taylor's series, for the approximate solution of numerical equations.

Newton's Interpolation Formula: A means of calculating intermediate values in a series of measurements.

Node: The point of intersection of two branches of a curve.

Partial Differential Equation: An equation containing a derivative and more than one independent variable.

Polar Coordinates: A system of locating points by giving (1) the distance from the origin, the radius vector, called r, and (2) the angle between the line and the initial line, the vectorial angle, called Θ.

Primitive: An equation from which a differential equation has been obtained by the elimination of constants.

Sturm's Theorem: A method for determining the number of real roots of an equation, and their approximate values. Infinity is substituted for x and the change of sign of the equation is noted.

Subtangent: The part of the X axis lying between the point of intersection of the tangent and the ordinate drawn from the given point.

Superior Limit: Upper limit of a definite integral (p. 123).

Triple Integral: $\int\int\int f(x, y, z)dx\ dy\ dz$. An integral involving three independent variables. Sometimes called the volume integral because useful in integrating in three dimensions when the equations are known. Three successive integrations are carried out (irrespective of order) and in each integration the other two quantities are kept constant.

Transcendental Function: A logarithmic or trigonometrical or other function as distinguished from an algebraic function.

Vector: A means of locating a point by specifying the direction and distance from a reference point.

Vector Analysis: Mathematical analysis making use of vectors and polar coordinates.

APPENDIX III

BIBLIOGRAPHY

Reference Books for Further Study

Algebra

Bocher, M.: Introduction to Advanced Algebra. The Macmillan Co., New York 1924.

Chrystal, G.: Algebra. 2 vols. A. and C. Black, London, 1906.

Weber, H.: Lehrbuch der Algebra. 3 vols. F. Vieweg und Sohn, Braunschweig, 1908.

Analytical Geometry

Slichter: Mathematical Analysis. McGraw-Hill Book Company, Inc., New York, 1921.

Goursat and Hedrick: A Course in Mathematical Analysis. Ginn & Co., Boston, 1917.

Calculus

Granville: Differential and Integral Calculus. Ginn & Co., Boston, 1911.

Wilson: Advanced Calculus. Ginn & Co., New York, 1912.

Osgood: Advanced Calculus. The MacMillan Co., New York, 1925.

Thompson: Calculus Made Easy. The MacMillan Co., London, 1919.

Chemical Applications of Mathematics

[1] Mellor: Higher Mathematics for Students of Physics and Chemistry. Longmans, Green, & Co., New York, 1919.

Nernst and Schönflies: Einführung in die Mathematische Behandlung der Naturwissenschaften. E. Wolff, Leipzig, 1895.

Hitchcock and Robinson: Differential Equations in Applied Chemistry. John Wiley and Sons, New York, 1923.

Michaelis: Einführung in Die Mathematik für Biologen und Chemiker. Julius Springer, Berlin, 1922.

Partington: Higher Mathematics for Chemical Students. D. Van Nostrand Company, New York, 1912.

[1] For a single reference book in mathematics, Mellor's "Higher Mathematics for Students of Physics and Chemistry," 641 pages, is unquestionably the best for chemists.

Differential Equations

FORSYTH: A Treatise on Differential Equations. University Press, Cambridge, 1906.

MURRAY: Introductory Course in Differential Equations. Longmans, Green and Co., New York, 1925.

COHEN: An Elementary Treatise on Differential Equations. D. C. Heath and Co., Boston, 1906.

Graphical Calculations

LIPKA: Graphical and Mechanical Computation. John Wiley and Sons, New York, 1921.

LELAND: Practical Least Squares. McGraw-Hill Book Company, Inc., New York, 1921.

RUNGE: Graphical Methods. Columbia University Press, New York, 1921.

RUNNING: Empirical Formulas. John Wiley and Sons, New York, 1917.

D'OCAGNE: Traité de nomagraphie. Gauthier-Villars, Paris, 1921.

Probability

COOLIDGE: An Introduction to Mathematical Probability. Oxford University Press, Oxford, 1925.

LAPLACE: A Philosophical Essay on Probabilities. John Wiley and Sons, New York, 1917.

POINCARÉ: Calcul des Probabilites. Gauthier-Villars, Paris, 1912.

WOODWARD: Probability and Theory of Errors. John Wiley and Sons, New York, 1906.

WRIGHT and HAYFORD: Adjustment of Observations. D. Van Nostrand Co., New York, 1907.

Miscellaneous

WHITTAKER and ROBINSON: A Short Course in Interpolation. D. Van Nostrand Co., New York, 1923.

LEWIS and RANDALL: Thermodynamics and the Free Energies of Chemical Substances. McGraw-Hill Book Company, Inc., New York, 1923.

RUNGE: Vector Analysis. Methuen and Co., London, 1923.

BRIDGMAN: "A Condensed Summary of Thermodynamic Formulas" Harvard University Press, Cambridge, 1925.

APPENDIX IV

THEOREMS OF ELEMENTARY MATHEMATICS

Algebra

$$(x + y)^2 = x^2 + 2xy + y^2$$
$$(x - y)^2 = x^2 - 2xy + y^2$$
$$(x \pm y)^3 = x^3 \pm 3x^2y + 3xy^2 \pm y^3$$
$$(x \pm y)^n = x^n \pm nx^{n-1}y + \frac{n(n-1)}{\underline{|2}}x^{n-2}y^2 \pm \frac{n(n-1)(n-2)}{\underline{|3}}x^{n-3}y^3 + \cdots$$
$$(x - y)^4 = x^4 - 4x^3y + \frac{4 \times 3}{1 \times 2}x^2y^2 - \frac{4 \times 3 \times 2}{1 \times 2 \times 3}xy^3 + y^4$$
$$\underline{|n} = 1 \times 2 \times 3 \times 4 \times \cdots (n-1) \times n$$
$$(x + y + z)^2 = x^2 + y^2 + z^2 + 2xy + 2xz + 2yz$$
$$(x + a)(x + b) = x^2 + (a + b)x + ab$$
$$(ax + b)(cx + d) = acx^2 + (bc + ad)x + bd$$
$$(a + b + c + \cdots)(x + y + z + \cdots) = ax + bx + cx + ay + by + cy + az + bz + cz$$
$$(x + y)(x - y) = x^2 - y^2$$

Factoring

Factoring is the process of finding two or more expressions, which will give back the original expression on multiplication.

$$x^2 - y^2 = (x + y)(x - y)$$
$$3x^3 + 6x^2y + 3xy^2 = 3x(x^2 + 2xy + y^2) = 3x(x + y)(x + y)$$
$$x^2 - x - 6 = (x - 3)(x + 2)$$

Exponents

$$a^x = a \times a \times a \times a \cdots \text{ to } x \text{ factors}$$
$$2^4 = 2 \times 2 \times 2 \times 2 = 16$$
$$a^{-x} = \frac{1}{a^x}$$
$$2^{-4} = \frac{1}{2^4} = \frac{1}{16}$$
$$a^{1/x} = \sqrt[x]{a}$$

$$9^{1/2} = \sqrt{9} = 3$$

$$a^x \times a^y = a^{x+y}$$

$$2^4 \times 2^2 = 2^6 = 64. \quad (2^4 \times 2^2 = 16 \times 4 = 64)$$

$$\frac{a^x}{a^y} = a^{x-y}$$

$$2^4 \div 2^2 = 2^{4-2} = 2^2 = 4. \quad (2^4 \div 2^2 = 16 \div 4 = 4)$$

$$(a^x)^y = a^{xy}$$

$$(2^4)^2 = 2^8 = 256. \quad ((2^4)^2 = 16^2 = 256)$$

$$a^\circ = 1$$

$$a^{x/y} = \sqrt[y]{a^x}$$

$$2^{2/3} = \sqrt[3]{2^2} = \sqrt[3]{4}$$

$$\sqrt{ab^2} = b\sqrt{a}$$

$$\sqrt{45} = \sqrt{5 \times 9} = \sqrt{5 \times 3^2} = 3\sqrt{5}$$

Proportion

If $a/b = c/d$, $bc = ad$

$$2/3 = 6/9. \quad 3 \times 6 = 2 \times 9. \quad 18 = 18$$

Solving Quadratic Equations

By formula,

$$\text{if } ax^2 + bx + c = 0$$

$$x = \frac{-b \pm \sqrt{b^2 - 4ac}}{2a}$$

when $b^2 - 4ac$ is negative the roots are imaginary

$$2x^2 + 5x - 40 = 0$$

$$x = \frac{-5 \pm \sqrt{25 + 320}}{4} = \frac{-5 \pm 18.57}{4} = -5.89 \text{ or } 3.39$$

By completing the square,

The equation $ax^2 + bx + c = 0$ is first divided by a, giving $x^2 + b'x + c' = 0$. Then c' is placed on the right hand side of the equation and the square of $\frac{1}{2}$ of b' is added to each side. This procedure gives a perfect square on the left. The square root of each side is taken.

$$x^2 - 12x = 13$$

$$x^2 - 12x + 36 = 13 + 36 = 49$$

$$x - 6 = \pm 7$$

$$x = \pm 7 + 6 = 13 \text{ or } -1$$

By factoring,

After factoring, a value of x, which makes one of the terms zero, causes the whole expression to become zero. Such a value is a root

$$x^2 - x + 2 = 0$$

$$(x + 1)(x - 2) = 0$$

When $x = 2$, $(x - 2) = 0$, and when $x = -1$, $(x + 1) = 0$. Hence $+2$ and -1 are the roots of the equation.

Solving Simultaneous Equations (Two or More Unknowns)

By addition or subtraction,

$$x - y = -2$$
$$3x + 2y = -1$$

$$3x - 3y = -6 \text{ (multiplying by 3)}$$
$$3x + 2y = -1$$

$$-5y = -5 \text{ (subtracting)}$$
$$y = 1 \text{ and } x - 1 = -2 \text{ or } x = -1.$$

By substitution,

$$x - y = -2$$
$$3x + 2y = -1$$
$$x = y - 2 \text{ (transposing } y\text{)}$$
$$3(y - 2) + 2y = -1 \text{ (substituting for } x\text{)}$$
$$3y - 6 + 2y = -1$$
$$5y = 6 - 1 = 5$$
$$y = 1 \text{ and } x = 1 - 2 = -1$$

By comparison,

$$x - y = -2$$
$$3x + 2y = -1$$
$$x = y - 2 \text{ (transposing first equation)}$$
$$x = \frac{-2y - 1}{3} \text{ (transposing second equation)}$$
$$y - 2 = \frac{-2y - 1}{3} \text{ (equating)}$$
$$3y - 6 = -2y - 1$$
$$5y = 5$$
$$y = 1 \text{ and } x = y - 2 = 1 - 2 = -1$$

Approximations

If x and y are very small quantities

$$(1 \pm x)^m = 1 \pm mx, \text{ approximately}$$
$$(1 \pm x)^m(1 \pm y)^n = 1 \pm mx \pm ny, \text{ approximately}$$

If n and m are nearly equal

$$mn = \frac{m^2 + n^2}{2}, \text{ approximately}$$

If θ is a very small angle expressed in radians,

$$\sin \theta = \theta, \cos \theta = 1, \text{ and } \tan \theta = 0, \text{ approximately}$$

Progressions

An arithmetical progression is a series of terms, each term of which is obtained from the preceding term by addition, or subtraction. When a

is the first term, l the last term, n the number of terms, and s the sum of the n terms and d the common difference, the following generalizations may be made:

$$l = a + (n - 1)d$$

$$s = \frac{n}{2}(a + l)$$

$$s = \frac{n}{2}[(2a + (n - 1)d]$$

A geometrical series is a series of terms, each term of which is obtained from the preceding term by multiplication or division.

When a, l, n, and s are defined as in the preceding paragraph and r is the common ratio, the following generalizations may be made.

$$l = ar^{n-1}$$

$$s = \frac{a(r^n - 1)}{r - 1}$$

$$s = \frac{a(1 - r^n)}{(1 - r)}$$

When n is infinity and r is a fraction, less than 1,

$$s = \frac{a}{1 - r}$$

Geometry

In a right triangle, the square of the hypotenuse = the sum of the squares of the other two sides.

Area of triangle = ½ (base × altitude).

Circumference of circle = $2\pi r$. $\pi = 3.1416$.

Area of circle = πr^2.

Arc of circle subtending angle $\theta = \frac{\theta \pi r}{180}$. ($\theta$ in degrees)

Length of a chord, subtending an angle $\theta = 2r \sin \frac{1}{2}\theta$.

Area of a sector (included angle = θ) = ½ arc × radius = $\frac{\pi r^2 \theta}{360}$.

Area of a spherical triangle = $(A + B + C - \pi)r^2$, where A, B, and C are the spherical angles of the triangle in radians.

Area of a polygon of n equal sides of length a, = $\frac{1}{4}na^2 \cot 180/n$.

Surface of a sphere = $4\pi r^2$.

Volume of a sphere = $\frac{4}{3}\pi r^3$.

Lateral surface of a right cylinder = $2\pi rh$.

Total surface of a right cylinder = $2\pi rh + 2\pi r^2$.

Volume of right cylinder = $\pi r^2 h$.

Lateral surface of a right cone (s = slant height) = πrs.

Total surface of a right cone (s = slant height) = $\pi rs + \pi r^2$.

Volume of a right cone = $\frac{1}{3}\pi r^2 h$.

TRIGONOMETRY

Definitions

$$\text{sine} = \frac{\text{ordinate}}{\text{hypotenuse}} = \frac{o}{h}$$

$$\text{cosine} = \frac{\text{abscissa}}{\text{hypotenuse}} = \frac{a}{h}$$

$$\text{tangent} = \frac{\text{ordinate}}{\text{abscissa}} = \frac{o}{a}$$

$$\text{cotangent} = \frac{\text{abscissa}}{\text{ordinate}} = \frac{a}{o}$$

$$\text{secant} = \frac{\text{hypotenuse}}{\text{abscissa}} = \frac{h}{a}$$

$$\text{cosecant} = \frac{\text{hypotenuse}}{\text{ordinate}} = \frac{h}{o}$$

SIGNS OF TRIGONOMETRICAL FUNCTIONS

Function	Quadrant I		Quadrant II		Quadrant III		Quadrant IV	
	Sign	Value	Sign	Value	Sign	Value	Sign	Value
sin....	+	0 to 1	+	1 to 0	−	0 to 1	−	1 to 0
cos....	+	1 to 0	−	0 to 1	−	1 to 0	+	0 to 1
tan....	+	0 to ∞	−	∞ to 0	+	0 to ∞	−	∞ to 0
cot....	+	∞ to 0	−	0 to ∞	+	∞ to 0	−	0 to ∞
sec....	+	1 to ∞	−	∞ to 1	−	1 to ∞	+	∞ to 1
cosec..	+	∞ to 1	+	1 to ∞	−	∞ to 1	−	1 to ∞

NUMERICAL VALUES FOR IMPORTANT ANGLES

degrees............	0°	30°	45°	60°	90°	180°	270°
radians............	0	0.5236	0.7854	1.0472	1.5708	3.1416	4.7124
sin................	0	½	$1/\sqrt{2}$	$\frac{1}{2}\sqrt{3}$	1	0	−1
cos................	1	$\frac{1}{2}\sqrt{3}$	$1/\sqrt{2}$	½	0	−1	0
tan................	0	$1/\sqrt{3}$	1	$\sqrt{3}$	∞	0	∞
cot................	∞	$\sqrt{3}$	1	$1/\sqrt{3}$	0	∞	0

Rules

Circumference $= 360° = 2\pi$ radians.

1 radian $= 57.295°$.

1 degree $= 0.0175$ radians.

$\sin^2 x + \cos^2 x = 1$.

$\sin (x + y) = \sin x \cos y + \cos x \sin y$.

$\sin (x - y) = \sin x \cos y - \cos x \sin y$.

$\cos (x + y) = \cos x \cos y - \sin x \sin y$.

$\cos (x - y) = \cos x \cos y + \sin x \sin y$.

$$\tan(x + y) = \frac{\tan x + \tan y}{1 - \tan x \tan y}.$$

$$\tan(x - y) = \frac{\tan x - \tan y}{1 + \tan x \tan y}.$$

$$\sin 2x = 2 \sin x \cos x.$$

$$\cos 2x = \cos^2 x - \sin^2 x = 2\cos^2 x - 1 = 1 - 2\sin^2 x.$$

$$\sin 3x = 3 \sin x - 4 \sin^3 x.$$

$$\cos 3x = 4 \cos^3 x - 3 \cos x.$$

$$\tan 2x = \frac{2 \tan x}{1 - \tan^2 x}.$$

$$\tan 3x = \frac{3 \tan x - \tan^3 x}{1 - \tan^2 x}.$$

$$\sin \frac{1}{2}x = \pm\sqrt{\frac{1 - \cos x}{2}}.$$

$$\cos \frac{1}{2}x = \pm\sqrt{\frac{1 + \cos x}{2}}.$$

$$\tan \frac{1}{2}x = \pm\sqrt{\frac{1 - \cos x}{1 + \cos x}}.$$

$$\sin x = \frac{1}{\operatorname{cosec} x}. \qquad \operatorname{cosec} x = \frac{1}{\sin x}.$$

$$\cos x = \frac{1}{\sec x}. \qquad \sec x = \frac{1}{\cos x}.$$

$$\tan x = \frac{1}{\cot x} = \frac{\sin x}{\cos x}.$$

$$\cot x = \frac{1}{\tan x} = \frac{\cos x}{\sin x}.$$

$$\sin x = \sqrt{1 - \cos^2 x}. \qquad \cos x = \sqrt{1 - \sin^2 x}.$$

$$\tan x = \sqrt{\sec^2 - 1}. \qquad \sec x = \sqrt{\tan^2 + 1}.$$

$$\cot x = \sqrt{\operatorname{cosec}^2 x - 1}. \qquad \operatorname{cosec} x = \sqrt{\cot^2 x + 1}.$$

$$\sin x = \cos(90 - x) = -\sin(180 - x).$$

$$\cos x = \sin(90 - x) = -\cos(180 - x).$$

$$\tan x = \cot(90 - x) = -\tan(180 - x).$$

$$\cot x = \tan(90 - x) = -\cot(180 - x).$$

$$\sin x + \sin y = 2 \sin \tfrac{1}{2}(x + y).\ \cos \tfrac{1}{2}(x - y)$$

$$\sin x - \sin y = 2 \cos \tfrac{1}{2}(x + y).\ \sin \tfrac{1}{2}(x - y)$$

APPENDIX V

TABLES

Chemical and Physical Constants

Electricity

1 ampere × 1 second = 1 coulomb.
1 volt × 1 coulomb = 1 joule.
1 volt × 1 ampere = 1 watt.
1 watt × 1 second = 1 joule.
1 ampere = 1 volt ÷ 1 ohm.
Charge on electron (e) = 1.59×10^{-19} coulomb.
1 faraday, (F) = 96,500 coulombs = quantity of electricity required to deposit one gram equivalent in electrolysis.

Gases

1 atmosphere = 760 mm. mercury.
Avogadro number (N) = 6.06×10^{23} molecules per gram molecule.
Gas constant (R)
joules per degree = 8.316
calories per degree = 1.9885
liter-atmospheres per degree = 0.08207
cubic centimeter − atmospheres per degree = 82.07
Gram molecular weight = molecular weight in grams.
Gram molecular volume = 22,412 cc. at 0° C. and 760 mm.

Heat

Absolute temperature (K.) = Centigrade temperature (C.) + 273.1.

Absolute zero = −273.1° C.
25° C. = 298.1° K.
298.1^2 = 88,863
298.1^3 = 26,490,000.

Fahrenheit Scale (F.)

0° C. = 32° F.
100° C. = 212° F.
(F.° − 32)$\frac{5}{9}$ = C.°
(C.°)$\frac{9}{5}$ + 32 = F.°
1 calorie (15° cal.) = 4.185 joules
1 joule = 0.2389 calorie

Light

1 ängstrom, (Å) = 1×10^{-8} cm.
1 milli mu ($m\mu$) = 1 mu mu ($\mu\mu$) = 10^{-6} mm. = 10^{-7} cm.
λ (lambda) = wave length of light (in centimeters, $m\mu$ or Å)
ν (nu) = frequency of light = velocity light (centimeters) ÷ wave length (centimeters)
velocity light = 3×10^{10} cm. per sec. = 186,000 miles per second
Planck's constant, h, = 6.56×10^{-27} erg-seconds
1 quantum (ϵ) = $h\nu$ ergs

Miscellaneous

Vapor pressure of water

0°	4.56 mm.	25°	23.71 mm.

Density water (absolute)

0°	0.99984	25°	0.99704

Density mercury

0°	13.5955	25°	13.5340

Metric Conversion Tables

(1 *inch* = 2.54 *centimeters*)
1/10 inch = 2.54001 millimeters
1 inch = 2.54001 centimeters
1 foot = 30.4801 centimeters
1 mile = 1.60935 kilometers
1 square inch = 6.452 square centi-centimeters
1 square foot = 929.0 square centimeters
1 cubic inch = 16.3872 cubic centimeters
1 cubic foot = .02832 cubic meter
1 grain = 0.06480 gram
1 ounce[1] = 28.3495 grams
1 pound[1] = 453.59 grams
1 liquid ounce (U. S.) = 29.574 milliliters
1 liquid quart (U. S.) = 0.94636 liter

(1 *pound* = 453.6 *grams*)
1 mm. = 0.0394 inch
1 cm. = 0.3937 inch
1 meter = 1.09361 yard
1 kilometer = 0.62137 mile
1 square mm. = 0.00155 square inch
1 square cm. = 0.1550 square inch
1 cubic millimeter = 0.000061 cubic inch
1 cubic cm. = 0.061 cubic inch
1 gram = 15.4324 grains
1 kilogram = 2.20462 pounds[1]
1 milliliter (c.c.) = 0.03381 liquid ounce (U. S.)
1 liter = 0.26417 gallon

[1] Avoirdupois.

Greek Alphabet

Greek letter	Greek name	English equivalent	Greek letter	Greek name	English equivalent
Α α	Alpha	a	Ν ν	Nu	n
Β β	Beta	b	Ξ ξ	Xi	x
Γ γ	Gamma	g	Ο ο	Omicron	ŏ
Δ δ	Delta	d	Π π	Pi	p
Ε ϵ	Epsilon	ĕ	Ρ ρ	Rho	r
Ζ ζ	Zeta	z	Σ σ	Sigma	s
Η η	Eta	ē	Τ τ	Tau	t
Θ θ	Theta	th	Υ υ	Upsilon	u
Ι ι	Iota	i	Φ φ	Phi	ph
Κ κ	Kappa	k	Χ χ	Chi	ch
Λ λ	Lambda	l	Ψ ψ	Psi	ps
Μ μ	Mu	m	Ω ω	Omega	ō

Atomic Weights

	Symbol	Atomic weight		Symbol	Atomic weight
Aluminum	Al	26.96	Molybdenum	Mo	96.0
Antimony	Sb	121.77	Neodymium	Nd	144.27
Argon	A	39.91	Neon	Ne	20.2
Arsenic	As	74.96	Nickel	Ni	58.69
Barium	Ba	137.37	Nitrogen	N	14.008
Beryllium	Be	9.02	Osmium	Os	190.8
Bismuth	Bi	209.00	**Oxygen**	**O**	**16.00**
Boron	B	10.82	Palladium	Pd	106.7
Bromine	Br	79.916	Phosphorus	P	31.03
Cadmium	Cd	112.40	Platinum	Pt	195.23
Cæsium	Cs	132.81	Potassium	K	39.096
Calcium	Ca	40.07	Praseodymium	Pr	140.92
Carbon	C	12.000	Radium	Ra	225.95
Cerium	Ce	140.25	Radon	Rn	222
Chlorine	Cl	35.46	Rhodium	Rh	102.91
Chromium	Cr	52.01	Rubidium	Rb	85.45
Cobalt	Co	58.97	Ruthenium	Ru	101.7
Columbium	Cb	93.1	Samarium	Sa	150.43
Copper	Cu	63.57	Scandium	Sc	45.1
Dysprosium	Dy	162.52	Selenium	Se	79.2
Erbium	Er	167.7	Silicon	Si	28.06
Europium	Eu	152.0	Silver	Ag	107.88
Fluorine	F	19.0	Sodium	Na	22.997
Gadolinium	Gd	157.26	Strontium	Sr	87.63
Gallium	Ga	69.72	Sulphur	S	32.06
Germanium	Ge	72.60	Tantalum	Ta	181.5
Gold	Au	197.2	Tellurium	Te	127.5
Helium	He	4.00	Terbium	Tb	159.2
Holmium	Ho	163.4	Thallium	Tl	204.4
Hydrogen	H	1.008	Thorium	Th	232.15
Indium	In	114.8	Thulium	Tm	169.4
Iodine	I	126.93	Tin	Sn	118.7
Iridium	Ir	193.1	Titanium	Ti	47.9
Iron	Fe	55.84	Tungsten	W	184.0
Krypton	Kr	82.9	Uranium	U	238.17
Lanthanum	La	138.90	Vanadium	V	50.96
Lead	Pb	207.20	Xenon	Xe	130.2
Lithium	Li	6.94	Ytterbium (Neoytterbi-		
Lutecium	Lu	175.0	um)	Yb	173.6
Magnesium	Mg	24.32	Yttrium	Yt	89.0
Manganese	Mn	54.93	Zinc	Zn	65.38
Mercury	Hg	200.61	Zirconium	Zr	91

Reciprocals, Squares, and Cubes

n	$1/n$	n^2	n^3	$\sqrt{n}$	$\sqrt[3]{n}$
1	1.	1	1	1	1.
2	0.50000	4	8	1.414	1.260
3	.33333	9	27	1.732	1.442
4	.25000	16	64	2.000	1.587
5	.20000	25	125	2.236	1.710
6	.16667	36	216	2.449	1.817
7	.14286	49	343	2.646	1.913
8	.12500	64	512	2.828	2.000
9	.11111	81	729	3.000	2.080
10	.10000	100	1,000	3.162	2.154
11	.09091	121	1,331	3.317	2.224
12	.08333	144	1,728	3.464	2.289
13	.07692	169	2,197	3.606	2.351
14	.07143	196	2,744	3.742	2.410
15	.06667	225	3,375	3.873	2.466
16	.06250	256	4,096	4.000	2.520
17	.05882	289	4,913	4.123	2.571
18	.05556	324	5,832	4.243	2.621
19	.05263	361	6,859	4.359	2.668
20	.05000	400	8,000	4.472	2.714
21	.04762	441	9,261	4.583	2.759
22	.04545	484	10,648	4.690	2.802
23	.04348	529	12,167	4.796	2.844
24	.04167	576	13,824	4.899	2.884
25	.04000	625	15,625	5.000	2.924
26	.03846	676	17,576	5.099	2.962
27	.03704	729	19,683	5.196	3.000
28	.03571	784	21,952	5.291	3.037
29	.03448	841	24,389	5.385	3.072
30	.03333	900	27,000	5.477	3.107
31	.03226	961	29,791	5.568	3.141
32	.03125	1,024	32,768	5.657	3.175
33	.03030	1,089	35,937	5.745	3.208
34	.02941	1,156	39,304	5.831	3.240
35	.02857	1,225	42,875	5.916	3.271
36	.02778	1,296	46,656	6.000	3.302
37	.02703	1,369	50,653	6.083	3.332
38	.02632	1,444	54,872	6.164	3.362
39	.02564	1,521	59,319	6.245	3.391
40	.02500	1,600	64,000	6.325	3.420
41	.02439	1,681	68,921	6.403	3.448
42	.02381	1,764	74,088	6.481	3.476
43	.02326	1,849	79,507	6.557	3.503
44	.02273	1,936	85,184	6.633	3.530
45	.02222	2,025	91,125	6.708	3.557
46	.02174	2,116	97,336	6.782	3.583
47	.02128	2,209	103,823	6.856	3.609
48	.02083	2,304	110,592	6.928	3.634
49	.02041	2,401	117,649	7.000	3.659
50	.02000	2,500	125,000	7.071	3.684

n	$1/n$	n^2	n^3	$\sqrt{n}$	$\sqrt[3]{n}$
51	.01961	2,601	132,651	7.141	3.708
52	.01923	2,704	140,608	7.211	3.733
53	.01887	2,809	148,877	7.280	3.756
54	.01852	2,916	157,464	7.348	3.780
55	.01818	3,025	166,375	7.416	3.803
56	.01786	3,136	175,616	7.483	3.826
57	.01754	3,249	185,193	7.550	3.849
58	.01724	3,364	195,112	7.616	3.871
59	.01695	3,481	205,379	7.681	3.893
60	.01667	3,600	216,000	7.746	3.915
61	.01639	3,721	226,981	7.810	3.936
62	.01613	3,844	238,328	7,874	3.958
63	.01587	3,969	250,047	7.937	3.979
64	.01563	4,096	262,144	8.000	4.000
65	.01538	4,225	274,625	8.062	4.021
66	.01515	4,356	287,496	8.124	4.041
67	.01493	4,489	300,763	8.185	4.062
68	.01471	4,624	314,432	8.246	4.082
69	.01449	4,761	328,509	8.307	4.102
70	.01429	4,900	343,000	8.367	4.121
71	.01408	5,041	357,911	8.426	4.141
72	.01389	5,184	373,248	8.485	4.160
73	.01370	5,329	389,017	8.544	4.179
74	.01351	5,476	405,224	8.602	4.198
75	.01333	5,625	421,875	8.660	4.217
76	.01316	5,776	438,976	8.718	4.236
77	.01299	5,929	456,533	8.775	4.254
78	.01282	6,084	474,552	8.832	4.273
79	.01266	6,241	493,039	8.888	4.291
80	.01250	6,400	512,000	8.944	4.309
81	.01235	6,561	531,441	9.000	4.327
82	.01220	6,724	551,368	9.055	4.344
83	.01205	6,889	571,787	9.110	4.362
84	.01191	7,056	592,704	9.165	4.380
85	.01177	7,225	614,125	9.220	4,397
86	.01163	7,396	636,056	9.274	4.414
87	.01149	7,569	658,503	9.327	4.431
88	.01136	7,744	681,472	9.381	4.448
89	.01124	7,921	704,969	9.434	4.465
90	.01111	8,100	729,000	9.487	4.481
91	.01099	8,281	753,571	9.539	4.498
92	.01087	8,464	778,688	9.592	4.514
93	.01075	8,649	804,357	9.644	4.531
94	.01064	8,836	830,584	9.695	4.547
95	.01053	9,025	857,375	9.747	4.563
96	.01042	9,216	884,736	9.798	4.579
97	.01031	9,409	912,673	9.849	4.595
98	.01020	9,604	941,192	9.899	4.610
99	.01010	9,801	970,299	9.950	4.626
100	.01000	10,000	1,000,000	10.000	4.642

Sines, Cosines, Tangents

Sine

°	0′	10′	20′	30′	40′	50′	60′	
0	0.0000	0029	0058	0087	0116	0145	0175	89
1	0175	0204	0233	0262	0291	0320	0349	88
2	0349	0378	0407	0436	0465	0494	0523	87
3	0523	0552	0581	0610	0640	0669	0698	86
4	0698	0727	0756	0785	0814	0843	0872	**85**
5	0.0872	0901	0929	0958	0987	1016	1045	84
6	1045	1074	1103	1132	1161	1190	1219	83
7	1219	1248	1276	1305	1334	1363	1392	82
8	1392	1421	1449	1478	1507	1536	1564	81
9	1564	1593	1622	1650	1679	1708	1736	**80**
10	0.1736	1765	1794	1822	1851	1880	1908	79
11	1908	1937	1965	1994	2022	2051	2079	78
12	2079	2108	2136	2164	2193	2221	2250	77
13	2250	2278	2306	2334	2363	2391	2419	76
14	2419	2447	2476	2504	2532	2560	2588	**75**
15	0.2588	2616	2644	2672	2700	2728	2756	74
16	2756	2784	2812	2840	2868	2896	2924	73
17	2924	2952	2979	3007	3035	3062	3090	72
18	3090	3118	3145	3173	3201	3228	3256	71
19	3256	3283	3311	3338	3365	3393	3420	**70**
20	0.3420	3448	3475	3502	3529	3557	3584	69
21	3584	3611	3638	3665	3692	3719	3746	68
22	3746	3773	3800	3827	3854	3881	3907	67
23	3907	3934	3961	3987	4014	4041	4067	66
24	4067	4094	4120	4147	4173	4200	4226	**65**
25	0.4226	4253	4279	4305	4331	4358	4384	64
26	4384	4410	4436	4462	4488	4514	4540	63
27	4540	4566	4592	4617	4643	4669	4695	62
28	4695	4720	4746	4772	4797	4823	4848	61
29	4848	4874	4899	4924	4950	4975	5000	**60**
30	0.5000	5025	5050	5075	5100	5125	5150	59
31	5150	5175	5200	5225	5250	5275	5299	58
32	5299	5324	5348	5373	5398	5422	5446	57
33	5446	5471	5495	5519	5544	5568	5592	56
34	5592	5616	5640	5664	5688	5712	5736	**55**
35	0.5736	5760	5783	5807	5831	5854	5878	54
36	5878	5901	5925	5948	5972	5995	6018	53
37	6018	6041	6065	6088	6111	6134	6157	52
38	6157	6180	6202	6225	6248	6271	6293	51
39	6293	6316	6338	6361	6383	6406	6428	**50**
40	0.6428	6450	6472	6494	6517	6539	6561	49
41	6561	6583	6604	6626	6648	6670	6691	48
42	6691	6713	6734	6756	6777	6799	6820	47
43	6820	6841	6862	6884	6905	6926	6947	46
44	6947	6967	6988	7009	7030	7050	7071	**45**
45	0.7071							
	60′	50′	40′	30′	20′	10′	0′	°

Sine

°	0′	10′	20′	30′	40′	50′	60′	
45	0.7071	7092	7112	7133	7153	7173	7193	44
46	7193	7214	7234	7254	7274	7294	7314	43
47	7314	7333	7353	7373	7392	7412	7431	42
48	7431	7451	7470	7490	7509	7528	7547	41
49	7547	7566	7585	7604	7623	7642	7660	**40**
50	0.7660	7679	7698	7716	7735	7753	7771	39
51	7771	7790	7808	7826	7844	7862	7880	38
52	7880	7898	7916	7934	7951	7969	7986	37
53	7986	8004	8021	8039	8056	8073	8090	36
54	8090	8107	8124	8141	8158	8175	8192	**35**
55	0.8192	8208	8225	8241	8258	8274	8290	34
56	8290	8307	8323	8339	8355	8371	8387	33
57	8387	8403	8418	8434	8450	8465	8480	32
58	8480	8496	8511	8526	8542	8557	8572	31
59	8572	8587	8601	8616	8631	8646	8660	**30**
60	0.8660	8675	8689	8704	8718	8732	8746	29
61	8746	8760	8774	8788	8802	8816	8829	28
62	8829	8843	8857	8870	8884	8897	8910	27
63	8910	8923	8936	8949	8962	8975	8988	26
64	8988	9001	9013	9026	9038	9051	9063	**25**
65	0.9063	9075	9088	9100	9112	9124	9135	24
66	9135	9147	9159	9171	9182	9194	9205	23
67	9205	9216	9228	9239	9250	9261	9272	22
68	9272	9283	9293	9304	9315	9325	9336	21
69	9336	9346	9356	9367	9377	9387	9397	**20**
70	0.9397	9407	9417	9426	9436	9446	9455	19
71	9455	9465	9474	9483	9492	9502	9511	18
72	9511	9520	9528	9537	9546	9555	9563	17
73	9563	9572	9580	9588	9596	9605	9613	16
74	9613	9621	9628	9636	9644	9652	9659	**15**
75	0.9659	9667	9674	9681	9689	9696	9703	14
76	9703	9710	9717	9724	9730	9737	9744	13
77	9744	9750	9757	9763	9769	9775	9781	12
78	9781	9787	9793	9799	9805	9811	9816	11
79	9816	9822	9827	9833	9838	9843	9848	**10**
80	0.9848	9853	9858	9863	9868	9872	9877	9
81	9877	9881	9886	9890	9894	9899	9903	8
82	9903	9907	9911	9914	9918	9922	9925	7
83	9925	9929	9932	9936	9939	9942	9945	6
84	9945	9948	9951	9954	9957	9959	9962	**5**
85	0.9962	9964	9967	9969	9971	9974	9976	4
86	9976	9978	9980	9981	9983	9985	9986	3
87	9986	9988	9989	9990	9992	9993	9994	2
88	9994	9995	9996	9997	9997	9998	9998	1
89	9998	9999	9999	*0000	*0000	*0000	*0000	**0**
90	1.0000							
	60′	50′	40′	30′	20′	10′	0′	°

Cosine

Tangent

°	0′	10′	20′	30′	40′	50′	60′	
0	0.0000	0029	0058	0087	0116	0145	0175	89
1	0175	0204	0233	0262	0291	0320	0349	88
2	0349	0378	0407	0437	0466	0495	0524	87
3	0524	0553	0582	0612	0641	0670	0699	86
4	0699	0729	0758	0787	0816	0846	0875	**85**
5	0.0875	0904	0934	0963	0992	1022	1051	84
6	1051	1080	1110	1139	1169	1198	1228	83
7	1228	1257	1287	1317	1346	1376	1405	82
8	1405	1435	1465	1495	1524	1554	1584	81
9	1584	1614	1644	1673	1703	1733	1763	**80**
10	0.1763	1793	1823	1853	1883	1914	1944	79
11	1944	1974	2004	2035	2065	2095	2126	78
12	2126	2156	2186	2217	2247	2278	2309	77
13	2309	2339	2370	2401	2432	2462	2493	76
14	2493	2524	2555	2586	2617	2648	2679	**75**
15	0.2679	2711	2742	2773	2805	2836	2867	74
16	2867	2899	2931	2962	2994	3026	3057	73
17	3057	3089	3121	3153	3185	3217	3249	72
18	3249	3281	3314	3346	3378	3411	3443	71
19	3443	3476	3508	3541	3574	3607	3640	**70**
20	0.3640	3673	3706	3739	3772	3805	3839	69
21	3839	3872	3906	3939	3973	4006	4040	68
22	4040	4074	4108	4142	4176	4210	4245	67
23	4245	4279	4314	4348	4383	4417	4452	66
24	4452	4487	4522	4557	4592	4628	4663	**65**
25	0.4663	4699	4734	4770	4806	4841	4877	64
26	4877	4913	4950	4986	5022	5059	5095	63
27	5095	5132	5169	5206	5243	5280	5317	62
28	5317	5354	5392	5430	5467	5505	5543	61
29	5543	5581	5619	5658	5696	5735	5774	**60**
30	0.5774	5812	5851	5890	5930	5969	6009	59
31	6009	6048	6088	6128	6168	6208	6249	58
32	6249	6289	6330	6371	6412	6453	6494	57
33	6494	6536	6577	6619	6661	6703	6745	56
34	6745	6787	6830	6873	6916	6959	7002	**55**
35	0.7002	7046	7089	7133	7177	7221	7265	54
36	7265	7310	7355	7400	7445	7490	7536	53
37	7536	7581	7627	7673	7720	7766	7813	52
38	7813	7860	7907	7954	8002	8050	8098	51
39	8098	8146	8195	8243	8292	8342	8391	**50**
40	0.8391	8441	8491	8541	8591	8642	8693	49
41	8693	8744	8796	8847	8899	8952	9004	48
42	9004	9057	9110	9163	9217	9271	9325	47
43	9325	9380	9435	9490	9545	9601	9657	46
44	9657	9713	9770	9827	9884	9942	*0000	**45**
45	1.0000							
	60′	50′	40′	30′	20′	10′	0′	°

Cotangent

Tangent

°	0′	10′	20′	30′	40′	50′	60′	
45	1.000	1.006	1.012	1.018	1.024	1.030	1.036	44
46	1.036	1.042	1.048	1.054	1.060	1.066	1.072	43
47	1.072	1.079	1.085	1.091	1.098	1.104	1.111	42
48	1.111	1.117	1.124	1.130	1.137	1.144	1.150	41
49	1.150	1.157	1.164	1.171	1.178	1.185	1.192	**40**
50	1.192	1.199	1.206	1.213	1.220	1.228	1.235	39
51	1.235	1.242	1.250	1.257	1.265	1.272	1.280	38
52	1.280	1.288	1.295	1.303	1.311	1.319	1.327	37
53	1.327	1.335	1.343	1.351	1.360	1.368	1.376	36
54	1.376	1.385	1.393	1.402	1.411	1.419	1.428	**35**
55	1.428	1.437	1.446	1.455	1.464	1.473	1.483	34
56	1.483	1.492	1.501	1.511	1.520	1.530	1.540	33
57	1.540	1.550	1.560	1.570	1.580	1.590	1.600	32
58	1.600	1.611	1.621	1.632	1.643	1.653	1.664	31
59	1.664	1.675	1.686	1.698	1.709	1.720	1.732	**30**
60	1.732	1.744	1.756	1.767	1.780	1.792	1.804	29
61	1.804	1.816	1.829	1.842	1.855	1.868	1.881	28
62	1.881	1.894	1.907	1.921	1.935	1.949	1.963	27
63	1.963	1.977	1.991	2.006	2.020	2.035	2.050	26
64	2.050	2.066	2.081	2.097	2.112	2.128	2.145	**25**
65	2.145	2.161	2.177	2.194	2.211	2.229	2.246	24
66	2.246	2.264	2.282	2.300	2.318	2.337	2.356	23
67	2.356	2.375	2.394	2.414	2.434	2.455	2.475	22
68	2.475	2.496	2.517	2.539	2.560	2.583	2.605	21
69	2.605	2.628	2.651	2.675	2.699	2.723	2.747	**20**
70	2.747	2.773	2.798	2.824	2.850	2.877	2.904	19
71	2.904	2.932	2.960	2.989	3.018	3.047	3.078	18
72	3.078	3.108	3.140	3.172	3.204	3.237	3.271	17
73	3.271	3.305	3.340	3.376	3.412	3.450	3.487	16
74	3.487	3.526	3.566	3.606	3.647	3.689	3.732	**15**
75	3.732	3.776	3.821	3.867	3.914	3.962	4.011	14
76	4.011	4.061	4.113	4.165	4.219	4.275	4.331	13
77	4.331	4.390	4.449	4.511	4.574	4.638	4.705	12
78	4.705	4.773	4.843	4.915	4.989	5.066	5.145	11
79	5.145	5.226	5.309	5.396	5.485	5.576	5.671	**10**
80	5.671	5.769	5.871	5.976	6.084	6.197	6.314	9
81	6.314	6.435	6.561	6.691	6.827	6.968	7.115	8
82	7.115	7.269	7.429	7.596	7.770	7.953	8.144	7
83	8.144	8.345	8.556	8.777	9.010	9.255	9.514	6
84	9.514	9.788	10.078	10.385	10.712	11.059	11.430	**5**
85	11.430	11.826	12.251	12.706	13.197	13.727	14.301	4
86	14.301	14.924	15.605	16.350	17.169	18.075	19.081	3
87	19.081	20.206	21.470	22.904	24.542	26.432	28.636	2
88	28.636	31.242	34.368	38.188	42.964	49.104	57.290	1
89	57.290	68.750	85.940	114.59	171.89	343.77	infinit.	**0**
90	infinit.							
	60′	50′	40′	30′	20′	10′	0′	°

Cotangent

$$\frac{x^2}{1-x}$$

$\frac{x^2}{1-x}$ from $x = 0.001$ to 0.0999 and from $x = 0.100$ to 0.999.

Position of the Decimal Point

x	$\frac{x^2}{1-x}$
0.0100	0.0001
0.0312	0.001
0.0952	0.01
0.271	0.1
0.619	1.0
0.917	10.
0.991	100.

x	0	1	2	3	4	5	6	7	8	9
0.010	1,010	1,030	1,051	1,072	1,093	1,114	1,136	1,157	1,179	1,201
11	1,223	1,246	1,268	1,291	1,315	1,337	1,361	1,385	1,408	1,433
12	1,457	1,482	1,507	1,532	1,557	1,582	1,608	1,633	1,659	1,686
13	1,712	1,739	1,765	1,792	1,820	1,847	1,875	1,903	1,931	1,959
14	1,987	2,016	2,045	2,074	2,104	2,133	2,163	2,193	2,223	2,253
15	2,284	2,314	2,345	2,376	2,408	2,440	2,473	2,505	2,537	2,569
16	2,602	2,635	2,668	2,706	2,734	2,768	2,802	2,836	2,871	2,905
17	2,940	2,975	3,010	3,046	3,081	3,118	3,154	3,190	3,226	3,262
18	3,299	3,336	3,373	3,411	3,449	3,487	3,525	3,563	3,602	3,641
19	3,680	3,719	3,758	3,798	3,838	3,878	3,918	3,958	3,999	4,040
0.020	4,082	4,123	4,164	4,206	4,248	4,290	4,333	4,376	4,418	4,461
21	4,505	4,548	4,591	4,635	4,680	4,724	4,759	4,813	4,858	4,903
22	4,949	4,994	5,041	5,087	5,133	5,179	5,226	5,273	5,320	5,367
23	5,415	5,462	5,510	5,558	5,607	5,655	5,704	5,753	5,802	5,852
24	5,902	5,952	6,002	6,052	6,103	6,154	6,204	6,256	6,307	6,358
25	6,410	6,462	6,514	6,567	6,619	6,672	6,725	6,778	6,832	6,886
26	6,940	6,995	7,049	7,104	7,159	7,213	7,269	7,324	7,380	7,436
27	7,492	7,548	7,605	7,662	7,719	7,777	7,834	7,892	7,949	8,007
28	8,066	8,124	8,183	8,242	8,301	8,360	8,420	8,478	8,538	8,599
29	8,661	8,721	8,782	8,844	8,905	8,966	9,028	9,090	9,152	9,215
0.030	9,278	9,341	9,404	9,467	9,531	9,595	9,659	9,723	9,788	9,852
31	9,917	9,982	1,005	1,011	1,017	1,025	1,031	1,038	1,044	1,051
32	1,057	1,063	1,070	1,077	1,084	1,091	1,098	1,104	1,111	1,118
33	1,125	1,132	1,138	1,146	1,153	1,160	1,167	1,174	1,181	1,188
34	1,196	1,204	1,212	1,219	1,226	1,233	1,241	1,248	1,255	1,263
35	1,270	1,277	1,285	1,292	1,300	1,307	1,314	1,322	1,330	1,337
36	1,345	1,352	1,360	1,368	1,375	1,383	1,391	1,398	1,406	1,414
37	1,422	1,430	1,438	1,446	1,454	1,462	1,470	1,478	1,486	1,494
38	1,502	1,510	1,518	1,526	1,534	1,543	1,551	1,559	1,567	1,575
39	1,583	1,592	1,600	1,608	1,616	1,625	1,633	1,642	1,650	1,658

$$\frac{x^2}{1-x}$$ *(Continued)*

x	0	1	2	3	4	5	6	7	8	9
0.040	1,667	1,675	1,684	1,692	1,701	1,710	1,718	1,727	1,736	1,744
41	1,753	1,762	1,770	1,779	1,788	1,797	1,805	1,814	1,823	1,832
42	1,841	1,850	1,859	1,868	1,877	1,886	1,895	1,904	1,913	1,922
43	1,932	1,941	1,950	1,959	1,968	1,978	1,987	1,996	2,005	2,015
44	2,024	2,034	2,043	2,053	2,062	2,071	2,081	2,090	2,100	2,110
45	2,119	2,129	2,139	2,149	2,159	2,168	2,178	2,188	2,198	2,208
46	2,217	2,227	2,237	2,247	2,257	2,267	2,277	2,287	2,297	2,307
47	2,317	2,327	2,337	2,347	2,357	2,368	2,379	2,389	2,399	2,409
48	2,420	2,430	2,440	2,450	2,461	2,471	2,482	2,492	2,503	2,513
49	2,524	2,534	2,545	2,555	2,566	2,577	2,587	2,599	2,610	2,620
0.050	2,631	2,642	2,653	2,663	2,674	2,685	2,696	2,707	2,718	2,729
51	2,741	2,752	2,763	2,774	2,785	2,796	2,807	2,818	2,829	2,840
52	2,852	2,863	2,874	2,885	2,897	2,908	2,919	2,931	2,942	2,953
53	2,965	2,977	2,989	3,000	3,012	3,023	3,035	3,047	3,058	3,070
54	3,081	3,093	3,105	3,116	3,128	3,140	3,152	3,164	3,176	3,187
55	3,199	3,211	3,223	3,235	3,248	3,260	3,272	3,284	3,296	3,308
56	3,321	3,333	3,345	3,357	3,370	3,383	3,395	3,407	3,419	3,432
57	3,444	3,457	3,469	3,481	3,494	3,507	3,520	3,532	3,545	3,558
58	3,570	3,583	3,595	3,608	3,621	3,634	3,647	3,660	3,673	3,686
59	3,699	3,711	3,724	3,737	3,751	3,764	3,777	3,790	3,803	3,816
0.060	3,830	3,843	3,856	3,870	3,883	3,896	3,910	3,923	3,936	3,950
61	3,963	3,977	3,990	4,004	4,017	4,030	4,044	4,057	4,071	4,084
62	4,098	4,111	4,125	4,139	4,153	4,166	4,180	4,194	4,208	4,222
63	4,236	4,250	4,264	4,278	4,292	4,306	4,320	4,334	4,348	4,362
64	4,376	4,391	4,405	4,419	4,434	4,448	4,462	4,477	4,491	4,505
65	4,519	4,534	4,548	4,563	4,577	4,592	4,606	4,621	4,635	4,650
66	4,664	4,679	4,694	4,708	4,723	4,738	4,752	4,767	4,782	4,796
67	4,811	4,826	4,841	4,856	4,871	4,886	4,901	4,916	4,931	4,946
68	4,961	4,976	4,992	5,007	5,023	5,036	5,054	5,069	5,085	5,100
69	5,115	5,130	5,146	5,161	5,177	5,192	5,208	5,223	5,239	5,254
0.070	5,269	5,284	5,300	5,316	5,331	5,347	5,362	5,378	5,394	5,410
71	5,426	5,442	5,458	5,474	5,490	5,506	5,522	5,538	5,554	5,570
72	5,586	5,602	5,619	5,636	5,652	5,668	5,685	5,701	5,717	5,733
73	5,749	5,766	5,782	5,799	5,815	5,832	5,848	5,865	5,881	5,898
74	5,914	5,931	5,947	5,964	5,981	5,997	6,014	6,031	6,047	6,064
75	6,081	6,098	6,115	6,132	6,149	6,166	6,183	6,200	6,217	6,234
76	6,251	6,268	6,286	6,303	6,320	6,338	6,355	6,372	6,390	6,407
77	6,424	6,442	6,459	6,477	6,494	6,512	6,529	6,547	6,564	6,582
78	6,599	6,617	6,634	6,652	6,670	6,687	6,705	6,723	6,740	6,758
79	6,776	6,794	6,812	6,829	6,847	6,865	6,883	6,901	6,919	6,937

$\frac{x^2}{1-x}$ (*Continued*)

x	0	1	2	3	4	5	6	7	8	9
0.080	6,955	6,973	6,992	7,010	7,029	7,047	7,066	7,084	7,103	7,121
81	7,139	7,158	7,176	7,197	7,215	7,234	7,252	7,270	7,288	7,307
82	7,325	7,344	7,362	7,381	7,400	7,418	7,437	7,456	7,475	7,494
83	7,513	7,532	7,551	7,570	7,589	7,608	7,627	7,646	7,665	7,684
84	7,703	7,722	7,741	7,761	7,780	7,799	7,819	7,838	7,857	7,876
85	7,896	7,916	7,935	7,955	7,975	7,994	8,014	8,033	8,053	8,072
86	8,092	8,112	8,131	8,151	8,171	8,190	8,210	8,230	8,250	8,270
87	8,290	8,310	8,330	8,350	8,370	8,391	8,411	8,431	8,451	8,471
88	8,491	8,511	8,532	8,552	8,572	8,593	8,613	8,633	8,654	8,674
89	8,695	8,715	8,736	8,757	8,777	8,798	8,819	8,839	8,860	8,881
0.090	8,901	8,922	8,942	8,963	8,984	9,005	9,026	9,047	9,068	9,089
91	9,110	9,131	9,152	9,173	9,195	9,216	9,237	9,258	9,280	9,301
92	9,322	9,343	9,365	9,386	9,408	9,429	9,451	9,472	9,494	9,515
93	9,536	9,557	9,579	9,601	9,622	9,644	9,666	9,687	9,709	9,731
94	9,753	9,775	9,796	9,818	9,840	9,862	9,884	9,906	9,928	9,950
95	9,972	9,994	1,002	1,004	1,006	1,008	1,011	1,013	1,015	1,017
96	1,020	1,022	1,024	1,027	1,029	1,031	1,033	1,036	1,038	1,040
97	1,042	1,044	1,047	1,049	1,051	1,054	1,056	1,058	1,060	1,063
98	1,065	1,067	1,069	1,072	1,074	1,076	1,079	1,081	1,083	1,086
99	1,088	1,090	1,092	1,095	1,097	1,099	1,101	1,104	1,106	1,109
0.10	1,111	1,135	1,159	1,183	1,207	1,232	1,257	1,282	1,308	1,333
11	1,360	1,386	1,413	1,440	1,467	1,494	1,522	1,550	1,579	1,607
12	1,636	1,666	1,695	1,725	1,755	1,786	1,817	1,848	1,879	1,911
13	1,943	1,975	2,007	2,040	2,073	2,107	2,141	2,175	2,209	2,244
14	2,279	2,314	2,350	2,386	2,422	2,459	2,496	2,533	2,571	2,609
15	2,647	2,686	2,725	2,764	2,803	2,843	2,883	2,924	2,965	3,006
16	3,048	3,090	3,132	3,174	3,217	3,261	3,304	3,348	3,392	3,437
17	3,482	3,527	3,573	3,619	3,665	3,712	3,759	3,807	3,855	3,903
18	3,951	4,000	4,049	4,099	4,149	4,199	4,250	4,301	4,353	4,403
19	4,457	4,509	4,562	4,616	4,670	4,724	4,778	4,833	4,888	4,944
0.20	5,000	5,056	5,113	5,171	5,228	5,286	5,345	5,403	5,463	5,522
21	5,582	5,643	5,704	5,765	5,826	5,889	5,951	6,014	6,077	6,141
22	6,205	6,270	6,335	6,400	6,466	6,532	6,599	6,666	6,734	6,802
23	6,870	6,939	7,008	7,078	7,148	7,219	7,290	7,362	7,434	7,506
24	7,579	7,652	7,726	7,800	7,875	7,950	8,026	8,102	8,179	8,256
25	8,333	8,411	8,490	8,569	8,648	8,728	8,809	8,890	8,971	9,053
26	9,135	9,218	9,301	9,385	9,470	9,554	9,640	9,726	9,812	9,899
27	9,986	1,007	1,016	1,025	1,034	1,043	1,052	1,061	1,070	1,080
28	1,089	1,099	1,108	1,117	1,127	1,136	1,146	1,155	1,165	1,175
29	1,185	1,194	1,204	1,214	1,224	1,234	1,245	1,255	1,265	1,275

$\frac{x^2}{1-x}$ *(Continued)*

x	0	1	2	3	4	5	6	7	8	9
0.30	1,286	1,296	1,307	1,317	1,328	1,339	1,349	1,360	1,371	1,382
31	1,393	1,404	1,415	1,426	1,437	1,449	1,460	1,471	1,483	1,494
32	1,506	1,518	1,529	1,541	1,553	1,565	1,577	1,589	1,601	1,613
33	1,625	1,638	1,650	1,663	1,675	1,688	1,700	1,713	1,726	1,739
34	1,752	1,765	1,778	1,791	1,804	1,817	1,831	1,844	1,857	1,871
35	1,885	1,898	1,912	1,926	1,940	1,954	1,968	1,982	1,996	2,011
36	2,025	2,040	2,054	2,068	2,083	2,098	2,113	2,128	2,143	2,158
37	2,173	2,188	2,203	2,219	2,234	2,250	2,266	2,281	2,297	2,313
38	2,329	2,345	2,361	2,378	2,394	2,410	2,427	2,443	2,460	2,477
39	2,493	2,510	2,527	2,545	2,562	2,579	2,596	2,614	2,631	2,649
0.40	2,667	2,685	2,702	2,720	2,739	2,757	2,775	2,793	2,812	2,830
41	2,849	2,868	2,887	2,906	2,925	2,944	2,963	2,983	3,002	3,022
42	3,041	3,061	3,081	3,101	3,121	3,141	3,162	3,182	3,203	3,223
43	3,244	3,265	3,286	3,307	3,328	3,349	3,371	3,392	3,414	3,435
44	3,457	3,479	3,501	3,523	3,546	3,568	3,591	3,613	3,636	3,659
45	3,682	3,705	3,728	3,752	3,775	3,799	3,822	3,846	3,870	3,894
46	3,919	3,943	3,967	3,992	4,017	4,042	4,067	4,092	4,117	4,142
47	4,168	4,194	4,219	4,245	4,271	4,298	4,324	4,351	4,377	4,404
48	4,431	4,458	4,485	4,512	4,540	4,568	4,595	4,613	4,651	4,680
49	4,708	4,736	4,765	4,794	4,823	4,852	4,881	4,911	4,940	4,970
0.50	5,000	5,030	5,060	5,091	5,121	5,152	5,183	5,214	5,245	5,277
51	5,308	5,340	5,372	5,404	5,436	5,469	5,501	5,534	5,567	5,600
52	5,633	5,667	5,701	5,734	5,768	5,803	5,837	5,871	5,906	5,941
53	5,977	6,012	6,048	6,083	6,119	6,155	6,192	6,228	6,265	6,302
54	6,339	6,377	6,414	6,452	6,490	6,528	6,566	6,605	6,644	6,683
55	6,722	6,762	6,801	6,841	6,882	6,922	6,963	7,003	7,044	7,086
56	7,127	7,169	7,211	7,253	7,296	7,339	7,382	7,425	7,468	7,512
57	7,556	7,600	7,645	7,689	7,734	7,779	7,825	7,871	7,917	7,963
58	8,010	8,056	8,103	8,151	8,199	8,246	8,295	8,343	8,392	8,441
59	8,490	8,540	8,590	8,640	8,691	8,741	8,792	8,844	8,896	8,948
0.60	9,000	9,053	9,106	9,159	9,213	9,267	9,321	9,375	9,430	9,485
61	9,541	9,597	9,653	9,710	9,767	9,824	9,882	9,940	9,998	1,006
62	1,012	1,018	1,024	1,030	1,036	1,042	1,048	1,054	1,060	1,066
63	1,073	1,079	1,085	1,092	1,098	1,105	1,111	1,118	1,124	1,131
64	1,138	1,145	1,151	1,158	1,165	1,172	1,179	1,186	1,193	1,200
65	1,207	1,214	1,222	1,229	1,236	1,244	1,251	1,258	1,266	1,274
66	1,281	1,289	1,297	1,304	1,312	1,320	1,328	1,336	1,344	1,352
67	1,360	1,369	1,377	1,385	1,393	1,402	1,410	1,419	1,428	1,436
68	1,445	1,454	1,463	1,473	1,482	1,491	1,499	1,508	1,517	1,526
69	1,536	1,545	1,555	1,564	1,574	1,583	1,593	1,603	1,613	1,623

$\frac{x^2}{1-x}$ *(Continued)*

x	0	1	2	3	4	5	6	7	8	9
0.70	1,633	1,643	1,654	1,664	1,674	1,685	1,695	1,706	1,717	1,727
71	1,738	1,749	1,760	1,771	1,783	1,794	1,805	1,817	1,828	1,840
72	1,851	1,863	1,875	1,887	1,899	1,911	1,924	1,936	1,949	1,961
73	1,974	1,987	1,999	2,012	2,025	2,039	2,052	2,065	2,079	2,092
74	2,106	2,120	2,134	2,148	2,162	2,177	2,191	2,206	2,220	2,235
75	2,250	2,265	2,280	2,296	2,311	2,327	2,342	2,358	2,374	2,390
76	2,407	2,423	2,440	2,456	2,473	2,490	2,508	2,525	2,542	2,560
77	2,578	2,596	2,614	2,632	2,651	2,669	2,688	2,707	2,727	2,746
78	2,766	2,785	2,805	2,825	2,846	2,866	2,887	2,908	2,929	2,950
79	2,972	2,994	3,016	3,038	3,060	3,083	3,106	3,129	3,153	3,176
0.80	3,200	3,224	3,249	3,273	3,298	3,323	3,348	3,374	3,400	3,427
81	3,453	3,480	3,507	3,535	3,562	3,590	3,619	3,648	3,677	3,706
82	3,736	3,766	3,796	3,827	3,858	3,889	3,921	3,953	3,986	4,019
83	4,052	4,086	4,120	4,155	4,190	4,225	4,262	4,298	4,335	4,372
84	4,410	4,448	4,487	4,526	4,566	4,606	4,648	4,689	4,731	4,773
85	4,816	4,860	4,905	4,950	4,995	5,042	5,088	5,136	5,184	5,233
86	5,283	5,333	5,384	5,436	5,489	5,542	5,597	5,652	5,708	5,765
87	5,822	5,881	5,941	6,001	6,063	6,125	6,189	6,253	6,319	6,386
88	6,453	6,522	6,593	6,664	6,737	6,811	6,886	6,963	7,041	7,120
89	7,201	7,283	7,367	7,453	7,540	7,629	7,719	7,812	7,906	8,002
0.90	8,100	8,200	8,302	8,406	8,513	8,621	8,732	8,846	8,962	9,080
91	9,201	9,324	9,452	9,581	9,714	9,850	9,989	1,013	1,028	1,043
92	1,058	1,074	1,090	1,107	1,123	1,141	1,158	1,177	1,196	1,215
93	1,236	1,256	1,277	1,299	1,321	1,345	1,369	1,393	1,419	1,445
94	1,473	1,501	1,530	1,560	1,592	1,624	1,658	1,692	1,728	1,766
95	1,805	1,846	1,888	1,933	1,979	2,027	2,077	2,130	2,185	2,244
96	2,304	2,368	2,436	2,506	2,582	2,660	2,744	2,833	2,928	3,029
97	3,136	3,251	3,374	3,507	3,649	3,803	3,970	4,150	4,347	4,564
98	4,802	5,005	5,358	5,684	6,052	6,468	6,945	7,493	8,134	8,892
99	9,801	1,091	1,230	1,409	1,647	1,980	2,480	3,313	4,980	9,980

e^x AND e^{-x}

VALUES OF e^x (p. 139)

x	0.00	0.01	0.02	0.03	0.04	0.05	0.06	0.07	0.08	0.09
0.0	1.000	1.010	1.020	1.030	1.041	1.051	1.062	1.073	1.083	1.094
0.1	1.105	1.116	1.127	1.139	1.150	1.162	1.174	1.185	1.197	1.209
0.2	1.221	1.234	1.246	1.259	1.271	1.284	1.297	1.310	1.323	1.336
0.3	1.350	1.363	1.377	1.391	1.405	1.419	1.433	1.448	1.462	1.477
0.4	1.492	1.507	1.522	1.537	1.553	1.568	1.584	1.600	1.616	1.632
0.5	1.649	1.665	1.682	1.699	1.716	1.733	1.751	1.768	1.786	1.804
0.6	1.822	1.840	1.859	1.878	1.896	1.916	1.935	1.954	1.974	1.994
0.7	2.014	2.034	2.054	2.075	2.096	2.117	2.138	2.160	2.181	2.203
0.8	2.226	2.248	2.270	2.293	2.316	2.340	2.363	2.387	2.411	2.435
0.9	2.460	2.484	2.509	2.535	2.560	2.586	2.612	2.638	2.664	2.691

x	0.0	0.1	0.2	0.3	0.4	0.5	0.6	0.7	0.8	0.9
1	2.718	3.004	3.320	3.669	4.055	4.482	4.953	5.474	6.050	6.686
2	7.389	8.166	9.025	9.974	11.02	12.18	13.46	14.88	16.44	18.17
3	20.09	22.20	24.53	27.11	29.96	33.12	36.60	40.45	44.70	49.40
4	54.60	60.34	66.69	73.70	81.45	90.02	99.48	110.0	121.5	134.3
5	148.4	164.0	181.3	200.3	221.4	244.7	270.4	298.9	330.3	365.0
6	403.4	445.9	492.8	544.6	601.9	665.1	735.1	812.4	897.9	992.3
7	1,097	1,212	1,339	1,480	1,636	1,808	1,998	2,208	2,441	2,697
8	2,981	3,295	3,641	4,024	4,447	4,915	5,432	6,003	6,634	7,332
9	8,103	8,955	9,897	10,938	12,088	13,360	14,765	16,318	18,034	19,930
10	22,026									

VALUES OF e^{-x} (p. 139)

x	0.00	0.01	0.02	0.03	0.04	0.05	0.06	0.07	0.08	0.09
0.0	1.000	0.990	0.980	0.970	0.961	0.951	0.942	0.932	0.923	0.914
0.1	0.905	0.896	0.887	0.878	0.869	0.861	0.852	0.844	0.835	0.827
0.2	0.819	0.811	0.803	0.795	0.787	0.779	0.771	0.763	0.756	0.748
0.3	0.741	0.733	0.726	0.719	0.712	0.705	0.698	0.691	0.684	0.677
0.4	0.670	0.664	0.657	0.651	0.644	0.638	0.631	0.625	0.619	0.613
0.5	0.607	0.600	0.595	0.589	0.583	0.577	0.571	0.566	0.560	0.554
0.6	0.549	0.543	0.538	0.533	0.527	0.522	0.517	0.512	0.507	0.502
0.7	0.497	0.492	0.487	0.482	0.477	0.472	0.468	0.463	0.458	0.454
0.8	0.449	0.445	0.440	0.436	0.432	0.427	0.423	0.419	0.415	0.411
0.9	0.407	0.403	0.399	0.395	0.391	0.387	0.383	0.379	0.375	0.372
1.0	0.368									

x	0.0	0.1	0.2	0.3	0.4	0.5	0.6	0.7	0.8	0.9
1.0	0.368	0.333	0.301	0.273	0.247	0.223	0.202	0.183	0.165	0.150
2.0	0.135	0.122	0.111	0.100	0.0907	0.0821	0.0743	0.0672	0.0608	0.0550
3.0	0.0498	0.0450	0.0408	0.0369	0.0334	0.0302	0.0273	0.0247	0.0224	0.0202
4.0	0.0183	0.0166	0.0150	0.0136	0.0123	0.0111	0.0101	0.00910	0.00823	0.00745
5.0	0.00674	0.00610	0.00552	0.00499	0.00452	0.00409	0.00370	0.00335	0.00303	0.00274
6.0	0.00248	0.00224	0.00203	0.00184	0.00166	0.00150	0.00136	0.00123	0.00111	0.00101
7.0	0.000912	0.000825	0.000747	0.000676	0.000611	0.000553	0.000500	0.000453	0.000410	0.000371
8.0	0.000335	0.000304	0.000275	0.000249	0.000225	0.000203	0.000184	0.000167	0.000151	0.000136
9.0	0.000123	0.000112	0.000101	0.000091	0.000083	0.000075	0.000068	0.000061	0.000055	0.000050
10.0	0.000045									

Rules of Calculus

The following rules have been discussed on previous pages of this book.

1. If $y = x^n$, $\dfrac{dy}{dx} = nx^{n-1}$.

2. If $y = f_1(x) + f_2(x)$, $\dfrac{dy}{dx} = \dfrac{df_1(x)}{dx} + \dfrac{df_2(x)}{dx}$.

3. If $y = k$, $\dfrac{dy}{dx} = 0$.

4. If $y = kf(x)$, $\dfrac{dy}{dx} = k\dfrac{df(x)}{dx}$.

5. If $y = uv$, $\dfrac{dy}{dx} = u\dfrac{dv}{dx} + v\dfrac{du}{dx}$.

6. If $y = \dfrac{u}{v}$, $\dfrac{dy}{dx} = \dfrac{v\dfrac{du}{dx} - u\dfrac{dv}{dx}}{v^2}$.

7. If $y = f_2(f_1x)$, $\dfrac{dy}{dx} = \dfrac{df_2(f_1x)}{d(f_1x)} \times \dfrac{d(f_1x)}{dx}$.

8. If $y = e^x$, $\dfrac{dy}{dx} = e^x$.

9. If $y = \log_e x$, $\dfrac{dy}{dx} = \dfrac{1}{x}$.

10. If $y = \log_{10} x$, $\dfrac{dy}{dx} = \dfrac{0.4343}{x}$.

11. If $y = \log_a x$, $\dfrac{dy}{dx} = \dfrac{\log_a \mathrm{e}}{x}$.

12. If $y = a^x$, $\dfrac{dy}{dx} = a^x \log_e a$.

13. For a maximum, or minimum or point of inflection, $\dfrac{dy}{dx} = 0$.

14. For a differential, $dy = \dfrac{dy}{dx}dx$.

15. $\int dx = x + C$.

16. $\int x dx = \dfrac{x^{n+1}}{n+1} + C$.

17. $\int kf(x)dx = k\int f(x)dx$.

18. $\int (f_1(x) + f_2(x) + f_3(x))dx = \int f_1(x)dx + \int f_2(x)dx + \int f_3(x)dx$.

19. $\int (k + f(x))dx = \int kdx + \int f(x)dx = kx + \int f(x)dx + C.$

20. $\int e^x dx = e^x + c.$

21. $\int \frac{dx}{x} = \log_e x + c.$

22. $\int a^x dx = \frac{a^x}{\log_e a} + c.$

23. $\int \log_e x dx = x(\log_e x - 1) + C.$

24. $\int \log_{10} x = 0.4343x(\log_e x - 1) + C.$

25. $\int \log_a x = (\log_a e)(x)(\log_e x - 1) + C.$

26. Integrating between limits, $\int_b^a x dx = \frac{a^2}{2} - \frac{b^2}{2}.$

27. If $y = \sin x$, $\frac{dy}{dx} = \cos x.$

28. If $y = \cos x$, $\frac{dy}{dx} = -\sin x.$

29. If $y = \tan x$, $\frac{dy}{dx} = \sec^2 x.$

30. If $y = \text{arc}\sin x$, $\frac{dy}{dx} = \frac{1}{\sqrt{1 - x^2}}.$

31. If $y = \text{arc}\tan x$, $\frac{dy}{dx} = \frac{1}{1 + x^2}.$

32. $\int \sin x dx = -\cos x + C.$

33. $\int \cos x dx = \sin x + C.$

34. $\int \tan x dx = -\log_e \cos x + C.$

35. $du = \left(\frac{\partial u}{\partial x}\right)_y dx + \left(\frac{\partial u}{\partial y}\right)_x dy.$

Table of Integrals[1]

In these formulas x and X represent variables, and all other letters represent constants. All angles are assumed to be in radians; and the base of logarithms is e unless otherwise indicated. An arbitrary constant may be added to the right-hand side of each formula. log refers to $\log_e$ in these tables. Throughout the rest of the book log refers to $\log_{10}$.

The use of these tables is explained on page 157.

[1] From Peirce and Carver's "Formulas and Tables," pages 33 to 46a. McGraw-Hill Book Company, Inc., New York.

Rational Algebraic Integrals

1. $\int dx = x$

2. $\int af(x)dx = a \int f(x)dx$

3. $\int [f(x) + \phi(x) - \psi(x)]dx = \int f(x)dx + \int \phi(x)dx - \int \psi(x)dx$

4. $\int x^m dx = \frac{x^{m+1}}{m+1}$ when $m \neq -1$

5. $\int \frac{dx}{x} = \log x$

6. $\int (ax+b)^m dx = \frac{(ax+b)^{m+1}}{a(m+1)}$ when $m \neq -1$

7. $\int \frac{dx}{ax+b} = \frac{1}{a} \log (ax+b)$

8. $\int \frac{xdx}{ax+b} = \frac{1}{a^2} \{ax + b - b \log (ax+b)\}$

9. $\int \frac{xdx}{(ax+b)^2} = \frac{1}{a^2} \left\{ \frac{b}{ax+b} + \log (ax+b) \right\}$

10. $\int \frac{x^2dx}{ax+b} = \frac{1}{a^3} \left\{ \frac{(ax+b)^2}{2} - 2b(ax+b) + b^2 \log (ax+b) \right\}$

11. $\int \frac{x^2dx}{(ax+b)^2} = \frac{1}{a^3} \left\{ ax + b - \frac{b^2}{ax+b} - 2b \log (ax+b) \right\}$

12. $\int \frac{dx}{x(ax+b)} = \frac{1}{b} \log \frac{x}{ax+b}$

13. $\int \frac{dx}{x(ax+b)^2} = \frac{1}{b(ax+b)} + \frac{1}{b^2} \log \frac{x}{ax+b}$

14. $\int \frac{dx}{x^2(ax+b)} = -\frac{1}{bx} + \frac{a}{b^2} \log \frac{ax+b}{x}$

15. $\int \frac{dx}{x^2(ax+b)^2} = -\frac{2ax+b}{b^2x(ax+b)} + \frac{2a}{b^3} \log \frac{ax+b}{x}$

16. $\int \frac{dx}{x^2+a^2} = \frac{1}{a} \tan^{-1} \frac{x}{a}$

17. $\int \frac{dx}{x^2-a^2} = \frac{1}{2a} \log \frac{x-a}{x+a} = -\frac{1}{a} \tanh^{-1} \frac{a}{x}$

$$\text{or } \frac{1}{2a} \log \frac{a-x}{a+x} = -\frac{1}{a} \tanh^{-1} \frac{x}{a}$$

The definition of the symbol $\tanh^{-1}$, is given on page 261.

$\int \frac{dx}{ax^2 + b}$ reduces to 16 or 17 by taking the factor $a/1$ outside the integral sign.

18. $$\int \frac{dx}{(ax^2 + b)^m} = \frac{x}{2(m - 1)b(ax^2 + b)^{m-1}} + \frac{2m - 3}{2(m - 1)b} \int \frac{dx}{(ax^2 + b)^{m-1}} \text{ when } m \neq 1$$

19. $$\int \frac{x\,dx}{(ax^2 + b)^m} = -\frac{1}{2(m - 1)a(ax^2 + b)^{m-1}} \text{ when } m \neq 1$$

20. $$\int \frac{x\,dx}{ax^2 + b} = \frac{1}{2a} \log (ax^2 + b)$$

21. $$\int \frac{x^2dx}{ax^2 + b} = \frac{x}{a} - \frac{b}{a} \int \frac{dx}{ax^2 + b}$$

22. $$\int \frac{x^2dx}{(ax^2 + b)^m} = -\frac{x}{2(m - 1)a(ax^2 + b)^{m-1}} + \frac{1}{2(m - 1)a} \int \frac{dx}{(ax^2 + b)^{m-1}} \text{ when } m \neq 1$$

23. $$\int \frac{dx}{ax^3 + b} = \frac{k}{3b} \left\{ \sqrt{3} \tan^{-1} \frac{2x - k}{k\sqrt{3}} + \log \frac{k + x}{\sqrt{k^2 - kx + x^2}} \right\}$$

24. $$\int \frac{x\,dx}{ax^3 + b} = \frac{1}{3ak} \left\{ \sqrt{3} \tan^{-1} \frac{2x - k}{k\sqrt{3}} - \log \frac{k + x}{\sqrt{k^2 - kx + x^2}} \right\}$$

where $k = \sqrt[3]{\frac{b}{a}}$ (for 23 and 24)

25. $$\int \frac{dx}{x(ax^n + b)} = \frac{1}{bn} \log \frac{x^n}{ax^n + b}$$

Let $X = ax^2 + bx + c$ and $q = b^2 - 4ac$.

26. $$\int \frac{dx}{X} = \frac{1}{\sqrt{q}} \log \frac{2ax + b - \sqrt{q}}{2ax + b + \sqrt{q}} \text{ when } q > 0$$

27. $$\int \frac{dx}{X} = \frac{2}{\sqrt{-q}} \tan^{-1} \frac{2ax + b}{\sqrt{-q}} \text{ when } q < 0$$

For the case $q = 0$, use formula 6 with $m = -2$.

28. $$\int \frac{dx}{X^n} = -\frac{2ax + b}{(n - 1)qX^{n-1}} - \frac{2(2n - 3)a}{q(n - 1)} \int \frac{dx}{X^{n-1}} \text{ when } n \neq 1$$

29. $$\int \frac{x\,dx}{X} = \frac{1}{2a} \log X - \frac{b}{2a} \int \frac{dx}{X}$$

30. $$\int \frac{(mx + n)dx}{X} = \frac{m}{2a} \log X + \frac{2an - bm}{2a} \int \frac{dx}{X}$$

31. $$\int \frac{x^2dx}{X} = \frac{x}{a} - \frac{b}{2a^2} \log X + \frac{b^2 - 2ac}{2a^2} \int \frac{dx}{X}$$

Integrals Involving $\sqrt{ax + b}$

$$\left.\begin{aligned} &\int \sqrt{ax + b}\, dx \\ &\int \frac{dx}{\sqrt{ax + b}} \\ &\int (ax + b)^n \sqrt{ax + b}\, dx \\ &\int \frac{dx}{(ax + b)^n \sqrt{ax + b}} \end{aligned}\right\} \text{These may all be integrated by formula 6.}$$

32. $$\int x\sqrt{ax + b}\, dx = \frac{2(3ax - 2b)\sqrt{(ax + b)^3}}{15a^2}$$

33. $$\int x^2\sqrt{ax + b}\, dx = \frac{2(15a^2x^2 - 12abx + 8b^2)\sqrt{(ax + b)^3}}{105a^3}$$

34. $$\int x^m\sqrt{ax + b}\, dx = \frac{2}{a(2m + 3)} \left\{ x^m\sqrt{(ax + b)^3} - mb \int x^{m-1}\sqrt{ax + b}\, dx \right\}$$

35. $$\int \frac{\sqrt{ax + b}\, dx}{x} = 2\sqrt{ax + b} + \sqrt{b} \log \frac{\sqrt{ax + b} - \sqrt{b}}{\sqrt{ax + b} + \sqrt{b}} \text{ when } b > 0$$

36. $$\int \frac{\sqrt{ax + b}\, dx}{x} = 2\sqrt{ax + b} - 2\sqrt{-b} \tan^{-1} \sqrt{\frac{ax + b}{-b}} \text{ when } b < 0$$

For the case $b = 0$, use formula 4

37. $$\int \frac{\sqrt{ax + b}\, dx}{x^m} = -\frac{1}{(m - 1)b} \left\{ \frac{\sqrt{(ax + b)^3}}{x^{m-1}} + \frac{(2m - 5)a}{2} \int \frac{\sqrt{ax + b}\, dx}{x^{m-1}} \right\} \text{ when } m \neq 1$$

38. $$\int \frac{xdx}{\sqrt{ax + b}} = \frac{2(ax - 2b)}{3a^2}\sqrt{ax + b}$$

39. $$\int \frac{x^2dx}{\sqrt{ax + b}} = \frac{2(3a^2x^2 - 4abx + 8b^2)}{15a^3}\sqrt{ax + b}$$

40. $$\int \frac{x^m dx}{\sqrt{ax+b}} = \frac{2}{a(2m+1)}\left\{x^m\sqrt{ax+b} - mb\int\frac{x^{m-1}dx}{\sqrt{ax+b}}\right\}$$ when $m \neq \frac{1}{2}$

41. $$\int \frac{dx}{x\sqrt{ax+b}} = \frac{1}{\sqrt{b}}\log\frac{\sqrt{ax+b}-\sqrt{b}}{\sqrt{ax+b}+\sqrt{b}}$$ when $b > 0$

42. $$\int \frac{dx}{x\sqrt{ax+b}} = \frac{2}{\sqrt{-b}}\tan^{-1}\sqrt{\frac{ax+b}{-b}}$$ when $b < 0$

For the case $b = 0$, use formula 4.

43. $$\int \frac{dx}{x^m\sqrt{ax+b}} = -\frac{\sqrt{ax+b}}{(m-1)bx^{m-1}} - \frac{(2m-3)a}{(2m-2)b}\int\frac{dx}{x^{m-1}\sqrt{ax+b}}$$ when $m \neq 1$

Integrals Involving $\sqrt{x^2 \pm a^2}$ *and* $\sqrt{a^2 - x^2}$

(These are special cases of the more general integrals given in the next section.)

44. $$\int\sqrt{x^2 \pm a^2}\,dx = \tfrac{1}{2}[x\sqrt{x^2 \pm a^2} \pm a^2\log(x + \sqrt{x^2 \pm a^2})]^*$$

45. $$\int\sqrt{a^2 - x^2}\,dx = \frac{1}{2}\left(x\sqrt{a^2 - x^2} + a^2\sin^{-1}\frac{x}{a}\right)$$

46. $$\int\frac{dx}{\sqrt{x^2 \pm a^2}} = \log(x + \sqrt{x^2 \pm a^2})^*$$

47. $$\int\frac{dx}{\sqrt{a^2 - x^2}} = \sin^{-1}\frac{x}{a}$$

48. $$\int x\sqrt{x^2 \pm a^2}\,dx = \frac{1}{3}\sqrt{(x^2 \pm a^2)^3}$$

49. $$\int x\sqrt{a^2 - x^2}\,dx = -\frac{1}{3}\sqrt{(a^2 - x^2)^3}$$

50. $$\int x^2\sqrt{x^2 \pm a^2}\,dx = \frac{x}{4}\sqrt{(x^2 \pm a^2)^3} \mp \frac{a^2}{8}\left(x\sqrt{x^2 \pm a^2} \pm a^2\log(x + \sqrt{x^2 \pm a^2})\right)^*$$

51. $$\int x^2\sqrt{a^2 - x^2}\,dx = -\frac{x}{4}\sqrt{(a^2 - x^2)^3} + \frac{a^2}{8}\left(x\sqrt{a^2 - x^2} + a^2\sin^{-1}\frac{x}{a}\right)$$

*** In these formulas we may replace**

$\log(x + \sqrt{x^2 + a^2})$ by $\sinh^{-1}\frac{x}{a}$

$\log(x + \sqrt{x^2 - a^2})$ by $\cosh^{-1}\frac{x}{a}$

$\log\frac{a + \sqrt{a^2 + x^2}}{x}$ by $\sinh^{-1}\frac{a}{x}$

$\log\frac{a + \sqrt{a^2 - x^2}}{x}$ by $\cosh^{-1}\frac{a}{x}$

page 261.

52. $\int \frac{\sqrt{a^2 \pm x^2}}{x}dx = \sqrt{a^2 \pm x^2} - a \log \frac{a + \sqrt{a^2 \pm x^2}}{x}$*

53. $\int \frac{\sqrt{x^2 - a^2}}{x}dx = \sqrt{x^2 - a^2} - a \cos^{-1} \frac{a}{x}$

54. $\int \frac{\sqrt{x^2 \pm a^2}dx}{x^2} = -\frac{\sqrt{x^2 \pm a^2}}{x} + \log (x + \sqrt{x^2 \pm a^2})$

55. $\int \frac{\sqrt{a^2 - x^2}}{x^2}dx = -\frac{\sqrt{a^2 - x^2}}{x} - \sin^{-1} \frac{x}{a}$

56. $\int \frac{xdx}{\sqrt{a^2 - x^2}} = -\sqrt{a^2 - x^2}$

57. $\int \frac{xdx}{\sqrt{x^2 \pm a^2}} = \sqrt{x^2 \pm a^2}$

58. $\int \frac{x^2dx}{\sqrt{x^2 \pm a^2}} = \frac{x}{2}\sqrt{x^2 \pm a^2} \mp \frac{a^2}{2} \log (x + \sqrt{x^2 \pm a^2})$*

59. $\int \frac{x^2dx}{\sqrt{a^2 - x^2}} = -\frac{x}{2}\sqrt{a^2 - x^2} + \frac{a^2}{2} \sin^{-1} \frac{x}{a}$

60. $\int \frac{dx}{x\sqrt{x^2 - a^2}} = \frac{1}{a} \cos^{-1} \frac{a}{x}$

61. $\int \frac{dx}{x\sqrt{a^2 \pm x^2}} = -\frac{1}{a} \log \left(\frac{a + \sqrt{a^2 \pm x^2}}{x}\right)$*

62. $\int \frac{dx}{x^2\sqrt{x^2 \pm a^2}} = \mp \frac{\sqrt{x^2 \pm a^2}}{a^2x}$

63. $\int \frac{dx}{x^2\sqrt{a^2 - x^2}} = -\frac{\sqrt{a^2 - x^2}}{a^2x}$

64. $\int \sqrt{(x^2 \pm a^2)^3}\, dx = \frac{1}{4}\left[x\sqrt{(x^2 \pm a^2)^3} \pm \frac{3a^2x}{2}\sqrt{x^2 \pm a^3}\right.$
$\left. + \frac{3a^4}{2} \log (x + \sqrt{x^2 \pm a^2})\right]$*

65. $\int \sqrt{(a^2 - x^2)^3}\, dx = \frac{1}{4}\left[x\sqrt{(a^2 - x^2)^3} + \frac{3a^2x}{2}\sqrt{a^2 - x^2}\right.$
$\left. + \frac{3a^4}{2} \sin^{-1} \frac{x}{a}\right]$

66. $\int \frac{dx}{\sqrt{(x^2 \pm a^2)^3}} = \frac{\pm x}{a^2\sqrt{x^2 \pm a^2}}$

67. $\int \frac{dx}{\sqrt{(a^2 - x^2)^3}} = \frac{x}{a^2\sqrt{a^2 - x^2}}$

Integrals Involving $\sqrt{ax^2 + bx + c}$

Let $X = ax^2 + bx + c$ and $q = b^2 - 4ac$

68. $$\int \frac{dx}{\sqrt{X}} = \frac{1}{\sqrt{a}} \log \left(\sqrt{X} + \frac{2ax + b}{2\sqrt{a}} \right) \text{ when } a > 0$$

69. $$\int \frac{dx}{\sqrt{X}} = \frac{1}{\sqrt{-a}} \sin^{-1} \left(\frac{-2ax - b}{\sqrt{q}} \right) \text{ when } a < 0$$

70. $$\int \frac{xdx}{\sqrt{X}} = \frac{\sqrt{X}}{a} - \frac{b}{2a} \int \frac{dx}{\sqrt{X}}$$

71. $$\int \frac{(mx + n)dx}{\sqrt{X}} = \frac{m\sqrt{X}}{a} + \frac{2an - bm}{2a} \int \frac{dx}{\sqrt{X}}$$

72. $$\int \frac{x^2dx}{\sqrt{X}} = \frac{(2ax - 3b)\sqrt{X}}{4a^2} + \frac{3b^2 - 4ac}{8a^2} \int \frac{dx}{\sqrt{X}}$$

73. $$\int \frac{dx}{x\sqrt{X}} = -\frac{1}{\sqrt{c}} \log \left(\frac{\sqrt{X} + \sqrt{c}}{x} + \frac{b}{2\sqrt{c}} \right) \text{ when } c > 0$$

74. $$\int \frac{dx}{x\sqrt{X}} = \frac{1}{\sqrt{-c}} \sin^{-1} \frac{bx + 2c}{x\sqrt{q}} \text{ when } c < 0$$

75. $$\int \frac{dx}{x\sqrt{X}} = -\frac{2\sqrt{X}}{bx} \text{ when } c = 0$$

where $k = an^2 - bmn + cm^2$ (applies to 76–78)

76. $$\int \frac{dx}{(mx + n)\sqrt{X}} = -\frac{1}{\sqrt{k}} \log \left(\frac{m\sqrt{X} + \sqrt{k}}{mx + n} + \frac{bm - 2an}{2\sqrt{k}} \right) \text{ when } k > 0$$

77. $$\int \frac{dx}{(mx + n)\sqrt{X}} = \frac{1}{\sqrt{-k}} \sin^{-1} \left\{ \frac{(bm - 2an)(mx + n) + 2k}{m(mx + n)\sqrt{q}} \right\} \text{ when } k < 0$$

78. $$\int \frac{dx}{(mx + n)\sqrt{X}} = -\frac{2m\sqrt{X}}{(bm - 2an)(mx + n)} \text{ when } k = 0$$

79. $$\int \frac{dx}{x^2\sqrt{X}} = -\frac{\sqrt{X}}{cx} - \frac{b}{2c} \int \frac{dx}{x\sqrt{X}}$$

80. $$\int \sqrt{X}\, dx = \frac{(2ax + b)\sqrt{X}}{4a} - \frac{q}{8a} \int \frac{dx}{\sqrt{X}}$$

81. $$\int x\sqrt{X}\, dx = \frac{X\sqrt{X}}{3a} - \frac{b(2ax + b)\sqrt{X}}{8a^2} + \frac{bq}{16a^2} \int \frac{dx}{\sqrt{X}}$$

82. $$\int x^2\sqrt{X}\, dx = \frac{(6ax - 5b)X\sqrt{X}}{24a^2} + \frac{(5b^2 - 4ac)(2ax + b)\sqrt{X}}{64a^3} - \frac{(5b^2 - 4ac)q}{128a^3} \int \frac{dx}{\sqrt{X}}$$

83. $$\int \frac{\sqrt{X}\,dx}{x} = \sqrt{X} + \frac{b}{2}\int \frac{dx}{\sqrt{X}} + c\int \frac{dx}{x\sqrt{X}}$$

84. $$\int \frac{\sqrt{X}\,dx}{mx + n} = \frac{\sqrt{X}}{m} + \frac{bm - 2an}{2m^2}\int \frac{dx}{\sqrt{X}} + \frac{an^2 - bmn + cm^2}{m^2}\int \frac{dx}{(mx + n)\sqrt{X}}$$

85. $$\int \frac{\sqrt{X}\,dx}{x^2} = -\frac{\sqrt{X}}{x} + \frac{b}{2}\int \frac{dx}{x\sqrt{X}} + a\int \frac{dx}{\sqrt{X}}$$

86. $$\int \frac{dx}{X\sqrt{X}} = -\frac{2(2ax + b)}{q\sqrt{X}}$$

87. $$\int X\sqrt{X}\,dx \quad \frac{(2ax + b)X\sqrt{X}}{8a} - \frac{3q(2ax + b)\sqrt{X}}{64a^2} + \frac{3q^2}{128a^2}\int \frac{dx}{\sqrt{X}}$$

Miscellaneous Irrational Integrals

88. $$\int \sqrt{2ax - x^2}\,dx = \frac{x - a}{2}\sqrt{2ax - x^2} + \frac{a^2}{2}\sin^{-1}\frac{x - a}{a}$$

89. $$\int \frac{dx}{\sqrt{2ax - x^2}} = \cos^{-1}\frac{a - x}{a}$$

90. $$\int \sqrt{\frac{mx + n}{ax + b}}\,dx = \int \frac{(mx + n)dx}{\sqrt{amx^2 + (bm + an)x + bn}},$$ then use formula 71.

Logarithmic Integrals

91. $$\int \log_a x dx = x \log_a \frac{x}{e}$$

92. $$\int \log x dx = x\,(\log x - 1)$$

93. $$\int x^m \log ax dx = x^{m+1}\left\{\frac{\log ax}{m + 1} - \frac{\log_a e}{(m + 1)^2}\right\}$$

94. $$\int x^m \log x dx = x^{m+1}\left\{\frac{\log x}{m + 1} - \frac{1}{(m + 1)^2}\right\}$$

Exponential Integrals

95. $$\int a^x dx = \frac{a^x}{\log a}$$

96. $$\int e^x dx = e^x$$

97. $\int xe^x dx = e^x(x - 1)$

98. $\int x^m e^x dx = x^m e^x - m \int x^{m-1} e^x dx$

Trigonometric Integrals

In these formulas m and n are *positive integers* unless otherwise indicated.

99. $\int \sin x dx = -\cos x$

100. $\int \sin^2 x dx = \frac{1}{2}(x - \sin x \cos x)$

101. $\int \sin^n x dx = \int (1 - \cos^2 x)^{\frac{n-1}{2}} \sin x dx$ when n is *odd*

Then expand $(1 - \cos^2 x)^{\frac{n-1}{2}}$ and use formula 112.

102. $\int \sin^n x dx = -\frac{\sin^{n-1} x \cos x}{n} + \frac{n-1}{n} \int \sin^{n-2} x dx$

when n is *even*

103. $\int \frac{dx}{\sin^n x} = -\frac{\cos x}{(n-1)\sin^{n-1} x} + \frac{n-2}{n-1} \int \frac{dx}{\sin^{n-2} x}$

when n is *odd*, $\neq 1$

104. $\int \frac{dx}{\sin^n x} = \int \csc^n x dx$ when n is *even*. Then use formula 131.

105. $\int \cos x dx = \sin x$

106. $\int \cos^2 x dx = \frac{1}{2}(x + \sin x \cos x)$

107. $\int \cos^n x dx = \int (1 - \sin^2 x)^{\frac{n-1}{2}} \cos x dx$ when n is *odd*

Then expand $(1 - \sin^2 x)^{\frac{n-1}{2}}$ and use formula 111.

108. $\int \cos^n x dx = \frac{\cos^{n-1} x \sin x}{n} + \frac{n-1}{n} \int \cos^{n-2} x dx$

when n is *even*

109. $$\int \frac{dx}{\cos^n x} = \frac{\sin x}{(n-1)\cos^{n-1} x} + \frac{n-2}{n-1}\int \frac{dx}{\cos^{n-2} x}$$ when n is *odd*, $\neq 1$

110. $$\int \frac{dx}{\cos^n x} = \int \sec^n x dx$$ when n is *even*. Then use formula 127.

111. $$\int \sin^n x \cos x dx = \frac{\sin^{n+1} x}{n+1}$$

112. $$\int \cos^n x \sin x dx = -\frac{\cos^{n+1} x}{n+1}$$

(n is *any constant* $\neq -1$)

113. $$\int \sin^2 x \cos^2 x dx = \frac{4x - \sin 4x}{32}$$

114. $$\int \frac{dx}{\sin x \cos x} = \log \tan x$$

$$\int \sin^m x \cos^n x dx, \quad \int \frac{\sin^m x dx}{\cos^n x}, \quad \int \frac{\cos^n x dx}{\sin^m x},$$

and $\int \frac{dx}{\sin^m x \cos^n x}$ may be reduced to integrals given above by the use of the following reduction formulas, in which r and s are any integers *positive* or *negative*.

115. $$\int \sin^r x \cos^s x dx = \frac{\cos^{s-1} x \sin^{r+1} x}{r+s} + \frac{s-1}{r+s}\int \sin^r x \cos^{s-2} x dx$$ when $r + s \neq 0$

116. $$\int \sin^r x \cos^s x dx = -\frac{\sin^{r-1} x \cos^{s+1} x}{r+s} + \frac{r-1}{r+s}\int \sin^{r-2} x \cos^s x dx$$ when $r + s \neq 0$

117. $$\int \sin^r x \cos^s x dx = \frac{\sin^{r+1} x \cos^{s+1} x}{r+1} + \frac{s+r+2}{r+1}\int \sin^{r+2} x \cos^s x dx$$ when $r \neq -1$

118. $$\int \sin^r x \cos^s x dx = -\frac{\sin^{r+1} x \cos^{s+1} x}{s+1} + \frac{s+r+2}{s+1}\int \sin^r x \cos^{s+2} x dx$$ when $s \neq -1$

119. $$\int \tan x dx = -\log \cos x$$

120. $\int \tan^n x\,dx = \frac{\tan^{n-1} x}{n-1} - \frac{\tan^{n-3} x}{n-3} + \frac{\tan^{n-5} x}{n-5} \cdots$
$\cdots \pm \tan x \mp x$ when n is *even*

121. $\int \tan^n x\,dx = \int (\sec^2 x - 1)^{\frac{n-1}{2}} \tan x\,dx$ when n is *odd*

Then expand $(\sec^2 x - 1)^{\frac{n-1}{2}}$ and use formula 133.

122. $\int \cot x\,dx = \log \sin x$

123. $\int \cot^n x\,dx = -\frac{\cot^{n-1} x}{n-1} + \frac{\cot^{n-3} x}{n-3} - \frac{\cot^{n-5} x}{n-5} \cdots$
$\cdots \pm \cot x \pm x$ when n is *even*

124. $\int \cot^n x\,dx = \int (\csc^2 x - 1)^{\frac{n-1}{2}} \cot x\,dx$ when n is *odd*

Then expand $(\csc^2 x - 1)^{\frac{n-1}{2}}$ and use formula 134.

125. $\int \sec x\,dx = \log(\sec x + \tan x)$ 126. $\int \sec^2 x\,dx = \tan x$

127. $\int \sec^n x\,dx = \int (\tan^2 x + 1)^{\frac{n-2}{2}} \sec^2 x\,dx$ when n is *even*

Then expand $(\tan^2 x + 1)^{\frac{n-2}{2}}$ and use formula 135.

128. $\int \sec^n x\,dx = \int \frac{dx}{\cos^n x}$ when n is odd, and use formula 109.

129. $\int \csc x\,dx = \log(\csc x - \cot x)$

130. $\int \csc^2 x\,dx = -\cot x$

131. $\int \csc^n x\,dx = \int (\cot^2 x + 1)^{\frac{n-2}{2}} \csc^2 x\,dx$ when n is *even*

Then expand $(\cot^2 x + 1)^{\frac{n-2}{2}}$ and use formula 136.

132. $\int \csc^n x\,dx = \int \frac{dx}{\sin^n x}$ when n is *odd*, and use formula 103.

133. $\int \sec^n x \tan x\,dx = \frac{\sec^n x}{n}$

134. $\int \csc^n x \cot x\,dx = -\frac{\csc^n x}{n}$

where n is *any constant* $\neq 0$ (applies to 133 and 134)

135. $\int \tan^n x \sec^2 x dx = \dfrac{\tan^{n+1} x}{n+1}$

136. $\int \cot^n x \csc^2 x dx = -\dfrac{\cot^{n+1} x}{n+1}$

where n is *any constant* $\neq -1$

137. $\int \dfrac{dx}{a + b\cos x} = \dfrac{-1}{\sqrt{a^2 - b^2}} \sin^{-1} \dfrac{b + a\cos x}{a + b\cos x}$ when $a^2 > b^2$

138. $\int \dfrac{dx}{a + b\cos x} = \dfrac{1}{\sqrt{b^2 - a^2}} \log \dfrac{b + a\cos x + \sqrt{b^2 - a^2}\sin x}{a + b\cos x}$ when $b^2 > a^2$

139. $\int \dfrac{dx}{a + b\sin x} = \dfrac{1}{\sqrt{a^2 - b^2}} \sin^{-1} \dfrac{b + a\sin x}{a + b\sin x}$ when $a^2 > b^2$

140. $\int \dfrac{dx}{a + b\sin x} = \dfrac{-1}{\sqrt{b^2 - a^2}} \log \dfrac{b + a\sin x + \sqrt{b^2 - a^2}\cos x}{a + b\sin x}$ when $b^2 > a^2$

Every integral involving only the trigonometric functions of x, and those only rationally, can be changed to an integral of a rational algebraic fraction by the substitution

$$z = \tan\frac{x}{2}.$$

For, if

$$\tan\frac{x}{2} = z,$$

$$\tan x = \frac{2z}{1 - z^2}$$

$$\sin x = \frac{2z}{1 + z^2}$$

$$\cos x = \frac{1 - z^2}{1 + z^2}$$

$$dx = \frac{2dz}{1 + z^2}$$

141. $\int \sqrt{1 - \cos x}\, dx = -2\sqrt{2}\cos\dfrac{x}{2}$

142. $\int \sqrt{(1 - \cos x)^3}\, dx = \dfrac{4\sqrt{2}}{3}\left(\cos^3\dfrac{x}{2} - 3\cos\dfrac{x}{2}\right)$

143. $\int x \sin x dx = \sin x - x\cos x$

144. $\int x^2 \sin x dx = 2x\sin x + (2 - x^2)\cos x$

145. $\int x \cos x dx = \cos x + x \sin x$

146. $\int x^2 \cos x dx = 2x \cos x + (x^2 - 2) \sin x$

Inverse Trigonometric Integrals

147. $\int \sin^{-1} x dx = x \sin^{-1} x + \sqrt{1 - x^2}$

148. $\int \cos^{-1} x dx = x \cos^{-1} x - \sqrt{1 - x^2}$

149. $\int \tan^{-1} x dx = x \tan^{-1} x - \log \sqrt{1 + x^2}$

150. $\int \cot^{-1} x dx = x \cot^{-1} x + \log \sqrt{1 + x^2}$

151. $\int \sec^{-1} x dx = x \sec^{-1} x - \log (x + \sqrt{x^2 - 1})$
$= x \sec^{-1} x - \cosh^{-1} x$

152. $\int \csc^{-1} x dx = x \csc^{-1} x + \log (x + \sqrt{x^2 - 1})$
$= x \csc^{-1} x + \cosh^{-1} x$

Exponential-trigonometric Integrals

153. $\int e^{ax} \sin x dx = \frac{e^{ax} (a \sin x - \cos x)}{a^2 + 1}$

154. $\int e^{ax} \cos x dx = \frac{e^{ax}(a \cos x + \sin x)}{a^2 + 1}$

Binomial Reduction Formulas

$$(A) \int x^m(ax^n + b)^p dx = \frac{x^{m-n+1}(ax^n + b)^{p+1}}{a(m + np + 1)} - \frac{b(m - n + 1)}{a(m + np + 1)} \int x^{m-n}(ax^n + b)^p dx,$$

$$(B) \int x^m(ax^n + b)^p dx = \frac{x^{m+1}(ax^n + b)^p}{m + np + 1} + \frac{bnp}{m + np + 1} \int x^m(ax^n + b)^{p-1} dx$$

(A) and (B): except when $m + np + 1 = 0$.

$$(C) \int x^m(ax^n + b)^p dx = \frac{x^{m+1}(ax^n + b)^{p+1}}{b(m + 1)} - \frac{a(m + np + n + 1)}{b(m + 1)} \int x^{m+n}(ax^n + b)^p dx, \text{ except when } m + 1 = 0.$$

$$(D) \int x^m(ax^n + b)^p dx = -\frac{x^{m+1}(ax^n + b)^{p+1}}{bn(p + 1)} + \frac{m + np + n + 1}{bn(p + 1)} \int x^m(ax^n + b)^{p+1} dx, \text{ except when } p + 1 = 0.$$

INDEX

P

Q

R

S

T

U

V

W

X

TABLE OF LOGARITHMS

Natural numbers	0	1	2	3	4	5	6	7	8	9	Proportional parts								
											1	2	3	4	5	6	7	8	9
10	0000	0043	0086	0128	0170	0212	0253	0294	0334	0374	4	8	12	17	21	25	29	33	37
11	0414	0453	0492	0531	0569	0607	0645	0682	0719	0755	4	8	11	15	19	23	26	30	34
12	0792	0828	0864	0899	0934	0969	1004	1038	1072	1106	3	7	10	14	17	21	24	28	31
13	1139	1173	1206	1239	1271	1303	1335	1367	1399	1430	3	6	10	13	16	19	23	26	29
14	1461	1492	1523	1553	1584	1614	1644	1673	1703	1732	3	6	9	12	15	18	21	24	27
15	1761	1790	1818	1847	1875	1903	1931	1959	1987	2014	3	6	8	11	14	17	20	22	25
16	2041	2068	2095	2122	2148	2175	2201	2227	2253	2279	3	5	8	11	13	16	18	21	24
17	2304	2330	2355	2380	2405	2430	2455	2480	2504	2529	2	5	7	10	12	15	17	20	22
18	2553	2577	2601	2625	2648	2672	2695	2718	2742	2765	2	5	7	9	12	14	16	19	21
19	2788	2810	2833	2856	2878	2900	2923	2945	2967	2989	2	4	7	9	11	13	16	18	20
20	3010	3032	3054	3075	3096	3118	3139	3160	3181	3201	2	4	6	8	11	13	15	17	19
21	3222	3243	3263	3284	3304	3324	3345	3365	3385	3404	2	4	6	8	10	12	14	16	18
22	3424	3444	3464	3483	3502	3522	3541	3560	3579	3598	2	4	6	8	10	12	14	15	17
23	3617	3636	3655	3674	3692	3711	3729	3747	3766	3784	2	4	6	7	9	11	13	15	17
24	3802	3820	3838	3856	3874	3892	3909	3927	3945	3962	2	4	5	7	9	11	12	14	16
25	3979	3997	4014	4031	4048	4065	4082	4099	4116	4133	2	3	5	7	9	10	12	14	15
26	4150	4166	4183	4200	4216	4232	4249	4265	4281	4298	2	3	5	7	8	10	11	13	15
27	4314	4330	4346	4362	4378	4393	4409	4425	4440	4456	2	3	5	6	8	9	11	13	14
28	4472	4487	4502	4518	4533	4548	4564	4579	4594	4609	2	3	5	6	8	9	11	12	14
29	4624	4639	4654	4669	4683	4698	4713	4728	4742	4757	1	3	4	6	7	9	10	12	13
30	4771	4786	4800	4814	4829	4843	4857	4871	4886	4900	1	3	4	6	7	9	10	11	13
31	4914	4928	4942	4955	4969	4983	4997	5011	5024	5038	1	3	4	6	7	8	10	11	12
32	5051	5065	5079	5092	5105	5119	5132	5145	5159	5172	1	3	4	5	7	8	9	11	12
33	5185	5198	5211	5224	5237	5250	5263	5276	5289	5302	1	3	4	5	6	8	9	10	12
34	5315	5328	5340	5353	5366	5378	5391	5403	5416	5428	1	3	4	5	6	8	9	10	11
35	5441	5453	5465	5478	5490	5502	5514	5527	5539	5551	1	2	4	5	6	7	9	10	11
36	5563	5575	5587	5599	5611	5623	5635	5647	5658	5670	1	2	4	5	6	7	8	10	11
37	5682	5694	5705	5717	5729	5740	5752	5763	5775	5786	1	2	3	5	6	7	8	9	10
38	5798	5809	5821	5832	5843	5855	5866	5877	5888	5899	1	2	3	5	6	7	8	9	10
39	5911	5922	5933	5944	5955	5966	5977	5988	5999	6010	1	2	3	4	5	7	8	9	10
40	6021	6031	6042	6053	6064	6075	6085	6096	6107	6117	1	2	3	4	5	6	8	9	10
41	6128	6138	6149	6160	6170	6180	6191	6201	6212	6222	1	2	3	4	5	6	7	8	9
42	6232	6243	6253	6263	6274	6284	6294	6304	6314	6325	1	2	3	4	5	6	7	8	9
43	6335	5345	6355	6365	6375	6385	6395	6405	6415	6425	1	2	3	4	5	6	7	8	9
44	6435	6444	6454	6464	6474	6484	6493	6503	6513	6522	1	2	3	4	5	6	7	8	9
45	6532	6542	6551	6561	6571	6580	6590	6599	6609	6618	1	2	3	4	5	6	7	8	9
46	6628	6637	6646	6656	6665	6675	6684	6693	6702	6712	1	2	3	4	5	6	7	7	8
47	6721	6730	6739	6749	6758	6767	6776	6785	6794	6803	1	2	3	4	5	5	6	7	8
48	6812	6821	6830	6839	6848	6857	6866	6875	6884	6893	1	2	3	4	4	5	6	7	8
49	6902	6911	6920	6928	6937	6946	6955	6964	6972	6981	1	2	3	4	4	5	6	7	8
50	6990	6998	7007	7016	7024	7033	7042	7050	7059	7067	1	2	3	3	4	5	6	7	8
51	7076	7084	7093	7101	7110	7118	7126	7135	7143	7152	1	2	3	3	4	5	6	7	8
52	7160	7168	7177	7185	7193	7202	7210	7218	7226	7235	1	2	2	3	4	5	6	7	7
53	7243	7251	7259	7267	7275	7284	7292	7300	7308	7316	1	2	2	3	4	5	6	6	7
54	7324	7332	7340	7348	7356	7364	7372	7380	7388	7396	1	2	2	3	4	5	6	6	7

TABLE OF LOGARITHMS

Natural numbers	0	1	2	3	4	5	6	7	8	9	Proportional parts								
											1	2	3	4	5	6	7	8	9
55	7404	7412	7419	7427	7435	7443	7451	7459	7466	7474	1	2	2	3	4	5	5	6	7
56	7482	7490	7497	7505	7513	7520	7528	7536	7543	7551	1	2	2	3	4	5	5	6	7
57	7559	7566	7574	7582	7589	7597	7604	7612	7619	7627	1	2	2	3	4	5	5	6	7
58	7634	7642	7649	7657	7664	7672	7679	7686	7694	7701	1	1	2	3	4	4	5	6	7
59	7709	7716	7723	7731	7738	7745	7752	7760	7767	7774	1	1	2	3	4	4	5	6	7
60	7782	7789	7796	7803	7810	7818	7825	7832	7839	7846	1	1	2	3	4	4	5	6	6
61	7853	7860	7868	7875	7882	7889	7896	7903	7910	7917	1	1	2	3	4	4	5	6	6
62	7924	7931	7938	7945	7952	7959	7966	7973	7980	7987	1	1	2	3	3	4	5	6	6
63	7993	8000	8007	8014	8021	8028	8035	8041	8048	8055	1	1	2	3	3	4	5	5	6
64	8062	8069	8075	8082	8089	8096	8102	8109	8116	8122	1	1	2	3	3	4	5	5	6
65	8129	8136	8142	8149	8156	8162	8169	8176	8182	8189	1	1	2	3	3	4	5	5	6
66	8195	8202	8209	8215	8222	8228	8235	8241	8248	8254	1	1	2	3	3	4	5	5	6
67	8261	8267	8274	8280	8287	8293	8299	8306	8312	8319	1	1	2	3	3	4	5	5	6
68	8325	8331	8338	8344	8351	8357	8363	8370	8376	8382	1	1	2	3	3	4	4	5	6
69	8388	8395	8401	8407	8414	8420	8426	8432	8439	8445	1	1	2	2	3	4	4	5	6
70	8451	8457	8463	8470	8476	8482	8488	8494	8500	8506	1	1	2	2	3	4	4	5	6
71	8513	8519	8525	8531	8537	8543	8549	8555	8561	8567	1	1	2	2	3	4	4	5	5
72	8573	8579	8585	8591	8597	8603	8609	8615	8621	8627	1	1	2	2	3	4	4	5	5
73	8633	8639	8645	8651	8657	8663	8669	8675	8681	8686	1	1	2	2	3	4	4	5	5
74	8692	8698	8704	8710	8716	8722	8727	8733	8739	8745	1	1	2	2	3	4	4	5	5
75	8751	8756	8762	8768	8774	8779	8785	8791	8797	8802	1	1	2	2	3	3	4	5	5
76	8808	8814	8820	8825	8831	8837	8842	8848	8854	8859	1	1	2	2	3	3	4	5	5
77	8865	8871	8876	8882	8887	8893	8899	8904	8910	8915	1	1	2	2	3	3	4	4	5
78	8921	8927	8932	8938	8943	8949	8954	8960	8965	8971	1	1	2	2	3	3	4	4	5
79	8976	8982	8987	8993	8998	9004	9009	9015	9020	9026	1	1	2	2	3	3	4	4	5
80	9031	9036	9042	9047	9053	9058	9063	9069	9074	9079	1	1	2	2	3	3	4	4	5
81	9085	9090	9096	9101	9106	9112	9117	9122	9128	9133	1	1	2	2	3	3	4	4	5
82	9138	9143	9149	9154	9159	9165	9170	9175	9180	9186	1	1	2	2	3	3	4	4	5
83	9191	9196	9201	9206	9212	9217	9222	9227	9232	9238	1	1	2	2	3	3	4	4	5
84	9243	9248	9253	9258	9263	9269	9274	9279	9284	9289	1	1	2	2	3	3	4	4	5
85	9294	9299	9304	9309	9315	9320	9325	9330	9335	9340	1	1	2	2	3	3	4	4	5
86	9345	9350	9355	9360	9365	9370	9375	9380	9385	9390	1	1	2	2	3	3	4	4	5
87	9395	9400	9405	9410	9415	9420	9425	9430	9435	9440	0	1	1	2	2	3	3	4	4
88	9445	9450	9455	9460	9465	9469	9474	9479	9484	9489	0	1	1	2	2	3	3	4	4
89	9494	9499	9504	9509	9513	9518	9523	9528	9533	9538	0	1	1	2	2	3	3	4	4
90	9542	9547	9552	9557	9562	9566	9571	9576	9581	9586	0	1	1	2	2	3	3	4	4
91	9590	9595	9600	9605	9609	9614	9619	9624	9628	9633	0	1	1	2	2	3	3	4	4
92	9638	9643	9647	9652	9657	9661	9666	9671	9675	9680	0	1	1	2	2	3	3	4	4
93	9685	9689	9694	9699	9703	9708	9713	9717	9722	9727	0	1	1	2	2	3	3	4	4
94	9731	9736	9741	9745	9750	9754	9759	9763	9768	9773	0	1	1	2	2	3	3	4	4
95	9777	9782	9786	9791	9795	9800	9805	9809	9814	9818	0	1	1	2	2	3	3	4	4
96	9823	9827	9832	9836	9841	9845	9850	9854	9859	9863	0	1	1	2	2	3	3	4	4
97	9868	9872	9877	9881	9886	9890	9894	9899	9903	9908	0	1	1	2	2	3	3	4	4
98	9912	9917	9921	9926	9930	9934	9939	9943	9948	9952	0	1	1	2	2	3	3	4	4
99	9956	9961	9965	9969	9974	9978	9983	9987	9991	9996	0	1	1	2	2	2	3	3	4